Human Microbiome

Sabu Thomas
Editor

Human Microbiome

Clinical Implications and Therapeutic Interventions

Springer

Editor
Sabu Thomas
Pathogen Biology
Rajiv Gandhi Centre for Biotechnology
Thiruvananthapuram, Kerala, India

ISBN 978-981-16-7671-0 ISBN 978-981-16-7672-7 (eBook)
https://doi.org/10.1007/978-981-16-7672-7

This Springer imprint is published by the registered company Springer Nature Singapore Pte Ltd.
The registered company address is: 152 Beach Road, #21-01/04 Gateway East, Singapore 189721, Singapore

Sabu Thomas
Editor

Human Microbiome

Clinical Implications and Therapeutic Interventions

Springer

Editor
Sabu Thomas
Pathogen Biology
Rajiv Gandhi Centre for Biotechnology
Thiruvananthapuram, Kerala, India

ISBN 978-981-16-7671-0 ISBN 978-981-16-7672-7 (eBook)
https://doi.org/10.1007/978-981-16-7672-7

This Springer imprint is published by the registered company Springer Nature Singapore Pte Ltd.
The registered company address is: 152 Beach Road, #21-01/04 Gateway East, Singapore 189721, Singapore

Foreword

The book *Human Microbiome: Clinical Implications and Therapeutic Interventions*, edited by Dr. Sabu Thomas of Rajiv Gandhi Centre for Biotechnology, is a collection of authentic and up-to-date information about the heaviest organ of our body the gut microbiota and its effect on the well-being of our body in keeping us healthy and in disease management. The editor has brought together eminent clinical practitioners, scientists, and academicians from the leading institutions in the world to expose the readers of this book to the latest in the field of the human microbiome. In the age of antibiotic resistance and difficulty in disease management, people have turned to understand the gut microbes and their role in neurological diseases, gastrointestinal disorders, depression, anxiety, diabetes, cancer, obesity, etc. The book has covered all aspects of the research going on in the field of the human microbiome and would act as a textbook material for the master's and PhD programs of many countries. The chapters of the book are organized in a way that it will also be a good reference and quick reading material for both beginners and established researchers in the area of microbiota. The editor is an established researcher with over two decades of research experience in the field of molecular biology and epidemiology of gut pathogens with special emphasis on cholera, probiotics and microbiomics, bacterial biofilm, and antimicrobial resistance. With his vast experience in this area, he has put together the best in the field to provide the readers a lucid description of the field and new challenges and open questions. The book also provides the future directions in this area and scope. The book is a great book for the uninitiated as well as established researchers and clinicians looking at gut microbes.

Chandrabhas Narayana
Rajiv Gandhi Centre for Biotechnology
Thiruvananthapuram, India

Preface

Human microbiome is the aggregate of all symbiotic microbes that reside on or within the human body colonizing various sites and niches. Hippocrates, the ancient Greek physician and father of modern medicine, stated "All diseases begin in the gut." From time immemorial, mankind has been aware of the importance of the microbial communities inhabiting our gut. It is not just the genes we inherit, but a large number of microbes as we pass through our mother's birth canal. At the time of childbirth, the newborn baby would have been blanketed by a plethora of diverse microorganisms. Further, these microbes are capable of impacting human physiology, both in health and disease, contributing to the enhancement or impairment of metabolic and immune functions. Any imbalance in the microbiome is central to elucidating the severity of a range of diseases including inflammatory bowel disease (IBD), autism spectrum disorder (ASD), diabetes, cancer, and obesity. Psychiatric illnesses and neurodegenerative diseases such as depression, anxiety, and ASD frequently co-occur with gastrointestinal pathology, as the host microbiome affects neurological functions through the gut–brain axis. Therefore, the microbiome has been duly proposed as an "essential organ" of the human body on account of its significant involvement in the host's well-being, in terms of nutritional requirements and immunomodulation. With the recent development in genomics and metagenome analysis, significant advances in our understanding of the human microbiome have been reported. Attempts to therapeutically manipulate microbiome are also rapidly progressing with respect to reversing a damaged, dysbiotic microbiome by probiotic therapy and fecal microbiota transplantation (FMT) for managing selected human diseases to attain a better standard of living. Today, human microbiome research is at a point where the statement of Hippocrates can be rephrased as "All diseases and its cure begin in the gut."

Through this book, we bring together eminent clinical practitioners, scientists, and academicians from different leading institutes to provide authentic and up-to-date information. The book focuses on various clinical implications of the human microbiome and opens by introducing the external and internal factors shaping the microbiome followed by a chapter on oral microbiome and its relation to systemic health. The next chapters in detail discuss the human gut microbiota and its association with functional gastrointestinal disorders, gastric cancer, and alcohol-associated liver disease. These chapters also provide up-to-date information on various diagnostic and therapeutic clinical application of the gut microbiota for the prevention and treatment of these diseases. Of special mention is the first-hand information on fecal microbiota transplantation (FMT) as an effective intervention shared by a team of practicing gastroenterologists. Furthermore, a chapter discussing the various genetic and epigenetic regulations of the gut microbiome in modulating metabolites and its association with chronic metabolic diseases has also been incorporated. In the next chapter, a team of neurologists investigates the impacts of the gut microbiome on neurological disorders which is governed primarily by vagal tone. The following chapter expounds the pros and cons of probiotics as an alternative to antimicrobials and the application of omics technologies in probiotic research. The subsequent chapter deals with the chronic conditions in geriatrics with respect to metabolic and infectious diseases with gut microbiota and polypharmacy. Another remarkable feature of the book is an exclusive chapter dedicated to virome, the less explored component of the human microbiome and its intricate relationships with the host. The book closes with a chapter very relevant to the current scenario, by exploring the influences of the human microbiome in the pathogenesis of the COVID-19 pandemic and potential therapies involving modulations of the microbiota.

I extend my sincere gratitude to all who have been part of this endeavor. Foremost, I would like to pen my sincere gratitude to Prof. Chandrabhas Narayana, Director, RGCB, Prof. M. Radhakrishna Pillai, former Director, and Dr. G. Balakrish Nair, Honorary Distinguished Professor, for their constant support and encouragement. I am extremely grateful to all the authors who munificently gave their time and energy to draft the book chapters, the editorial team for their critical and thoughtful comments, and Springer Nature for their excellent assistance throughout the process. I also acknowledge the Department of Biotechnology of the Government of India. I sincerely hope this volume sparks the interest of young researchers in extending the knowledge on the realm of the human microbiome that will further assist in the discovery of robust microbial biomarkers underlying various diseases. Widespread applications of effective microbiome-based therapeutic strategies will reduce the dependency on conventional medicines and thereby decelerate the pace of antimicrobial resistance.

Thiruvananthapuram, India Sabu Thomas

Acknowledgments

I would like to thank the editorial team Aparna Shankar, Lekshmi Narendrakumar, Karthika Suryaletha, and Merin Paul who have given their valuable time to sincerely assist in reviewing and editing the book chapters. Their diligence in propagating scientific knowledge is much appreciated.

Contents

About the Editor

Sabu Thomas M.Sc., Ph.D. is a senior faculty scientist heading the Cholera and Biofilm Research Group at Rajiv Gandhi Centre for Biotechnology (National Institute under the Department of Biotechnology, Govt. of India). Dr. Thomas has been working in the area of environmental and clinical pathogenic bacteria for more than two decades, with a special focus on gut and chronic wound infected pathogens, antimicrobial resistance (AMR), and alternative strategies to curb AMR. Dr. Thomas's team has published 13 book chapters and more than 80 research articles in prestigious journals at the national and international level, and edited two books. He has received various honors, including membership in the World Health Organization's Global Task Force on Cholera Control. Currently, he is a member of the State Working Committee on Antimicrobial Resistance, and research coordinator for AMR activities in Kerala. He was part of the Second Indian Arctic Scientific Expedition team, formed by the Govt. of India to assess the bioprospecting potential of psychrophilic bacteria in the Polar region. He is also affiliated with various reputed organizations around the globe, including the Global Foodborne Infections Network, CHOLDInet—Global Laboratory Network for Cholera and other Diarrhoeal Infections, International Society for Infectious Diseases, and Freshwater Action Network South Asia.

Chapter 1
Human Microbiome: Implication of Age and External Factors

Hilal Bashir, Anchal Bawa, and Rashmi Kumar

1 Introduction

The human body is colonized by a range of microorganisms, including bacteria, fungi, archaea, viruses, and unicellular eukaryotes, which play an important role in the development of their host in profound ways. The collective term for all these organisms on a particular niche is called microbiota, and the combined genome of constituent microbes is termed as microbiome (Hsiao et al. 2013). The microbiota residing in the human gut outnumber human cells by a factor of 10 (Sender et al. 2016), out of which bacteria predominate all other microbial forms. According to an estimate, almost 500–1000 different species of bacteria reside in the human gut (Sommer and Backhed 2013). Two predominant bacterial phyla Bacteroidetes and Firmicutes constitute more than 90% of the total gut bacterial community, with the remaining 10% comprising other phyla including Proteobacteria, Actinobacteria, and Verrucomicrobia (Qin et al. 2010). This composition basically remains stable and unaffected by temporal disturbance once established early in life (as early as within 3 years of birth) after a dynamic nature of gut microbiota in neonatal life (Rodriguez et al. 2015). The commensal microbial communities remain stable while the relative abundance of bacteria and their diversity undergo dynamic changes depending upon the genetic and environmental factors of the host. Even though inter-individual variation in gut microbiota profile exists, there remains a shared array of bacterial genes, constituting core microbiome provided by individual

H. Bashir
CSIR-Institute of Microbial Technology, Chandigarh, India

A. Bawa · R. Kumar (✉)
CSIR-Institute of Microbial Technology, Chandigarh, India

Academy of Scientific and Innovative Research (AcSIR), Ghaziabad, India
e-mail: rashmi@imtech.res.in

S. Thomas (ed.), *Human Microbiome*,
https://doi.org/10.1007/978-981-16-7672-7_1

bacterial taxa having functional redundancy (Lozupone et al. 2012). This stability starts losing with age-associated dysbiosis and varies from individual to individual, and in general, the beta diversity of gut microbiota increases and becomes more variable with advancement in age (Claesson et al. 2011). Apart from the gut, complex aggregates of microbiome reside at various body sites, including oral cavity, skin, respiratory tract, urinary tract, and reproductive tract (Sender et al. 2016), where they perform various metabolic functions, provide colonization resistance to pathogens, and regulate the immune system. Several studies have identified variables of external and environmental factors and age-related microbial dysbiosis associated with health complications of old age. However, the identification of causal relationship is still at large. It has been demonstrated that altering gut microbiota with the intervention of diet, probiotics, prebiotics, and synbiotics is a promising strategy for healthy aging and has a beneficial effect on age-related health consequences. In this chapter, we have reviewed and summarized all relevant studies describing the association of microbiome with human host at all stages of life. Microbial acquisition, their succession with chronological aging of host, role and impact of acquired microbiota on the host, and reciprocal effect of host-related factors on microbial abundance are summarized.

2 Acquisition of Microbiota in Early Life

Neonates acquire microbiota quite early in life during the prenatal period, and the composition of pioneering microbial communities influences early metabolic and immunologic development of infants. The initial microbiota is highly dynamic in nature and is acquired both prenatally and postnatally. Infants gain microbiota from their mother and surrounding environment, and their composition is structured by many covariates including mode of delivery, type of feeding, early exposure to antibiotics, and mother's health during pregnancy. The prenatal microbial acquisition is primarily achieved via vertical transmission from the mother through the placenta during pregnancy, as demonstrated by the presence of *Lactobacillus* and *Bifidobacterium* DNA in the placenta of both vaginal and cesarean born infants (Satokari et al. 2009). *Fusobacterium nucleatum*, prominent oral bacteria associated with intrauterine infections, penetrate the vascular endothelial cells, disseminate systemically, colonize placenta, and subsequently reach the amniotic fluid and the fetus through the attachment of FadA adhesin with its endothelial receptor, VE-cadherin (Fardini et al. 2011). Mode of delivery is another deciding factor for the structuring of the neonatal microbiome. Vaginally delivered and cesarean section born infants differ in their normal gut microbial consortia and immunity. Vaginally delivered neonatal microbiota resemble maternal vaginal microbial communities, predominantly *Bifidobacterium*, *Lactobacillus*, *Prevotella*, and *Sneathia* spp. while the caesarean born have bacterial communities similar to the mother's skin, predominated by *Staphylococcus*, *Corynebacterium*, and *Propionibacterium* spp. (Sampaio-Maia and Monteiro-Silva 2014). The presence of *Bifidobacterium*,

Enterobacteriaceae, Enterococcaceae, and *Prevotella* in first-pass meconium also suggests in utero acquisition of bacterial communities (Hansen et al. 2015; Martin et al. 2016). *Bifidobacterium* predominance has been shown to provide augmented vaccine response while the dominance of *Clostridiales*, *Enterobacterales*, and *Pseudomonadales* associates with neutrophilia and weak immune response against vaccines (Huda et al. 2014). Additionally, although quite controversial, the place of birth can be a covariate having an impact on the early bacterial colonization of neonates (Chong et al. 2018). The gut microbiota of neonates has been reported to resemble that of the hospital environment, which can lead to the spread of pathogens and vulnerability toward early atopic manifestations (Brooks et al. 2014). *Clostridium difficile* was found associated with vaginally delivered hospital born infants in comparison to home born infants and was linked with wheeze, eczema, and ultimately the procurement of asthma during 6–7 years of life (van Nimwegen et al. 2011). Alongside the place of birth, geographical location of birth also has a role in the build-up of healthy gut microbial consortia. In a study, comparing gut microbiota of 6-week infants from five distinct geographical locations of Europe has shown a high prevalence of *Bifidobacterium* in the fecal samples of infants from Northern European countries while those belonging to Southern Europe possess much diverse microbiota, mainly dominated by *Bacteroides* and *Lactobacillus* (Fallani et al. 2010).

After birth, the acquisition of microbiota is governed by nutrition. Breastfed and formula-fed infants differ in their gut microbial composition, which is simultaneously linked with their immune, neurological, and metabolic development (Martin et al. 2016). Mother's milk is the chief source of microbial seeding into infants after birth. Besides being a rich source of nutrients, it possesses a much complex and diverse microbial ecosystem that lays the foundation for an infant's immune development (de Andres et al. 2017). Bacterial colonization in mammary glands and consequent prevalence in the milk is supposed to be due to the translocation of bacteria from the gut to the extra-intestinal regions during pregnancy, as demonstrated by the presence of orally administered lactic acid bacteria strains—*Lactococcus lactis* and *L. salivarius* in the mammary gland of pregnant dams (de Andres et al. 2017). *Streptococcus*, *Lactobacillus*, and *Bifidobacteria* are the most frequent genera present in breast milk coinciding with their presence in the infant's gut (Solis et al. 2010). Breastfeeding also contributes to the acquisition of *Staphylococcus* spp. from the areolar skin of the mother (Stewart et al. 2018). Colostrum, the initial mother's milk, is unique and different in its microbial composition from the mature milk (Toscano et al. 2017). Bacterial communities in the colostrum belong to the genera *Staphylococcus* dominated by *S. epidermidis*, *Streptococcus* dominated by *S. salivaris*, *Pseudomonas*, *Prevotella*, and *Bacteroides* (Toscano et al. 2017; Obermajer et al. 2014). The gut ecosystem of a healthy breastfed infant as compared to a formula-fed infant shows the initial prevalence of facultative anaerobes including *Streptococcus* and *Enterococcus* and later on the obligate anaerobes like *Bifidobacterium* and *Lactobacillus* from day 10 until 3 months of age (Martin et al. 2016; Solis et al. 2010). On the other hand, formula-fed infants possess *Clostridium* XVIII, *Lachnospiraceae incertae sedis*, *Enterococcus*, *Veillonella*, and *Streptococcus* in abundance (Wang et al. 2015). *Propionibacterium* serves as an important tool for distinguishing breastfed and formula-fed infants, being significantly abundant in the fecal samples of breastfed infants (Wang et al. 2020). Transition to solid food leads to the intrusion by another class of microbes that resembles adult-like microbiota with high efficiency to digest solid food. This is mainly composed of *Bacteroidetes* and *Clostridium* cluster IV and XIV (Martin et al. 2016; Koenig et al. 2011).

Initial acquisition of microbiota is markedly influenced by the usage of antibiotics by mothers and infants (Penders et al. 2006). Maternal or neonatal antibiotic exposure has been well documented to induce ecological imbalance in an infant's gut characterized by decreased microbial diversity, which impairs appropriate immune development and predisposes infants to immune disorders quite early in life (Eck et al. 2020). Antibiotic usage by pregnant mothers during the perinatal period leads to microbial dysbiosis in the maternal vagina and neonatal meconium, which further makes infants susceptible to early-onset sepsis (Zhou et al. 2020). Increased colonization by multi-drug resistant *Enterococcus* and *Enterobacteriaceae* and abatement of commensals like *Bifidobacterium*, *Eubacterium rectale,* and *Bacteroidetes* were reported in infant gut after administration of broad-spectrum antibiotics in the first 4 days of life (Tanaka et al. 2009; Eck et al. 2020). This antibiotic-induced dysbiosis has been shown to normalize within 15 days upon breastfeeding (Eck et al. 2020). Breastfeeding along with probiotic supplementation to mother and to the infant has been shown to ameliorate the effect of antibiotics and birth mode displayed by increased *Bifidobacterium* and decreased *Clostridia* and *Proteobacteria* abundance in treated infants (Korpela et al. 2018). However, alteration of dysbiosis is transient and lasts only until treatment continues. Durable changes in gut microbiota are an area of active research, and in this regard, probiotic inoculation of *Bifidobacterium longum* subsp. *infantis* EVC001 has shown significant alteration of gut microbiota and colonization of the gut with healthy microbiota (Jacobsen et al. 1999), even after the cessation of therapy (Frese et al. 2017).

Mother's BMI, weight, and weight gain during pregnancy are decisive factors for the establishment of a healthy gut ecosystem in neonates. Kids of overweight mothers or the mothers that gain excess weight during their pregnancies show a preponderance of *Staphylococcus*, *Akkermansia muciniphila*, *Bacteroides*, and *Clostridium difficile* group and paucity of *Bifidobacterium* as compared to those from the normal weight mothers, and this is concurrently linked with the risk of obesity. The relative proportion of *Bifidobacterium* and *Clostridium coccoides* regulates higher and lesser body weight, respectively (Collado et al. 2010). Mother's weight also alters breast milk composition in respect to cytokines and microbiota, which in turn exerts its influence on an infant's development. Overweight mothers possess decreased levels of transforming growth factor-β2 (TGF-β2) and soluble CD14 (sCD14) and increased levels of IL-6 in milk. A higher abundance of *Akkermansia muciniphila* regulates the level of IL-6 in colostrum of overweight mothers, which in turn relates to the lower count of *Bifidobacteria* in infants, making them at risk for obesity (Collado et al. 2012). Although maternal obesity impinges an infant's gut microbial ecosystem during the exclusive breastfeeding period, the same does not seem to hold importance during the complementary feeding period, and diet starts regulating gut microbial diversity during this phase (Laursen et al. 2016).

The bacterial communities at different body sites of a newborn, namely oral, intestinal, skin, and nasopharyngeal, are very similar to each other (Dominguez-Bello et al. 2010). Pioneering oral microbial communities including *Streptococcus* and *Staphylococcus* are acquired in the first 24 hrs of birth through breathing, breastfeeding, and contact with surrounding people (Sampaio-Maia and Monteiro-Silva 2014; Li et al. 2017), and these early colonizers further condition the subsequent colonization leading to the established and complex oral ecosystem. A complex oral microbial community with a predominance of *Streptococcus*, *Haemophilus*, *Neisseria*, and *Veillonella* establishes at around 5 months of age. Microbial diversity in the oral cavity is influenced by exposure to environmental factors such as feeding patterns, solid food introduction, and dentition, and this maturation process continues until adulthood (Cephas et al. 2011).

The skin of the human body provides an extensive interface with the environment and is also considered to be dominated by diverse commensal microbiota having a crucial role in the development of the host's immune system against foreign pathogens (Grice et al. 2009). Neonatal skin is structurally similar to adult skin but with a distinct metabolic profile, environmental exposure, and immune activity. Early colonizers of neonatal skin are maternal strains that colonize the skin soon after birth (Casterline and Paller 2020). A shift in microbiota composition was observed as early as 72 h after birth, and acquisition of antibiotic resistance genes (ARGs) was reported in postpartum hospitalization settings in newborns (Klassert et al. 2020).

The initial microbiota of infants is highly dynamic in nature and varies within hours, days, months, and years of life (Rodriguez et al. 2015). The variations are observed not only in the microbial diversity but also in the relative richness of different taxa. These fluctuations are majorly observed up to the age of 3 years, and afterward, an adult-like microbial framework is achieved, which gradually gets stabilized with time (Derrien et al. 2019).

3 Site-Specific Succession of Microbiota with Age

The ecologically stable healthy microbiome of different body niches includes both bacterial and nonbacterial communities residing in almost all the body parts including gastrointestinal tract, skin, vagina, oral cavity, etc. (Backhed et al. 2012). Human gut microbiota abundance changes along with host aging and manifests in either expansion or abatement of certain microbial species that are known to be simultaneously linked with age-related health implications (Xu et al. 2019). Considering the gut microbial communities, an infant's gut is primarily colonized by the populations of facultative anaerobes belonging to *Enterobacteria* including *E. coli*, *Klebsiella*, and *Enterobacter* and coagulase-negative *Staphylococcus* for the first 2 months of age. As the infant ages toward adulthood, the population size of these facultative anaerobes gradually decreases and is soon outnumbered by obligate anaerobes pioneered by *Bifidobacterium* and *Clostridium* (Adlerberth et al. 2006). The major factor reasonable for the shift in the microbial composition in the fundamental years of life is the change in the dietary substrates from milk to solid food containing indigestible carbohydrates. The milk diet favors the growth and proliferation of bacteria that can efficiently metabolize milk oligosaccharides including soluble and conjugated glycans. Thus, infants exhibit a preponderance of *Lactobacillus*, *Bifidobacterium longum* subsp. *infantis* ATCC15697, *Bacteroides fragilis*, and *Bacteroides vulgatus* (Marcobal et al. 2010; Sela and Mills 2014). With the introduction of solid food, an adult-like microenvironment starts developing, favoring the growth of *Bacteroidetes* and *Firmicutes* (Koenig et al. 2011). A fibrous and prudent-style diet facilitates the growth of beneficial microbes mainly enriched by *Prevotella* and *Xylanibacter*, which hydrolyze the dietary cellulose and xylan and produce short-chain fatty acids including acetate, propionate, and butyrate, which provide protection against inflammation and noninfectious colonic diseases, whereas a high-fat and high-sugar westernized diet causes intestinal impairment and endotoxemia (Koenig et al. 2011; Pendyala et al. 2012).

In the case of a rare population of extremely long-lived individuals, i.e., centenarians and super-centenarians, diverse microbiota with decreased relative abundance compared to healthy individuals was observed where alpha diversity can predict longevity, and microbial diversity is modifiable with dietary interventions (Kong et al. 2016). A universal healthy microbial signature of longevity was also identified, which consists of *Clostridium* cluster XIVa, *Lachnospiraceae*, *Akkermansia*, and *Ruminococcaceace* by comparing two super-centenarian cohorts from distinct geographical locations (Biagi et al. 2016; Kong et al. 2016). Very long-lived frail individuals were represented with less diverse microbial signatures (Jackson et al. 2016). A substantial decrease in a "potent probiotic" species like *Ruminococcaceae*, *Erysipelotrichaceae*, *Lachnospiraceae*, and *Faecalibacterium prausnitzii* and an increase in certain species like *Eubacterium dolichum* and Coriobacteriaceae (e.g., *Eggerthellalenta*) have been reported (Jackson et al. 2016; Rampelli et al. 2013). The extent of age-associated conditions like sarcopenia, systemic inflammation, mental health, and frailty have been demonstrated to correlate with frail microbiota (Amato et al. 2019). Factors like reduced exercise, dietary changes, antibiotic exposure, and medication are claimed to be the reason for the loss of diversity (Claesson et al. 2012). The acquisition and succession of microbiota from birth to old age are depicted in Fig. 1.1.

Skin is the largest body organ with miscellaneous niches and harbors a plethora of diverse microorganisms with a predominance of Gram-positive bacteria such as *Staphylococcus*, *Propionibacterium*, and *Corynebacterium* species and fungal genus *Malassezia* with under-representation of Gram-negative bacteria (Wu et al. 2020). An age-associated shift in skin bacterial communities reported by recent studies acknowledges the influence of environmental factors and chronological and physiological skin aging (Shibagaki et al. 2017). *Propionibacterium* spp. were enriched in the skin of young when compared with elderly and centenarians and correlated with the activity of the sebaceous gland, while potential pathogenic

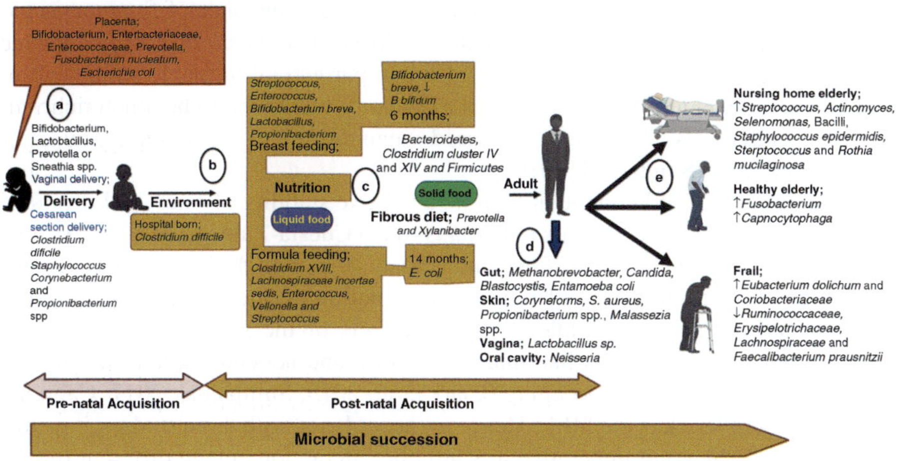

Fig. 1.1 Microbial succession from early life to late life: (**a**) different microbial stratum with varied modes of delivery like vaginal and cesarean section, (**b**) location of delivery resulting in the predominance of many microbial species, (**c**) postnatal factors like food-liquid/solid resulting in different microbial structure, (**d**) site-specific microbial abundance through the adult stage, and (**e**) microbial divergence in different elderly stages like healthy, nursing home, and frail

strains of *Staphylococcus* and *Streptococcus* spp. were detected in centenarians (Wu et al. 2020).

Microbial communities of the vagina play an important role in promoting homeostasis and prevention from pathogenic bacteria. The vagina is predominantly occupied by the lactic acid-producing bacteria including *Lactobacillus iners*, *L. crispatus*, *L. gasseri*, and *L. jensenii*. They provide protection by lowering the vaginal pH through lactic acid, producing hydrogen peroxide and bacteriocins, and giving competition to pathogenic strains for nutrients and space (Ravel et al. 2011). Acquisition of vaginal microbiome occurs shortly after birth, and as of other body sites, it is also influenced by the mode of birth and resembles maternal microbiota. Composition of the vaginal microbiome changes with hormonal shifts associated with age, which makes the environment from neutral/slightly alkaline environment in childhood to acidic at the attainment of puberty and again alkaline at menopause (Fettweis et al. 2019). In young children, the vaginal microbiome consists of a mixed population of aerobes, anaerobes, and enteric organisms; this changes to the predominance of *Lactobacillus* during adulthood (Hickey et al. 2012). Menopause causes physiological changes in the vaginal environment and is accompanied by depletion of *Lactobacillus* and increase in biodiversity with enrichment of *Prevotella*, *Porphyromonas*, *Peptonipjilus*, and *Bacillus* (Gliniewicz et al. 2019; Hummelen et al. 2011).

In comparison to other body sites, studies related to age-associated alteration in the oral microbiome are still scarce (Garcia-Pena et al. 2017). With age, a shift in the oral microbiome occurs from core microbiota to periodontal pathogens (Dewhirst et al. 2010). Analysis of oral microbiome in the elderly, living in nursing homes, has depicted less diversity at phyla level as compared to those living independently. The nursing home elderly are also found to have a higher abundance of *Streptococcus*, *Actinomyces*, *Selenomonas*, *Bacilli* with a reduced abundance of *Campylobacter* and *Fusobacterium* (Ogawa et al. 2018). On comparing oral microbiome in healthy and non-healthy elderly individuals, alpha diversity was found to be much richer in healthy individuals with an increased abundance of *Fusobacterium* and *Capnocytophaga* (Singh et al. 2019).

Although it is well known that there is a relative stability of microbial diversity at middle age approx. at 40 years in humans (de la Cuesta-Zuluaga et al. 2019), loss of stability occurs at extremes of age and is associated with dysbiosis (O'Toole and Jeffery 2015). Determination of relative contribution of age, medication, co-morbidities, and diet is quite challenging, but effects of the said factors have been reported in model organisms like mice, fruit flies, and nematodes, thus suggesting dysbiosis to be the attribute of aging (Clark et al. 2015; Smith et al. 2017; Cabreiro et al. 2013; Broderick et al. 2014). Hence, in accordance with these studies, it would be quite interesting and imperative to study the interaction among different tissue/organ-associated microbial diversities with age to prevent associated diseases.

4 Host and Environmental Factors Influencing Microbiota with Age

The microbiome of different body niches, as discussed above, are essential to human health; their abundance and composition are influenced by various environmental and host-related factors. Here, in this section, we will summarize the influence of various factors such as host physiology, diet, nutrition, and medication on microbial composition with age.

4.1 Host Physiology

Host physiology drastically changes with age and is one of the key factors that shape microbiome structure (Amato et al. 2019). Life span can be extended by preventing age-associated changes in physiology, which dampens dysbiosis at old age (H. Li et al. 2016). Mucin provides protective lining to the gastrointestinal tract (GIT) and prevents microbial infiltration through epithelial cells (Dieterich et al. 2018). A decrease in mucin production with age leads to chronic inflammatory response due to unregulated infiltration of microbes (Elderman et al. 2017). Members of *Clostridiaceae*, *Akkermansiaceae*, *Bifidobacteriaceae*, and *Bacteroidaceae* families show alteration with age, which uses mucin as a nutrient source (Derrien et al. 2010). Administration of *Akkermansia muciniphila*, a potent inducer of mucin production, has been found to mitigate age-associated loss of mucin, resulting in improved immune and health status with increased lifespan in animal models (Barcena et al. 2019; Depommier et al. 2019).

Increased intestinal permeability is an important cause of intestinal dysbiosis (Dumic et al. 2019). Loss of tight junction proteins like zonulin and claudins during uncontrolled pro-inflammatory response results in loss of many microbial niches (Mabbott 2015). Low-grade inflammation with increased levels of inflammatory cytokines like IL-6 and TNF-α contribute to frailty and chronic inflammatory disorders (Franceschi et al. 2000). This leads to impairment of monocytes, neutrophils, and tissue macrophage development, which are important for intestinal integrity maintenance in response to dysbiosis (Zhang et al. 2015). Although studies related to the role of myeloid immunosenescence in barrier or microbial dysfunction are in paucity, it has been shown that retarding chronic inflammation alleviates myeloid dysfunction and dysbiosis (Thevaranjan et al. 2017).

4.2 Diet

Components of diet, such as macronutrients; carbohydrates, proteins, and fat, play a quintessential role in regulating intestinal microbiome diversity and shift in human microbiota (David et al. 2014; O'Keefe et al. 2015). Factors like sex, medication, age, and ethnicity corroborate with diet in shaping the microbiota, which results in further complexity to decipher the collective responsiveness (Sanz et al. 2018).

Richness and diversity of human microbiota decrease on the consumption of dietary saturated fatty acids in adults and infants (Wolters et al. 2019). High-fat diet intervention in healthy adults results in elevated levels of *Bacteroides* and *Alistipes*, which is inversely proportional to *Faecalibacterium* species and is in concurrence with cardiovascular and metabolic disorder-associated metabolites like p-cresol and indole (Wan et al. 2019). Consumption of omega-3 polyunsaturated fatty acids (PUFAs), which are known for anti-cancer and anti-inflammatory effects, increases certain butyrate-producing bacteria (Watson et al. 2018). Dietary proteins similar to fats are also very potent in altering the normal microbiota structure with considerable inter-individual variation in microbial abundance and composition. In addition to the type of proteins, the source of the same, i.e., animal or plant derived, also plays a crucial role in determining the response (Zhu et al. 2015). Consistent consumption of protein-rich diet results in enrichment in bile-tolerant species like *Bacteroides*, *Alistipes*, and *Bilophila* and reduction of saccharolytic species like *Roseburia*, *Ruminococcus bromii*, and *Eubacterium rectale* (David et al. 2014). Consumption of plant-based protein diet like glycated pea proteins significantly increases *Lactobacilli* and *Bifidobacteria* with concurrent production of short-chain fatty acids (SCFAs) in humans (Swiatecka et al. 2011). Carbohydrate-rich diet selectively flourishes microbial species having the ability to digest them. Carbohydrates consumed in the form of fibers found in fruits, grains, vegetables, and milk products affect gut microbiota composition, which is restricted to the type and duration of intake. Long duration of complex carbohydrate intake increases

Prevotella abundance in humans (Lang et al. 2018), while arabinoxylan-degrading *Bifidobacteria* were absent in hunter-gatherer populations with reduced grain diet (Schnorr et al. 2014). Non-digestible carbohydrates favor a significant increase in phylum like *Firmicutes* with species including *Ruminococci*, *Eubacterium*, and *Roseburia* (Walker et al. 2011), while low intake of carbohydrates leads to a significant decrease in butyrate-producing *Firmicutes* and *Bifidobacteria* along with low levels of fecal butyrate in obese subjects (Duncan et al. 2007). Diet has an important role in defining frail microbiota signature. Low-fiber/vegetable and high-saturated fat/sugar diet contribute substantially to the frail microbiome (Claesson et al. 2012). Changes in specific metabolites on nutrient acquisition and inflammatory modulation are known to be the common feature of extreme age with frailty (Krishnan et al. 2018). Short-chain fatty acid (SCFA)-producing species have been reported by many studies to decrease with age, but a concordant decrease in the fiber-rich diet makes it quite difficult to determine the exact cause. At the population level, personalized microbial signature has been found associated in response to specific dietary fiber- and carbohydrate-containing prebiotics (Korem et al. 2017), and such baseline microbial diversity plays an important role in predicting microbial response to dietary carbohydrates (Salonen et al. 2014).

These studies collectively affirm that beyond genetic make-up, factors like nutritional content, meal timing, gut microbiome, and host metabolism are some key determinants for the dietary response.

4.3 Nutrition

Nutrition is recognized as a spearhead in regulating human microbiome structure via modulating numerous individual species and their functionality (David et al. 2014). Combinatorial effects of host and microbial features direct the observed interpersonal variation in a population in response to nutritional components (Rothschild et al. 2018). Various chemical molecules of nutrients regulate host metabolism by promoting the secretion of enzymes and other regulatory molecules, directing immune response toward bacterial colonization (Zmora et al. 2019).

SCFAs like butyrate, propionate, acetate, lactic acid, and acetic acid have an imperative role in promoting the proliferation of epithelial cells and enhancing the expression of tight junction proteins (Makki et al. 2018). Apart from these core molecular functions, SCFAs have a quite considerable role as an energy source for resident microbes, thus maintaining the diversity of the same (Van den Abbeele et al. 2013). However, how age and frailty-associated decrease in SCFA-producing species like *Lachnospiraceae* and *Ruminococcus* spp. directly impact extreme age conditions and barrier integrity is still elusive.

Vitamin D is another covariate that regulates intestinal microbial composition and tissue homeostasis by governing adaptive immune system components: Treg, Th17, Th1, and B cells (Yamamoto and Jorgensen 2019). It acts on vitamin D receptor (VDR), a nuclear receptor, expressed in immune cells and colonic epithelial cells (Bizzaro et al. 2017; Thomas et al. 2020). Increased vitamin D activation ratio, i.e., hormone to prohormone ratio, augments the microbial diversity with a preponderance of butyrate-producing bacteria, including *Firmicutes* in the gut. These butyrate-producing bacteria synthesize SCFAs, which are used by the colonic immune cells and enterocytes for energy production and further 1,25-dihydroxycholecalciferol synthesis (Thomas et al. 2020). This bidirectional interplay maintains both tissue and immune homeostasis. Any perturbation in their interaction, either due to impaired vitamin D biosynthesis or absence of VDR or both, results in intestinal dysbiosis with increased levels of *Bacteroides* and *Proteobacteria* and decreased levels of *Lachnospiraceae*, *Lachnobacillaceae*, and *Ruminococcaceae* linked with the elevated risks of developing autoimmune diseases (Thomas et al. 2020; Bizzaro et al. 2017). Additionally, vitamin D also promotes microbial vitamin B synthesis and simultaneously maintains normal levels of Actinobacteria, Firmicutes, Bacteroidetes, and Proteobacteria and averts the development of atherosclerosis and autoimmunity (Gominak 2016). We may interpret that a comprehensive study about the role and mechanism of action of vitamin D in regulating intestinal microbial diversity can help avert the trailing ailments that appear with its deficiency.

4.4 Medication

Antibiotics are consumed to prevent pathogenic infections, but they affect commensals as well and their long-term consumption can severely disrupt a healthy microbiome (Langdon et al. 2016). Administration of ciprofloxacin in healthy adults has been reported to affect about one-third of the total gut bacterial taxa, thereby influencing the richness, diversity, and evenness of residential communities (Dethlefsen et al. 2008). Similarly, clindamycin, which is used against Gram-positive and anaerobic bacterial infections and is prescribed to patients allergic to β-lactam antibiotics, causes loss of intestinal microbial diversity with about 90% reduction in cecal microbial taxa, thus enhancing intestinal pathogenic bacterium *Clostridum difficile* and predisposing individuals toward *C. difficile*-induced colitis (Bulloch et al. 2016; Buffie et al. 2015). Similarly, the combined action of streptomycin and vancomycin causes alteration of gut microbial composition culminating in host's susceptibility to pathogenic infection by *Salmonella enterica* serovar *Typhimurium* (Sekirov et al. 2008). Antibiotics can have a long-term influence on indigenous gut microbiota even after cessation of treatment, and further acquisition of antibiotic-resistant genes was also reported. The use of clarithromycin and metronidazole for therapy against *Helicobacter pylori*-induced peptic ulcers and gastric cancer resulted in a reduction of the abundance of *Actinobacter* in both throat and feces along with the expression of antibiotic-resistant genes, which make subsequent antibiotic treatment more challenging (Jakobsson et al. 2010). The gut microbiome was found to be more susceptible to several antibiotic classes than the oral microbiome, and the knowledge about the mechanism for resilience toward antibiotic-induced dysbiosis in the oral microbiome can help combat dysbiosis in other body niches after antibiotics treatment (Zaura et al. 2015). Hence, antibiotics must be consumed with high precision, considering their effects on the innocuous microbes in different body sites.

Rapamycin, an immunosuppressant drug and inhibitor of mammalian target of rapamycin and metformin, an anti-diabetic drug, possesses both anti-aging and anti-cancer properties. It prevents immune cell activation of macrophages and monocytes and impedes the age-related autoimmunity and inflammatory responses (Xu et al. 2020). However, it is a potent modulator of the gut microbiota (Blagosklonny 2019). A recent finding has shown that transient administration of rapamycin for 3 months in middle-aged mice remodels the small intestinal microbiota with a significant upsurge in the population of Gram-positive segmented filamentous bacteria, as confirmed from their fecal DNA real-time PCR analysis and increases the life expectancy by up to 60% (Bitto et al. 2016). When studying the role of rapamycin in the resolution of multiple sclerosis, an autoimmune disease, Xu et al. found that rapamycin administration in the experimental autoimmune encephalomyelitis (EAE) model can restore the gut microbial abundance to normal. They found that the EAE group shows a decrease in Bacteroidetes and an increase in Firmicutes in their fecal sample analysis with a simultaneous appearance of Enterobacter, a pathogenic form of Proteobacteria. Moreover, both rapamycin and MCC950 (a specific small-molecule inhibitor of NLRP3 inflammasome) treatment, individually or synergistically, can revive their normal microbial composition, alleviate the associated symptoms, promote autophagy, and narrow down the autoimmune responses (Xu et al. 2020).

Skin microbiota that stably colonizes over time gets destabilized with topical use of antibiotics, pre-operative antiseptics, and temporal bleach bath, resulting in temporal depletion of skin microbiota. These treatments typically target pathogenic skin colonizers like group A *Streptococci* or methicillin-resistant *Staphylococcus aureus*, but due to their nonspecificity, they often lead to collateral damage and reduced microbial diversity (Claesen 2018); however, recovery occurs quickly within 6–12 hrs (SanMiguel et al. 2018). Frequent use of soap and skin cleansers alters the diversity and richness of skin microbiota and also reduces the abundance of antimicrobial peptides (Yu et al. 2018; Two et al. 2016). Similarly, oral microbiota gets influenced by lifestyle and behavior. Dietary choice, consumption of tobacco and alcohol, and oral hygiene practices influence oral microbiome composition. Oral microbiota undergo dysbiosis in response to external stimuli such as cigarette smoke and alcohol (Camelo-Castillo et al. 2015; Fan et al. 2018). Mouth washes, which are effective for oral conditions like gingival inflammation, bleeding, and dental plaque, impact the oral microbiome. A seven-day use of chlorohexidine mouth wash led to an increase in the abundance of *Neisseria*, *Streptococcus*, and *Granulicatella* while lowering the abundance of *Actinomyces* (Bescos et al. 2020).

5 Age-Associated Diseases and Their Influence on Gut Microbiota

The physiological process of aging contributes to the onset of many diseases, and a decrease in gut microbiota diversity with age has been shown to be a potent risk factor. Age-associated ailments like inflammatory bowel disease (DeJong et al. 2020), cancer (O'Keefe et al. 2009), neurological diseases (Parkinson's disease, Alzheimer's disease, multiple sclerosis) (Sampson et al. 2016; Hu et al. 2016), metabolic diseases (obesity, diabetes) (Barlow et al. 2015; Komaroff 2017), musculoskeletal conditions (frailty, osteoporosis, rheumatoid arthritis, gout) (Jackson et al. 2016; Vieira et al. 2015; Britton et al. 2014; Scher et al. 2013), and tuberculosis (Negi et al. 2019) have been reported to be associated with microbial dysbiosis. However, there is still a causality dilemma, whether alteration in microbiome causes age-associated inflammation and senescence, which results in various diseases, or the latter is involved in microbial dysbiosis. Human studies have shown a substantial difference in long-lived and frail individuals' microbiome, but their impact on the promotion or prevention of later life diseases is still in paucity (DeJong et al. 2020).

Chronic and low-grade systemic inflammation known as inflammaging, related to changes in pro-inflammatory cytokines (IL-6, IL-8, C-reactive proteins and TNFα), has a key role in the pathogenesis of various age-associated diseases (Ferrucci and Fabbri 2018; Castaneda-Delgado et al. 2017; Sarkar and Fisher 2006). Concurrently, numerous studies have shown a substantial role of microbiota in these age-related inflammatory conditions (Buford 2017), along with key geriatric syndromes like dementia and physical frailty (Jackson et al. 2016; Cattaneo et al. 2017). A gradual reduction in microbial components, especially in the gut, results in retarded resilience and homeostasis, making individuals susceptible to diseases (Elinav et al. 2011; Kranich et al. 2011). A decrease in biodiversity weakens host defense toward invading pathogens, as in the case of *Clostridium difficile*-associated diarrhea, the major complication in hospitalized elderly (Rea et al. 2012). Microbiota composition in older people is highly variable and differs from core phyla with respect to young (Claesson et al. 2011). They regulate the homeostasis and activation of the innate and adaptive immune system and induce age-associated immunosenescence, evident with persistent NF-kB-mediated inflammation and loss of naive $CD4^+$ T cell pool (Garrett et al. 2010). Increasing focus on the possible involvement of microbiota on the pathophysiology of physical frailty and sarcopenia with age has led to many findings regarding alteration of microbiota involved in reduced muscle size and function with adverse clinical outcomes (Grosicki et al. 2018; Ticinesi et al. 2017). Increased gut mucosal permeability with age results in uncontrolled intake of bacterial products like lipopolysaccharides, leading to enhanced inflammatory response through increased circulatory levels of pro-inflammatory cytokines tumor necrosis factor (TNF) and interleukin-6 (IL-6) (Thevaranjan et al. 2017). Increased abundance of opportunistic pro-inflammatory bacteria, notably pathobionts, and reduced prevalence of anti-inflammatory symbiotic species like clostridial clusters such as *Faecalibacterium prauznitzii* and related species in centenarians have been linked with systemic inflammation (Santoro et al. 2018). Levels of circulatory pro-inflammatory cytokines like IL-6 and IL-8 correlate with enrichment of Proteobacteria phylum and decrease in anti-inflammatory butyrate-producing Firmicutes phylum such as clostridial clusters (Biagi et al. 2010). This clearly depicts that microbiota in older individuals are a key risk factor for inflammaging (Franceschi et al. 2018). Hence, exploring the role of microbial alteration on age-associated diseases with emphasis on inflammatory conditions would be quite envisaging for future therapeutics. As the onset of these conditions is mostly evident with age progression, it is very much conducive that alteration in microbiota with age may stem the associated pathologies.

The estimated rate of cancer, mostly known as the disease of old age, is predicted to correspondingly increase with life expectancy. In the United States, by 2030, individuals above 65 years are thought to contribute around 70% of all cancer types (White et al. 2014). A plethora of clinical studies from the past decade has enhanced our understanding of the role of microbiota in the development of cancer and associated therapies. Considering their roles, microbiota seem to influence many aspects of immune and cancer development (Fulbright et al. 2017). Microbes can induce the onset and progression of cancer directly by producing genotoxins or indirectly by modulation of antitumor immune response and cancer therapy (Gopalakrishnan et al. 2018; Boleij et al. 2015). Native microbial species are reported to affect cancer therapy by altering drug metabolism, immune response, and toxicity across a range of diverse cancer treatments, including chemotherapy, immunotherapy, radiotherapy, and immune checkpoint therapy (Geller et al. 2017; Al-Dasooqi et al. 2011; Singh et al. 2021). Therefore, it is essential to understand the role of microbiota interaction in age-associated cancer progression for the development of suitable antitumor therapies.

Age-associated gut dysbiosis can also lead to a decline in cognitive function. Peripheral inflammation and brain amyloidosis among cognition-impaired elders like in Alzheimer's disease (AD) have been reported to have an increased abundance of pro-inflammatory microbiota. Pro-inflammatory microbial taxa such as *Escherichia/Shigella*, *Pseudomonas aeruginosa*, *Eubacterium rectale*, and *Faecalibacterium prausnitzii* with concurrent increase in pro-inflammatory circulatory cytokines, viz., IL-1β, IL-6, IL-18, IL-8, and TNFα were found to be abundant in stool samples of cognitively impaired patients (Cattaneo et al. 2017). Numerous studies, using models for age-related cognitive declines like Parkinson's disease and multiple sclerosis, have demonstrated a well-entrenched gut–brain axis leading to age-related dementias (Proctor et al. 2017).

The risk of cardiovascular disorders and atherosclerosis is associated with the gradual advancement of age (Benjamin et al. 2017; Gregory et al. 2015) owing to chronic low-grade inflammation and increased oxidative stress. Reduced bioavailability of nitric oxide, a potent vasodilatory molecule and deteriorative modifications in extracellular matrix components, results in pathophysiological events like stiffening of elastic arteries and vascular endothelial dysfunction (El Assar et al. 2012). Although there is a paucity of mechanistic details behind age-driven chronic inflammation, vascular oxidative stress, and dysfunction, the host microbiome is considered to be one of the driving factors in these pathophysiological adversities (Clemente et al. 2012). In experimental studies, gut dysbiosis promoted by a high-fat diet has been found to impair arterial function (Battson et al. 2018) while antibiotic-mediated microbial suppression ameliorated age-related arterial dysfunction. Similarly, increased oxidative stress due to reduced abundance of genus *Desulfovibrio* was shown to be involved in the synthesis of trimethylamine N-oxide (TMAO) (Brunt et al. 2019). Metabolites like TMAO and pre-cresyl sulfate, which are produced via microbial conversion of dietary precursors, viz., L-carnitine, choline, tyrosine, and tryptophan are known to increase CVD risk by promoting atherosclerosis (Tang et al. 2013; Wu et al. 2015). Several interesting studies have found the DNA of intestinal bacteria such as Proteobacteria, *Cryseomonas* spp., *Staphylococcus* spp., *Propionibacterium* spp., and *Chlamydia* spp. in coronary plaques (Ott et al. 2006). *Enterobacteriaceae* and *Streptococcus* spp. have been found to be in high abundance in atherosclerotic CVD patients as compared to healthy controls (Jie et al. 2017). Differential abundance of bacteria was reported from feces of atherosclerotic and chronic heart failure (CHF) patients wherein patients harbored elevated levels of *Collinsella* while *Eubacterium* and *Roseburia* were reported in controls (Karlsson et al. 2012). Moreover, pathogenic bacteria *Campylobacter, Candida*, and *Shigella* were found positively correlated with the severity of the disease (Pasini et al. 2016). Age-related disorders reported having an association with gut microbial dysbiosis are depicted in Fig. 1.2.

Thus, in accordance with current findings, it will be very intriguing and quintessential to decipher the significance of intertwining interactions of microbiota with

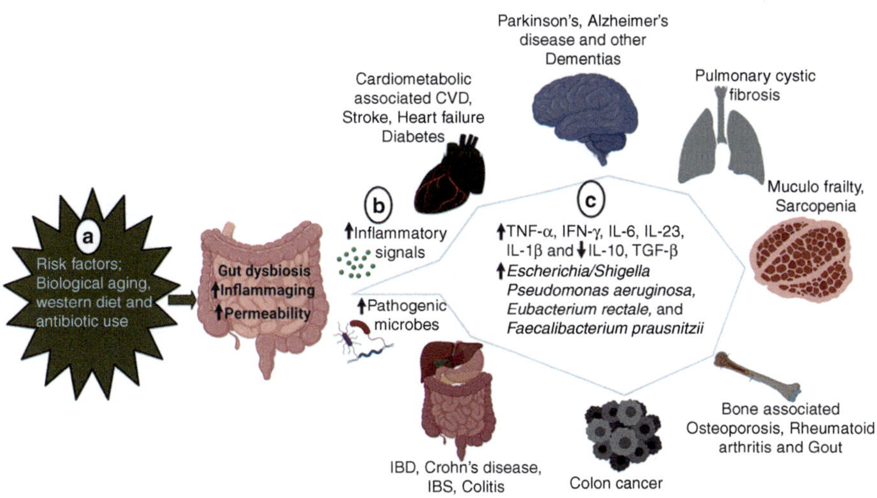

Fig. 1.2 Gut microbial dysbiosis in age-related disorders: (**a**) risk factors involved in gut dysbiosis, (**b**) increased inflammaging and permeability resulting in enhanced inflammatory and pathogenic signals, and (**c**) increased pro-inflammatory cytokines and pathogenic microbes promoting different age-associated diseases

age-related diseases for the development of the same as a therapeutic tool for healthy aging. In corroboration with this, from the past decade, a multitude of seminal studies on microbiota and its imperative role in human health and diseases have prompted the utilization of this interaction to develop microbiota-based therapeutics. Among the various microbiome-based therapeutic interventions, a vast number of studies have demonstrated that fecal microbiota transplant (FMT) probiotic and prebiotic interventions play a substantial role in the management of many microbiome-associated diseases.

6 Conclusions

In recent years, advancements in high-throughput sequencing technologies and other culture-independent techniques for the identification of microorganisms have enabled researchers to identify and assign functional attributes to complex microbial niches. Microbial association with health and disease has been studied in great detail, and now a determinant role has been assigned to microbiota for host health. As discussed, a plethora of studies have characterized and identified microbiota, specific to different body sites, with a succession of age and their dependence on host-related factors. Aging microbiota is distinct from early age microbiota and is often less diverse. These studies have explored the changes not only in gut microbiota but other body sites with chronological age. However, more exploration in terms of inter-microbiota interaction at different ecological niches is warranted to understand the exact mechanism of microbiota shift with age. Mechanistic details underlying the beneficial effect of microbial manipulation, especially through prebiotics, probiotics, diet, and nutrition, hold great promise for future personalized diet-mediated therapy. Inflammation-prone environment of the elderly gut set the baseline for various age-associated disorders, especially metabolic disorders. In this case, precise molecular detail of microbial community-mediated alteration of host physiology at the cellular level can improve human health with age. Studies related to age-associated disease management, medication, and response of complex microbiota are an intensive area of research that will establish personalized microbial therapy as an alternative therapeutic option. A unified approach for data acquisition, analysis, and extensive amalgamation with clinical research across the globe will help in further characterization of the microbiome with age in health and disease and will aid in the discovery of age-associated preventive therapy.

References

Adlerberth I, Lindberg E, Aberg N, Hesselmar B, Saalman R, Strannegard IL et al (2006) Reduced enterobacterial and increased staphylococcal colonization of the infantile bowel: an effect of hygienic lifestyle? Pediatr Res 59(1):96–101. https://doi.org/10.1203/01.pdr.0000191137.12774.b2

Al-Dasooqi N, Bowen JM, Gibson RJ, Logan RM, Stringer AM, Keefe DM (2011) Irinotecan-induced alterations in intestinal cell kinetics and extracellular matrix component expression in the dark agouti rat. Int J Exp Pathol 92(5):357–365. https://doi.org/10.1111/j.1365-2613.2011.00771.x

Amato KR, Sanders JG, Song SJ, Nute M, Metcalf JL, Thompson LR et al (2019) Evolutionary trends in host physiology outweigh dietary niche in structuring primate gut microbiomes. ISME J 13(3):576–587. https://doi.org/10.1038/s41396-018-0175-0

Backhed F, Fraser CM, Ringel Y, Sanders ME, Sartor RB, Sherman PM et al (2012) Defining a healthy human gut microbiome: current concepts, future directions, and clinical applications. Cell Host Microbe 12(5):611–622. https://doi.org/10.1016/j.chom.2012.10.012

Barcena C, Valdes-Mas R, Mayoral P, Garabaya C, Durand S, Rodriguez F et al (2019) Healthspan and lifespan extension by fecal microbiota transplantation into progeroid mice. Nat Med 25(8):1234–1242. https://doi.org/10.1038/s41591-019-0504-5

Barlow GM, Yu A, Mathur R (2015) Role of the gut microbiome in obesity and diabetes mellitus. Nutr Clin Pract 30(6):787–797. https://doi.org/10.1177/0884533615609896

Battson ML, Lee DM, Jarrell DK, Hou S, Ecton KE, Weir TL et al (2018) Suppression of gut dysbiosis reverses Western diet-induced vascular dysfunction. Am J Physiol Endocrinol Metab 314(5):E468–E477. https://doi.org/10.1152/ajpendo.00187.2017

Benjamin EJ, Blaha MJ, Chiuve SE, Cushman M, Das SR, Deo R et al (2017) Heart disease and stroke Statistics-2017 update: a report from the American Heart Association. Circulation 135(10):e146–e603. https://doi.org/10.1161/CIR.0000000000000485

Bescos R, Ashworth A, Cutler C, Brookes ZL, Belfield L, Rodiles A et al (2020) Effects of chlorhexidine mouthwash on the oral microbiome. Sci Rep 10(1):5254. https://doi.org/10.1038/s41598-020-61912-4

Biagi E, Franceschi C, Rampelli S, Severgnini M, Ostan R, Turroni S et al (2016) Gut microbiota and extreme longevity. Curr Biol 26(11):1480–1485. https://doi.org/10.1016/j.cub.2016.04.016

Biagi E, Nylund L, Candela M, Ostan R, Bucci L, Pini E et al (2010) Through ageing, and beyond: gut microbiota and inflammatory status in seniors and centenarians. PLoS One 5(5):e10667. https://doi.org/10.1371/journal.pone.0010667

Bitto A, Ito TK, Pineda VV, LeTexier NJ, Huang HZ, Sutlief E et al (2016) Transient rapamycin treatment can increase lifespan and healthspan in middle-aged mice. elife 5. https://doi.org/10.7554/eLife.16351

Bizzaro G, Antico A, Fortunato A, Bizzaro N (2017) Vitamin D and autoimmune diseases: is vitamin D receptor (VDR) polymorphism the culprit? Isr Med Assoc J 19(7):438–443

Blagosklonny MV (2019) Rapamycin for longevity: opinion article. Aging (Albany NY) 11(19):8048–8067. https://doi.org/10.18632/aging.102355

Boleij A, Hechenbleikner EM, Goodwin AC, Badani R, Stein EM, Lazarev MG et al (2015) The Bacteroides fragilis toxin gene is prevalent in the colon mucosa of colorectal cancer patients. Clin Infect Dis 60(2):208–215. https://doi.org/10.1093/cid/ciu787

Britton RA, Irwin R, Quach D, Schaefer L, Zhang J, Lee T et al (2014) Probiotic L. reuteri treatment prevents bone loss in a menopausal ovariectomized mouse model. J Cell Physiol 229(11):1822–1830. https://doi.org/10.1002/jcp.24636

Broderick NA, Buchon N, Lemaitre B (2014) Microbiota-induced changes in drosophila melanogaster host gene expression and gut morphology. mBio 5(3):e01117–e01114. https://doi.org/10.1128/mBio.01117-14

Brooks B, Firek BA, Miller CS, Sharon I, Thomas BC, Baker R et al (2014) Microbes in the neonatal intensive care unit resemble those found in the gut of premature infants. Microbiome 2(1):1. https://doi.org/10.1186/2049-2618-2-1

Brunt VE, Gioscia-Ryan RA, Richey JJ, Zigler MC, Cuevas LM, Gonzalez A et al (2019) Suppression of the gut microbiome ameliorates age-related arterial dysfunction and oxidative stress in mice. J Physiol 597(9):2361–2378. https://doi.org/10.1113/JP277336
Buffie CG, Bucci V, Stein RR, McKenney PT, Ling L, Gobourne A et al (2015) Precision microbiome reconstitution restores bile acid mediated resistance to Clostridium difficile. Nature 517(7533):205–208. https://doi.org/10.1038/nature13828
Buford TW (2017) (Dis)Trust your gut: the gut microbiome in age-related inflammation, health, and disease. Microbiome 5(1):80. https://doi.org/10.1186/s40168-017-0296-0
Bulloch MN, Baccas JT, Arnold S (2016) Clindamycin-induced hypersensitivity reaction. Infection 44(3):357–359. https://doi.org/10.1007/s15010-015-0826-2
Cabreiro F, Au C, Leung KY, Vergara-Irigaray N, Cocheme HM, Noori T et al (2013) Metformin retards aging in C. elegans by altering microbial folate and methionine metabolism. Cell 153(1):228–239. https://doi.org/10.1016/j.cell.2013.02.035
Camelo-Castillo AJ, Mira A, Pico A, Nibali L, Henderson B, Donos N et al (2015) Subgingival microbiota in health compared to periodontitis and the influence of smoking. Front Microbiol 6:119. https://doi.org/10.3389/fmicb.2015.00119
Castaneda-Delgado JE, Frausto-Lujan I, Gonzalez-Curiel I, Montoya-Rosales A, Serrano CJ, Torres-Juarez F et al (2017) Differences in cytokine production during aging and its relationship with antimicrobial peptides production. Immunol Investig 46(1):48–58. https://doi.org/10.1080/08820139.2016.1212873
Casterline BW, Paller AS (2020) Early development of the skin microbiome: therapeutic opportunities. Pediatr Res. https://doi.org/10.1038/s41390-020-01146-2
Cattaneo A, Cattane N, Galluzzi S, Provasi S, Lopizzo N, Festari C et al (2017) Association of brain amyloidosis with pro-inflammatory gut bacterial taxa and peripheral inflammation markers in cognitively impaired elderly. Neurobiol Aging 49:60–68. https://doi.org/10.1016/j.neurobiolaging.2016.08.019
Cephas KD, Kim J, Mathai RA, Barry KA, Dowd SE, Meline BS et al (2011) Comparative analysis of salivary bacterial microbiome diversity in edentulous infants and their mothers or primary care givers using pyrosequencing. PLoS One 6(8):e23503. https://doi.org/10.1371/journal.pone.0023503
Chong CYL, Bloomfield FH, O'Sullivan JM (2018) Factors affecting gastrointestinal microbiome development in neonates. Nutrients 10(3). https://doi.org/10.3390/nu10030274
Claesen J (2018) Topical antiseptics and the skin microbiota. J Invest Dermatol 138(10):2106–2107. https://doi.org/10.1016/j.jid.2018.06.001
Claesson MJ, Cusack S, O'Sullivan O, Greene-Diniz R, de Weerd H, Flannery E et al (2011) Composition, variability, and temporal stability of the intestinal microbiota of the elderly. Proc Natl Acad Sci U S A 108(Suppl 1):4586–4591. https://doi.org/10.1073/pnas.1000097107
Claesson MJ, Jeffery IB, Conde S, Power SE, O'Connor EM, Cusack S et al (2012) Gut microbiota composition correlates with diet and health in the elderly. Nature 488(7410):178–184. https://doi.org/10.1038/nature11319
Clark RI, Salazar A, Yamada R, Fitz-Gibbon S, Morselli M, Alcaraz J et al (2015) Distinct shifts in microbiota composition during drosophila aging impair intestinal function and drive mortality. Cell Rep 12(10):1656–1667. https://doi.org/10.1016/j.celrep.2015.08.004
Clemente JC, Ursell LK, Parfrey LW, Knight R (2012) The impact of the gut microbiota on human health: an integrative view. Cell 148(6):1258–1270. https://doi.org/10.1016/j.cell.2012.01.035
Collado MC, Isolauri E, Laitinen K, Salminen S (2010) Effect of mother's weight on infant's microbiota acquisition, composition, and activity during early infancy: a prospective follow-up study initiated in early pregnancy. Am J Clin Nutr 92(5):1023–1030. https://doi.org/10.3945/ajcn.2010.29877
Collado MC, Laitinen K, Salminen S, Isolauri E (2012) Maternal weight and excessive weight gain during pregnancy modify the immunomodulatory potential of breast milk. Pediatr Res 72(1):77–85. https://doi.org/10.1038/pr.2012.42

David LA, Maurice CF, Carmody RN, Gootenberg DB, Button JE, Wolfe BE et al (2014) Diet rapidly and reproducibly alters the human gut microbiome. Nature 505(7484):559–563. https://doi.org/10.1038/nature12820

de Andres J, Jimenez E, Chico-Calero I, Fresno M, Fernandez L, Rodriguez JM (2017) Physiological translocation of lactic acid bacteria during pregnancy contributes to the composition of the Milk microbiota in mice. Nutrients 10(1). https://doi.org/10.3390/nu10010014

de la Cuesta-Zuluaga J, Kelley ST, Chen Y, Escobar JS, Mueller NT, Ley RE et al (2019) Age- and sex-dependent patterns of gut microbial diversity in human adults. mSystems 4(4). https://doi.org/10.1128/mSystems.00261-19

DeJong EN, Surette MG, Bowdish DME (2020) The gut microbiota and unhealthy aging: disentangling cause from consequence. Cell Host Microbe 28(2):180–189. https://doi.org/10.1016/j.chom.2020.07.013

Depommier C, Everard A, Druart C, Plovier H, Van Hul M, Vieira-Silva S et al (2019) Supplementation with Akkermansia muciniphila in overweight and obese human volunteers: a proof-of-concept exploratory study. Nat Med 25(7):1096–1103. https://doi.org/10.1038/s41591-019-0495-2

Derrien M, Alvarez AS, de Vos WM (2019) The gut microbiota in the first decade of life. Trends Microbiol 27(12):997–1010. https://doi.org/10.1016/j.tim.2019.08.001

Derrien M, van Passel MW, van de Bovenkamp JH, Schipper RG, de Vos WM, Dekker J (2010) Mucin-bacterial interactions in the human oral cavity and digestive tract. Gut Microbes 1(4):254–268. https://doi.org/10.4161/gmic.1.4.12778

Dethlefsen L, Huse S, Sogin ML, Relman DA (2008) The pervasive effects of an antibiotic on the human gut microbiota, as revealed by deep 16S rRNA sequencing. PLoS Biol 6(11):e280. https://doi.org/10.1371/journal.pbio.0060280

Dewhirst FE, Chen T, Izard J, Paster BJ, Tanner AC, Yu WH et al (2010) The human oral microbiome. J Bacteriol 192(19):5002–5017. https://doi.org/10.1128/JB.00542-10

Dieterich W, Schink M, Zopf Y (2018) Microbiota in the gastrointestinal tract. Med Sci (Basel) 6(4). https://doi.org/10.3390/medsci6040116

Dominguez-Bello MG, Costello EK, Contreras M, Magris M, Hidalgo G, Fierer N et al (2010) Delivery mode shapes the acquisition and structure of the initial microbiota across multiple body habitats in newborns. Proc Natl Acad Sci U S A 107(26):11971–11975. https://doi.org/10.1073/pnas.1002601107

Dumic I, Nordin T, Jecmenica M, Stojkovic Lalosevic M, Milosavljevic T, Milovanovic T (2019) Gastrointestinal tract disorders in older age. Can J Gastroenterol Hepatol 2019:6757524. https://doi.org/10.1155/2019/6757524

Duncan SH, Belenguer A, Holtrop G, Johnstone AM, Flint HJ, Lobley GE (2007) Reduced dietary intake of carbohydrates by obese subjects results in decreased concentrations of butyrate and butyrate-producing bacteria in feces. Appl Environ Microbiol 73(4):1073–1078. https://doi.org/10.1128/AEM.02340-06

Eck A, Rutten N, Singendonk MMJ, Rijkers GT, Savelkoul PHM, Meijssen CB et al (2020) Neonatal microbiota development and the effect of early life antibiotics are determined by two distinct settler types. PLoS One 15(2):e0228133. https://doi.org/10.1371/journal.pone.0228133

El Assar M, Angulo J, Vallejo S, Peiro C, Sanchez-Ferrer CF, Rodriguez-Manas L (2012) Mechanisms involved in the aging-induced vascular dysfunction. Front Physiol 3:132. https://doi.org/10.3389/fphys.2012.00132

Elderman M, Sovran B, Hugenholtz F, Graversen K, Huijskes M, Houtsma E et al (2017) The effect of age on the intestinal mucus thickness, microbiota composition and immunity in relation to sex in mice. PLoS One 12(9):e0184274. https://doi.org/10.1371/journal.pone.0184274

Elinav E, Strowig T, Kau AL, Henao-Mejia J, Thaiss CA, Booth CJ et al (2011) NLRP6 inflammasome regulates colonic microbial ecology and risk for colitis. Cell 145(5):745–757. https://doi.org/10.1016/j.cell.2011.04.022

Fallani M, Young D, Scott J, Norin E, Amarri S, Adam R et al (2010) Intestinal microbiota of 6-week-old infants across Europe: geographic influence beyond delivery mode, breast-feeding,

and antibiotics. J Pediatr Gastroenterol Nutr 51(1):77–84. https://doi.org/10.1097/MPG.0b013e3181d1b11e
Fan X, Peters BA, Jacobs EJ, Gapstur SM, Purdue MP, Freedman ND et al (2018) Drinking alcohol is associated with variation in the human oral microbiome in a large study of American adults. Microbiome 6(1):59. https://doi.org/10.1186/s40168-018-0448-x
Fardini Y, Wang X, Temoin S, Nithianantham S, Lee D, Shoham M et al (2011) Fusobacterium nucleatum adhesin FadA binds vascular endothelial cadherin and alters endothelial integrity. Mol Microbiol 82(6):1468–1480. https://doi.org/10.1111/j.1365-2958.2011.07905.x
Ferrucci L, Fabbri E (2018) Inflammageing: chronic inflammation in ageing, cardiovascular disease, and frailty. Nat Rev Cardiol 15(9):505–522. https://doi.org/10.1038/s41569-018-0064-2
Fettweis JM, Serrano MG, Brooks JP, Edwards DJ, Girerd PH, Parikh HI et al (2019) The vaginal microbiome and preterm birth. Nat Med 25(6):1012–1021. https://doi.org/10.1038/s41591-019-0450-2
Franceschi C, Bonafe M, Valensin S, Olivieri F, De Luca M, Ottaviani E et al (2000) Inflamm-aging. An evolutionary perspective on immunosenescence. Ann N Y Acad Sci 908:244–254. https://doi.org/10.1111/j.1749-6632.2000.tb06651.x
Franceschi C, Garagnani P, Parini P, Giuliani C, Santoro A (2018) Inflammaging: a new immune-metabolic viewpoint for age-related diseases. Nat Rev Endocrinol 14(10):576–590. https://doi.org/10.1038/s41574-018-0059-4
Frese SA, Hutton AA, Contreras LN, Shaw CA, Palumbo MC, Casaburi G et al (2017) Persistence of supplemented Bifidobacterium longum subsp. infantis EVC001 in breastfed infants. mSphere 2(6). https://doi.org/10.1128/mSphere.00501-17
Fulbright LE, Ellermann M, Arthur JC (2017) The microbiome and the hallmarks of cancer. PLoS Pathog 13(9):e1006480. https://doi.org/10.1371/journal.ppat.1006480
Garcia-Pena C, Alvarez-Cisneros T, Quiroz-Baez R, Friedland RP (2017) Microbiota and aging. A review and commentary. Arch Med Res 48(8):681–689. https://doi.org/10.1016/j.arcmed.2017.11.005
Garrett WS, Gordon JI, Glimcher LH (2010) Homeostasis and inflammation in the intestine. Cell 140(6):859–870. https://doi.org/10.1016/j.cell.2010.01.023
Geller LT, Barzily-Rokni M, Danino T, Jonas OH, Shental N, Nejman D et al (2017) Potential role of intratumor bacteria in mediating tumor resistance to the chemotherapeutic drug gemcitabine. Science 357(6356):1156–1160. https://doi.org/10.1126/science.aah5043
Gliniewicz K, Schneider GM, Ridenhour BJ, Williams CJ, Song Y, Farage MA et al (2019) Comparison of the vaginal microbiomes of premenopausal and postmenopausal women. Front Microbiol 10:193. https://doi.org/10.3389/fmicb.2019.00193
Gominak SC (2016) Vitamin D deficiency changes the intestinal microbiome reducing B vitamin production in the gut. The resulting lack of pantothenic acid adversely affects the immune system, producing a "pro-inflammatory" state associated with atherosclerosis and autoimmunity. Med Hypotheses 94:103–107. https://doi.org/10.1016/j.mehy.2016.07.007
Gopalakrishnan V, Spencer CN, Nezi L, Reuben A, Andrews MC, Karpinets TV et al (2018) Gut microbiome modulates response to anti-PD-1 immunotherapy in melanoma patients. Science 359(6371):97–103. https://doi.org/10.1126/science.aan4236
Gregory JC, Buffa JA, Org E, Wang Z, Levison BS, Zhu W et al (2015) Transmission of atherosclerosis susceptibility with gut microbial transplantation. J Biol Chem 290(9):5647–5660. https://doi.org/10.1074/jbc.M114.618249
Grice EA, Kong HH, Conlan S, Deming CB, Davis J, Young AC et al (2009) Topographical and temporal diversity of the human skin microbiome. Science 324(5931):1190–1192. https://doi.org/10.1126/science.1171700
Grosicki GJ, Fielding RA, Lustgarten MS (2018) Gut microbiota contribute to age-related changes in skeletal muscle size, composition, and function: biological basis for a gut-muscle Axis. Calcif Tissue Int 102(4):433–442. https://doi.org/10.1007/s00223-017-0345-5

Hansen R, Scott KP, Khan S, Martin JC, Berry SH, Stevenson M et al (2015) First-pass meconium samples from healthy term vaginally-delivered neonates: an analysis of the microbiota. PLoS One 10(7):e0133320. https://doi.org/10.1371/journal.pone.0133320

Hickey RJ, Zhou X, Pierson JD, Ravel J, Forney LJ (2012) Understanding vaginal microbiome complexity from an ecological perspective. Transl Res 160(4):267–282. https://doi.org/10.1016/j.trsl.2012.02.008

Hsiao EY, McBride SW, Hsien S, Sharon G, Hyde ER, McCue T et al (2013) Microbiota modulate behavioral and physiological abnormalities associated with neurodevelopmental disorders. Cell 155(7):1451–1463. https://doi.org/10.1016/j.cell.2013.11.024

Hu X, Wang T, Jin F (2016) Alzheimer's disease and gut microbiota. Sci China Life Sci 59(10):1006–1023. https://doi.org/10.1007/s11427-016-5083-9

Huda MN, Lewis Z, Kalanetra KM, Rashid M, Ahmad SM, Raqib R et al (2014) Stool microbiota and vaccine responses of infants. Pediatrics 134(2):e362–e372. https://doi.org/10.1542/peds.2013-3937

Hummelen R, Macklaim JM, Bisanz JE, Hammond JA, McMillan A, Vongsa R et al (2011) Vaginal microbiome and epithelial gene array in post-menopausal women with moderate to severe dryness. PLoS One 6(11):e26602. https://doi.org/10.1371/journal.pone.0026602

Jackson MA, Jeffery IB, Beaumont M, Bell JT, Clark AG, Ley RE et al (2016) Erratum to: signatures of early frailty in the gut microbiota. Genome Med 8(1):21. https://doi.org/10.1186/s13073-016-0275-2

Jacobsen CN, Rosenfeldt Nielsen V, Hayford AE, Moller PL, Michaelsen KF, Paerregaard A et al (1999) Screening of probiotic activities of forty-seven strains of lactobacillus spp. by in vitro techniques and evaluation of the colonization ability of five selected strains in humans. Appl Environ Microbiol 65(11):4949–4956. https://doi.org/10.1128/AEM.65.11.4949-4956.1999

Jakobsson HE, Jernberg C, Andersson AF, Sjolund-Karlsson M, Jansson JK, Engstrand L (2010) Short-term antibiotic treatment has differing long-term impacts on the human throat and gut microbiome. PLoS One 5(3):e9836. https://doi.org/10.1371/journal.pone.0009836

Jie Z, Xia H, Zhong SL, Feng Q, Li S, Liang S et al (2017) The gut microbiome in atherosclerotic cardiovascular disease. Nat Commun 8(1):845. https://doi.org/10.1038/s41467-017-00900-1

Karlsson FH, Fak F, Nookaew I, Tremaroli V, Fagerberg B, Petranovic D et al (2012) Symptomatic atherosclerosis is associated with an altered gut metagenome. Nat Commun 3:1245. https://doi.org/10.1038/ncomms2266

Klassert TE, Zubiria-Barrera C, Kankel S, Stock M, Neubert R, Lorenzo-Diaz F et al (2020) Early bacterial colonization and antibiotic resistance Gene Acquisition in newborns. Front Cell Infect Microbiol 10:332. https://doi.org/10.3389/fcimb.2020.00332

Koenig JE, Spor A, Scalfone N, Fricker AD, Stombaugh J, Knight R et al (2011) Succession of microbial consortia in the developing infant gut microbiome. Proc Natl Acad Sci U S A 108(Suppl 1):4578–4585. https://doi.org/10.1073/pnas.1000081107

Komaroff AL (2017) The microbiome and risk for obesity and diabetes. JAMA 317(4):355–356. https://doi.org/10.1001/jama.2016.20099

Kong F, Hua Y, Zeng B, Ning R, Li Y, Zhao J (2016) Gut microbiota signatures of longevity. Curr Biol 26(18):R832–R833. https://doi.org/10.1016/j.cub.2016.08.015

Korem T, Zeevi D, Zmora N, Weissbrod O, Bar N, Lotan-Pompan M et al (2017) Bread affects clinical parameters and induces gut microbiome-associated personal glycemic responses. Cell Metab 25(6):1243–1253 e5. https://doi.org/10.1016/j.cmet.2017.05.002

Korpela K, Salonen A, Vepsalainen O, Suomalainen M, Kolmeder C, Varjosalo M et al (2018) Probiotic supplementation restores normal microbiota composition and function in antibiotic-treated and in caesarean-born infants. Microbiome 6(1):182. https://doi.org/10.1186/s40168-018-0567-4

Kranich J, Maslowski KM, Mackay CR (2011) Commensal flora and the regulation of inflammatory and autoimmune responses. Semin Immunol 23(2):139–145. https://doi.org/10.1016/j.smim.2011.01.011

Krishnan S, Ding Y, Saedi N, Choi M, Sridharan GV, Sherr DH et al (2018) Gut microbiota-derived tryptophan metabolites modulate inflammatory response in hepatocytes and macrophages. Cell Rep 23(4):1099–1111. https://doi.org/10.1016/j.celrep.2018.03.109

Lang JM, Pan C, Cantor RM, Tang WHW, Garcia-Garcia JC, Kurtz I et al (2018) Impact of individual traits, saturated fat, and protein source on the gut microbiome. MBio 9(6). https://doi.org/10.1128/mBio.01604-18

Langdon A, Crook N, Dantas G (2016) The effects of antibiotics on the microbiome throughout development and alternative approaches for therapeutic modulation. Genome Med 8(1):39. https://doi.org/10.1186/s13073-016-0294-z

Laursen MF, Andersen LB, Michaelsen KF, Molgaard C, Trolle E, Bahl MI et al (2016) Infant gut microbiota development is driven by transition to family foods independent of maternal obesity. mSphere 1(1). https://doi.org/10.1128/mSphere.00069-15

Li H, Qi Y, Jasper H (2016) Preventing age-related decline of gut compartmentalization limits microbiota Dysbiosis and extends lifespan. Cell Host Microbe 19(2):240–253. https://doi.org/10.1016/j.chom.2016.01.008

Li SW, Watanabe K, Hsu CC, Chao SH, Yang ZH, Lin YJ et al (2017) Bacterial composition and diversity in breast Milk samples from mothers living in Taiwan and mainland China. Front Microbiol 8:965. https://doi.org/10.3389/fmicb.2017.00965

Lozupone CA, Stombaugh JI, Gordon JI, Jansson JK, Knight R (2012) Diversity, stability and resilience of the human gut microbiota. Nature 489(7415):220–230. https://doi.org/10.1038/nature11550

Mabbott NA (2015) A breakdown in communication? Understanding the effects of aging on the human small intestine epithelium. Clin Sci (Lond) 129(7):529–531. https://doi.org/10.1042/CS20150364

Makki K, Deehan EC, Walter J, Backhed F (2018) The impact of dietary fiber on gut microbiota in host health and disease. Cell Host Microbe 23(6):705–715. https://doi.org/10.1016/j.chom.2018.05.012

Marcobal A, Barboza M, Froehlich JW, Block DE, German JB, Lebrilla CB et al (2010) Consumption of human milk oligosaccharides by gut-related microbes. J Agric Food Chem 58(9):5334–5340. https://doi.org/10.1021/jf9044205

Martin R, Makino H, Cetinyurek Yavuz A, Ben-Amor K, Roelofs M, Ishikawa E et al (2016) Early-life events, including mode of delivery and type of feeding, siblings and gender, shape the developing gut microbiota. PLoS One 11(6):e0158498. https://doi.org/10.1371/journal.pone.0158498

Negi S, Pahari S, Bashir H, Agrewala JN (2019) Gut microbiota regulates Mincle mediated activation of lung dendritic cells to protect against mycobacterium tuberculosis. Front Immunol 10:1142. https://doi.org/10.3389/fimmu.2019.01142

O'Keefe SJ, Li JV, Lahti L, Ou J, Carbonero F, Mohammed K et al (2015) Fat, fibre and cancer risk in African Americans and rural Africans. Nat Commun 6:6342. https://doi.org/10.1038/ncomms7342

O'Keefe SJ, Ou J, Aufreiter S, O'Connor D, Sharma S, Sepulveda J et al (2009) Products of the colonic microbiota mediate the effects of diet on colon cancer risk. J Nutr 139(11):2044–2048. https://doi.org/10.3945/jn.109.104380

O'Toole PW, Jeffery IB (2015) Gut microbiota and aging. Science 350(6265):1214–1215. https://doi.org/10.1126/science.aac8469

Obermajer T, Lipoglavsek L, Tompa G, Treven P, Lorbeg PM, Matijasic BB et al (2014) Colostrum of healthy Slovenian mothers: microbiota composition and bacteriocin gene prevalence. PLoS One 10(4):e0123324. https://doi.org/10.1371/journal.pone.0123324

Ogawa T, Hirose Y, Honda-Ogawa M, Sugimoto M, Sasaki S, Kibi M et al (2018) Composition of salivary microbiota in elderly subjects. Sci Rep 8(1):414. https://doi.org/10.1038/s41598-017-18677-0

Ott SJ, El Mokhtari NE, Musfeldt M, Hellmig S, Freitag S, Rehman A et al (2006) Detection of diverse bacterial signatures in atherosclerotic lesions of patients with coronary heart disease. Circulation 113(7):929–937. https://doi.org/10.1161/CIRCULATIONAHA.105.579979

Pasini E, Aquilani R, Testa C, Baiardi P, Angioletti S, Boschi F et al (2016) Pathogenic gut flora in patients with chronic heart failure. JACC Heart Fail 4(3):220–227. https://doi.org/10.1016/j.jchf.2015.10.009

Penders J, Thijs C, Vink C, Stelma FF, Snijders B, Kummeling I et al (2006) Factors influencing the composition of the intestinal microbiota in early infancy. Pediatrics 118(2):511–521. https://doi.org/10.1542/peds.2005-2824

Pendyala S, Walker JM, Holt PR (2012) A high-fat diet is associated with endotoxemia that originates from the gut. Gastroenterology 142(5):1100–1101 e2. https://doi.org/10.1053/j.gastro.2012.01.034

Proctor C, Thiennimitr P, Chattipakorn N, Chattipakorn SC (2017) Diet, gut microbiota and cognition. Metab Brain Dis 32(1):1–17. https://doi.org/10.1007/s11011-016-9917-8

Qin J, Li R, Raes J, Arumugam M, Burgdorf KS, Manichanh C et al (2010) A human gut microbial gene catalogue established by metagenomic sequencing. Nature 464(7285):59–65. https://doi.org/10.1038/nature08821

Rampelli S, Candela M, Turroni S, Biagi E, Collino S, Franceschi C et al (2013) Functional metagenomic profiling of intestinal microbiome in extreme ageing. Aging (Albany NY) 5(12):902–912. https://doi.org/10.18632/aging.100623

Ravel J, Gajer P, Abdo Z, Schneider GM, Koenig SS, McCulle SL et al (2011) Vaginal microbiome of reproductive-age women. Proc Natl Acad Sci U S A 108(Suppl 1):4680–4687. https://doi.org/10.1073/pnas.1002611107

Rea MC, O'Sullivan O, Shanahan F, O'Toole PW, Stanton C, Ross RP et al (2012) Clostridium difficile carriage in elderly subjects and associated changes in the intestinal microbiota. J Clin Microbiol 50(3):867–875. https://doi.org/10.1128/JCM.05176-11

Rodriguez JM, Murphy K, Stanton C, Ross RP, Kober OI, Juge N et al (2015) The composition of the gut microbiota throughout life, with an emphasis on early life. Microb Ecol Health Dis 26:26050. https://doi.org/10.3402/mehd.v26.26050

Rothschild D, Weissbrod O, Barkan E, Kurilshikov A, Korem T, Zeevi D et al (2018) Environment dominates over host genetics in shaping human gut microbiota. Nature 555(7695):210–215. https://doi.org/10.1038/nature25973

Salonen A, Lahti L, Salojarvi J, Holtrop G, Korpela K, Duncan SH et al (2014) Impact of diet and individual variation on intestinal microbiota composition and fermentation products in obese men. ISME J 8(11):2218–2230. https://doi.org/10.1038/ismej.2014.63

Sampaio-Maia B, Monteiro Silva F (2014) Acquisition and maturation of oral microbiome throughout childhood: an update. Dent Res J (Isfahan) 11(3):291–301

Sampson TR, Debelius JW, Thron T, Janssen S, Shastri GG, Ilhan ZE et al (2016) Gut microbiota regulate motor deficits and Neuroinflammation in a model of Parkinson's disease. Cell 167(6):1469–1480 e12. https://doi.org/10.1016/j.cell.2016.11.018

SanMiguel AJ, Meisel JS, Horwinski J, Zheng Q, Bradley CW, Grice EA (2018) Antiseptic agents elicit short-term, personalized, and body site-specific shifts in resident skin bacterial communities. J Invest Dermatol 138(10):2234–2243. https://doi.org/10.1016/j.jid.2018.04.022

Santoro A, Ostan R, Candela M, Biagi E, Brigidi P, Capri M et al (2018) Gut microbiota changes in the extreme decades of human life: a focus on centenarians. Cell Mol Life Sci 75(1):129–148. https://doi.org/10.1007/s00018-017-2674-y

Sanz Y, Romani-Perez M, Benitez-Paez A, Portune KJ, Brigidi P, Rampelli S et al (2018) Towards microbiome-informed dietary recommendations for promoting metabolic and mental health: opinion papers of the MyNewGut project. Clin Nutr 37(6 Pt A):2191–2197. https://doi.org/10.1016/j.clnu.2018.07.007

Sarkar D, Fisher PB (2006) Molecular mechanisms of aging-associated inflammation. Cancer Lett 236(1):13–23. https://doi.org/10.1016/j.canlet.2005.04.009

Satokari R, Gronroos T, Laitinen K, Salminen S, Isolauri E (2009) Bifidobacterium and lactobacillus DNA in the human placenta. Lett Appl Microbiol 48(1):8–12. https://doi.org/10.1111/j.1472-765X.2008.02475.x

Scher JU, Sczesnak A, Longman RS, Segata N, Ubeda C, Bielski C et al (2013) Expansion of intestinal Prevotella copri correlates with enhanced susceptibility to arthritis. elife 2:e01202. https://doi.org/10.7554/eLife.01202

Schnorr SL, Candela M, Rampelli S, Centanni M, Consolandi C, Basaglia G et al (2014) Gut microbiome of the Hadza hunter-gatherers. Nat Commun 5:3654. https://doi.org/10.1038/ncomms4654

Sekirov I, Tam NM, Jogova M, Robertson ML, Li Y, Lupp C et al (2008) Antibiotic-induced perturbations of the intestinal microbiota alter host susceptibility to enteric infection. Infect Immun 76(10):4726–4736. https://doi.org/10.1128/IAI.00319-08

Sela DA, Mills DA (2014) The marriage of nutrigenomics with the microbiome: the case of infant-associated Bifidobacteria and milk. Am J Clin Nutr 99(3):697S–703S. https://doi.org/10.3945/ajcn.113.071795

Sender R, Fuchs S, Milo R (2016) Revised estimates for the number of human and bacteria cells in the body. PLoS Biol 14(8):e1002533. https://doi.org/10.1371/journal.pbio.1002533

Shibagaki N, Suda W, Clavaud C, Bastien P, Takayasu L, Iioka E et al (2017) Aging-related changes in the diversity of women's skin microbiomes associated with oral bacteria. Sci Rep 7(1):10567. https://doi.org/10.1038/s41598-017-10834-9

Singh H, Torralba MG, Moncera KJ, DiLello L, Petrini J, Nelson KE et al (2019) Gastro-intestinal and oral microbiome signatures associated with healthy aging. Geroscience 41(6):907–921. https://doi.org/10.1007/s11357-019-00098-8

Singh RP, Bashir H, Kumar R (2021) Emerging role of microbiota in immunomodulation and cancer immunotherapy. Semin Cancer Biol 70:37–52. https://doi.org/10.1016/j.semcancer.2020.06.008

Smith P, Willemsen D, Popkes M, Metge F, Gandiwa E, Reichard M et al (2017) Regulation of life span by the gut microbiota in the short-lived African turquoise killifish. elife 6. https://doi.org/10.7554/eLife.27014

Solis G, de Los Reyes-Gavilan CG, Fernandez N, Margolles A, Gueimonde M (2010) Establishment and development of lactic acid bacteria and bifidobacteria microbiota in breast-milk and the infant gut. Anaerobe 16(3):307–310. https://doi.org/10.1016/j.anaerobe.2010.02.004

Sommer F, Backhed F (2013) The gut microbiota–masters of host development and physiology. Nat Rev Microbiol 11(4):227–238. https://doi.org/10.1038/nrmicro2974

Stewart CJ, Ajami NJ, O'Brien JL, Hutchinson DS, Smith DP, Wong MC et al (2018) Temporal development of the gut microbiome in early childhood from the TEDDY study. Nature 562(7728):583–588. https://doi.org/10.1038/s41586-018-0617-x

Swiatecka D, Narbad A, Ridgway KP, Kostyra H (2011) The study on the impact of glycated pea proteins on human intestinal bacteria. Int J Food Microbiol 145(1):267–272. https://doi.org/10.1016/j.ijfoodmicro.2011.01.002

Tanaka S, Kobayashi T, Songjinda P, Tateyama A, Tsubouchi M, Kiyohara C et al (2009) Influence of antibiotic exposure in the early postnatal period on the development of intestinal microbiota. FEMS Immunol Med Microbiol 56(1):80–87. https://doi.org/10.1111/j.1574-695X.2009.00553.x

Tang WH, Wang Z, Levison BS, Koeth RA, Britt EB, Fu X et al (2013) Intestinal microbial metabolism of phosphatidylcholine and cardiovascular risk. N Engl J Med 368(17):1575–1584. https://doi.org/10.1056/NEJMoa1109400

Thevaranjan N, Puchta A, Schulz C, Naidoo A, Szamosi JC, Verschoor CP et al (2017) Age-associated microbial Dysbiosis promotes intestinal permeability, systemic inflammation, and macrophage dysfunction. Cell Host Microbe 21(4):455–466 e4. https://doi.org/10.1016/j.chom.2017.03.002

Thomas RL, Jiang L, Adams JS, Xu ZZ, Shen J, Janssen S et al (2020) Vitamin D metabolites and the gut microbiome in older men. Nat Commun 11(1):5997. https://doi.org/10.1038/s41467-020-19793-8

Ticinesi A, Lauretani F, Milani C, Nouvenne A, Tana C, Del Rio D et al (2017) Aging gut microbiota at the cross-road between nutrition, physical frailty, and sarcopenia: is there a gut-muscle Axis? Nutrients 9(12). https://doi.org/10.3390/nu9121303

Toscano M, De Grandi R, Peroni DG, Grossi E, Facchin V, Comberiati P et al (2017) Impact of delivery mode on the colostrum microbiota composition. BMC Microbiol 17(1):205. https://doi.org/10.1186/s12866-017-1109-0

Two AM, Nakatsuji T, Kotol PF, Arvanitidou E, Du-Thumm L, Hata TR et al (2016) The cutaneous microbiome and aspects of skin antimicrobial defense system resist acute treatment with topical skin cleansers. J Invest Dermatol 136(10):1950–1954. https://doi.org/10.1016/j.jid.2016.06.612

Van den Abbeele P, Belzer C, Goossens M, Kleerebezem M, De Vos WM, Thas O et al (2013) Butyrate-producing clostridium cluster XIVa species specifically colonize mucins in an in vitro gut model. ISME J 7(5):949–961. https://doi.org/10.1038/ismej.2012.158

Van Nimwegen FA, Penders J, Stobberingh EE, Postma DS, Koppelman GH, Kerkhof M et al (2011) Mode and place of delivery, gastrointestinal microbiota, and their influence on asthma and atopy. J Allergy Clin Immunol 128(5):948-55 e1-3. https://doi.org/10.1016/j.jaci.2011.07.027

Vieira AT, Macia L, Galvao I, Martins FS, Canesso MC, Amaral FA et al (2015) A role for gut microbiota and the metabolite-sensing receptor GPR43 in a murine model of gout. Arthritis Rheumatol 67(6):1646–1656. https://doi.org/10.1002/art.39107

Walker AW, Ince J, Duncan SH, Webster LM, Holtrop G, Ze X et al (2011) Dominant and diet-responsive groups of bacteria within the human colonic microbiota. ISME J 5(2):220–230. https://doi.org/10.1038/ismej.2010.118

Wan Y, Wang F, Yuan J, Li J, Jiang D, Zhang J et al (2019) Effects of dietary fat on gut microbiota and faecal metabolites, and their relationship with cardiometabolic risk factors: a 6-month randomised controlled-feeding trial. Gut 68(8):1417–1429. https://doi.org/10.1136/gutjnl-2018-317609

Wang M, Li M, Wu S, Lebrilla CB, Chapkin RS, Ivanov I et al (2015) Fecal microbiota composition of breast-fed infants is correlated with human milk oligosaccharides consumed. J Pediatr Gastroenterol Nutr 60(6):825–833. https://doi.org/10.1097/MPG.0000000000000752

Wang Z, Neupane A, Vo R, White J, Wang X, Marzano SL (2020) Comparing gut microbiome in Mothers' own breast Milk- and formula-fed moderate-late preterm infants. Front Microbiol 11:891. https://doi.org/10.3389/fmicb.2020.00891

Watson H, Mitra S, Croden FC, Taylor M, Wood HM, Perry SL et al (2018) A randomised trial of the effect of omega-3 polyunsaturated fatty acid supplements on the human intestinal microbiota. Gut 67(11):1974–1983. https://doi.org/10.1136/gutjnl-2017-314968

White MC, Holman DM, Boehm JE, Peipins LA, Grossman M, Henley SJ (2014) Age and cancer risk: a potentially modifiable relationship. Am J Prev Med 46(3 Suppl 1):S7–S15. https://doi.org/10.1016/j.amepre.2013.10.029

Wolters M, Ahrens J, Romani-Perez M, Watkins C, Sanz Y, Benitez-Paez A et al (2019) Dietary fat, the gut microbiota, and metabolic health–a systematic review conducted within the MyNewGut project. Clin Nutr 38(6):2504–2520. https://doi.org/10.1016/j.clnu.2018.12.024

Wu CY, Hu HY, Chou YJ, Huang N, Chou YC, Li CP (2015) High blood pressure and all-cause and cardiovascular disease mortalities in community-dwelling older adults. Medicine (Baltimore) 94(47):e2160. https://doi.org/10.1097/MD.0000000000002160

Wu L, Zeng T, Deligios M, Milanesi L, Langille MGI, Zinellu A et al (2020) Age-related variation of bacterial and fungal communities in different body habitats across the Young, elderly, and centenarians in Sardinia. mSphere 5(1). https://doi.org/10.1128/mSphere.00558-19

Xu C, Zhu H, Qiu P (2019) Aging progression of human gut microbiota. BMC Microbiol 19(1):236. https://doi.org/10.1186/s12866-019-1616-2

Xu L, Zhang C, He D, Jiang N, Bai Y, Xin Y (2020) Rapamycin and MCC950 modified gut microbiota in experimental autoimmune encephalomyelitis mouse by brain gut axis. Life Sci 253:117747. https://doi.org/10.1016/j.lfs.2020.117747

Yamamoto EA, Jorgensen TN (2019) Relationships between vitamin D, gut microbiome, and systemic autoimmunity. Front Immunol 10:3141. https://doi.org/10.3389/fimmu.2019.03141

Yu JJ, Manus MB, Mueller O, Windsor SC, Horvath JE, Nunn CL (2018) Antibacterial soap use impacts skin microbial communities in rural Madagascar. PLoS One 13(8):e0199899. https://doi.org/10.1371/journal.pone.0199899

Zaura E, Brandt BW, Teixeira de Mattos MJ, Buijs MJ, Caspers MP, Rashid MU et al (2015) Same exposure but Two radically different responses to antibiotics: resilience of the salivary microbiome versus long-term microbial shifts in feces. mBio 6(6):e01693–e01615. https://doi.org/10.1128/mBio.01693-15

Zhang D, Chen G, Manwani D, Mortha A, Xu C, Faith JJ et al (2015) Neutrophil ageing is regulated by the microbiome. Nature 525(7570):528–532. https://doi.org/10.1038/nature15367

Zhou P, Zhou Y, Liu B, Jin Z, Zhuang X, Dai W et al (2020) Perinatal antibiotic exposure affects the transmission between maternal and neonatal microbiota and is associated with early-onset sepsis. mSphere 5(1). https://doi.org/10.1128/mSphere.00984-19

Zhu Y, Lin X, Zhao F, Shi X, Li H, Li Y et al (2015) Meat, dairy and plant proteins alter bacterial composition of rat gut bacteria. Sci Rep 5:15220. https://doi.org/10.1038/srep15220

Zmora N, Suez J, Elinav E (2019) You are what you eat: diet, health and the gut microbiota. Nat Rev Gastroenterol Hepatol 16(1):35–56. https://doi.org/10.1038/s41575-018-0061-2

Chapter 2
Oral Microbiome: An Opening to Healthy Possibilities

Sahana Vasudevan, Prasanna Neelakantan, and Adline Princy Solomon

1 Introduction

The association of the human and microbiome hosted by humans is referred to as "supraorganism" where the core and variable microbiome define the fine balance between human health and disease states (Turnbaugh et al. 2007; Sonnenburg and Fischbach 2011). However, each microhabitat maintains an inter/intra species and cross-kingdom-level interaction with their co-partners in an ecosystem having varied environmental and nutritional compositions (Sonnenburg and Fischbach 2011). Interestingly, the diverse oral microbiome is observed as an ecologically balanced ecosystem as individual members practice commensalism with their co-habitants to allow each one of them to flourish in the same ecosystem at no expense, as it is crucial for human health (Ruby and Goldner 2007; Filoche et al. 2010). Most importantly, dysbiosis in the oral microbiome may disturb the ecosystem as it is a primary gateway for oral inhabitants to spread to non-oral sites of the human body and cause diseases (Dewhirst et al. 2010). This chapter offers a glimpse into the biogeography of the oral microbiota that predominate healthy oral cavities and the various factors that cause ecological shifts in the oral microbiome to cause systemic disorders.

S. Vasudevan · A. P. Solomon (✉)
Quorum Sensing Laboratory, Centre for Research in Infectious Diseases (CRID), School of Chemical and Biotechnology, SASTRA Deemed to be University, Thanjavur, India
e-mail: adlineprinzy@biotech.sastra.edu

P. Neelakantan
Faculty of Dentistry, The University of Hong Kong, Pok Fu Lam, Hong Kong
e-mail: prasanna@hku.hk

S. Thomas (ed.), *Human Microbiome*,
https://doi.org/10.1007/978-981-16-7672-7_2

2 Oral Microbiome: The Site Specialist in Natural Niches

Biogeography aims to explain the spatial and temporal distribution and patterns of diversity in a certain region to understand variation and the important compositional microbes present. Knowing the biogeography of the buccal cavity is necessary to understand the transition of a microbiome from a healthy state to a diseased state. Biogeography of a niche is characterized by culture-independent methods of sequencing such as whole genome sequencing or 16S rRNA amplicon (16S) sequencing. The oral cavity contains different niches, which are the tongue, teeth, palate, gingiva (supra- and sub-gingival), periodontal pockets, buccal mucosa, and saliva (Proctor et al. 2020). The microbiome differs from niche to niche as it depends on factors including moisture and oxygen content, salivary flow, temperature, and pH. Other factors that affect microbiome composition are metabolites and molecules of inhibitory activity, which can lead to inter-species co-existence and co-exclusion (Wilbert et al. 2020). It has been shown that the oral microbiome, much like other microbiomes, is dominated by one-to-many genera while also containing multiple unidentified microorganisms (Welch et al. 2019). The different niches have varied features and provide unique habitable areas for microbes. For instance, the chewing surfaces of teeth provide a permanent surface for adhesion thanks to the numerous grooves and fossae, while the pockets and crevices between teeth and gums provide a nutrient-rich region. On the other hand, the palate and the gingiva provide a sturdier surface while the mucosa and saliva provide a fluid, flexible surface on the cheeks, tongue, and the further regions of the oral cavity, and the tongue has a rough surface due to the papillae (Welch et al. 2014). The oral environment has low oxygen levels leading to the development of predominantly anaerobic microbiota (Bernardi 2019). Besides these physical features, chemical features also vary across these regions. We will go about each region individually.

The teeth provide multiple surfaces for growth as each tooth has five surfaces: occlusal surface or the biting/chewing surface, the proximal surfaces (mesial and distal), the buccal surface or the surface facing the cheek, and the lingual/palatal surface or the surface facing the tongue/palate of the mouth. However, it is difficult to estimate the composition of the community at such a small level. Before microbes bind to the surface of the tooth, a pellicle is formed by the adhesive proteins secreted by the microbes. The microbial community attaches to the solid surface when the planktonic cells embed in an exopolysaccharide matrix. The initial colonizers belong to the *Actinomyces* sp., *Streptococcus* sp., *Lactobacillus* sp., and *Candida* sp. Variations in this colonizing community will affect the formation of the biofilms over the basal layer of the supragingival biome (Liu et al. 2012; Hannig et al. 2017). Studies focus on the supragingival plaque and the subgingival plaque as the two regions depend on the tooth and have varied microbiota due to the location. Gram-positive anaerobes dominate the supragingival plaque while Gram-negative microbes dominate the subgingival plaque. The plaque microbiota is strongly organized into many multi-genus consortia. The genus *Corynebacterium* has been found to be specific to supragingival and subgingival plaque alone and is thus a biomarker for dental plaque (Welch et al. 2019). *Corynebacterium* spp. have been found to have specific interactions with cocci. Other species that are specific to this region belong to the *Capnocytophaga* and *Lautropia* genera (Welch et al. 2014, 2016).

In a study done to summarize the subgingival microbial architecture, it was shown to have four layers: outside layer, top layer, intermediate layer, and basal layer. The outside layer had no characteristic organization but was primarily inhabited by the Sphirochaetes and other bacterial aggregates. Initial colonization is achieved at the basal layer by *Streptococcus* sp. and *Actinomyces* sp. With time, the population density proportion between the *Streptococcus* sp. and *Fusobacterium nucleatum* is inverted. The top and intermediate layers contain bacteria under the *Cytophaga–Flavobacterium–Bacteroides* group (CFB-cluster) and the *Synergistetes* group. The intermediate layer contains spindle-shaped bacteria, including *Tannerella forsythia* and other species belonging to the same genus, and rod-shaped bacteria, such as *Prevotella* sp. and the *Bacteroidetes* sp. (Zijnge et al. 2010). The supragingival microbiota has two general layers: the basal layer and the layer above it. The basal layer plays its role as a foundation for the four different kinds of biofilms that form over it. These types of biofilms have compositions as follows: type one has *Actinomyces* sp., type two has some cocci in addition to *Actinomyces* sp., type three has *Streptococcus* sp. and yeast that form unique colonies with the yeast at the center of streptococcal colonies, and type four has coexistence of *Lactobacillus* sp. and *Streptococcus* sp. The layer above this basal biofilm layer comprises *Streptococcus* sp. varyingly. There are also scattered colonies of *Lactobacillus* sp. and bacterial cells belonging to the CFB-cluster. Corncobs can also be found as the *Streptococcus* sp. adhere to a *Candida* cell or hyphae, and test-tube brushes are observed when filaments from the CFB-cluster and *T. forsythia* and *F. nucleatum* adhere perpendicularly to the Lactobacilli in the supragingival plaque (Zijnge et al. 2010). The corncob structures (single or double layered) are commonly found and indicate commensalistic, mutualistic, and competitive interspecies interactions. Some single-layered corncobs comprised either *Porphyromonas* or *Streptococcus*; double-layered corncobs had an inner layer of *Streptococcus* and an outer layer of *Aggregatibacter/Haemophilus* (Welch et al. 2016). In a situation where the environmental conditions of the subgingival and the supragingival regions are the same, the microbial composition does not vary due to the absence of different dietary factors, oxygen concentration, saliva, and gingival crevicular fluids (Zijnge et al. 2010; Welch et al. 2019).

The dorsum of the tongue has a unique microbial community too. It is in direct contact with the mucosa and is in indirect contact with all other parts of the mouth through saliva. Although the regions in closest proximity and in constant contact are the teeth, the species that exist predominantly in the mucosal regions are not found on the teeth. There are around 17 genera with more than 100 species in total that prevail on the tongue. *Actinomyces* spp. are found on the tongue dorsum as well, but the two species found in abundance here are not the same as the two species found in abundance in the dental region. *Streptococcus mitis* are found in abundance in the mucosal and nonmucosal surfaces (Wilbert et al. 2020). The surface of the tongue is lined with mucosal papillae that provide niches on their hairy surfaces and in the crevices. The varied geography of the tongue leads to the formation of two layers of biofilms with the lower, thinner layer being anaerobic as the top layer of biofilm is thick. Mechanical stress does not alter the anaerobicity of the lower layer as the papillae protect the layers from being disengaged (Bernardi 2019). The most prevalent species belonged to the genera *Streptococcus*, *Veillonella*, *Prevotella*, *Neisseria*, *Actinomyces*, *Fusobacterium*, *Haemophilus*, and *Rothia*. *Veillonella* spp. are good biomarkers for the tongue dorsum microbiota (Welch et al. 2019). The three genera that are present in all individuals and in all the consortiums are *Actinomyces*, *Rothia*, and *Streptococcus*. *Actinomyces* spp., *A. odontolyticus* and *A. graevenitzii*, occupy a region closer to the core of the tongue, while *Streptococcus* spp., *S. mitis*, *S. salivarius*, and *S. parasanguinis*, inhabited the outlying regions, as a crust, and as stripes. *Rothia mucilaginosa* formed large patches with interspersed cultures of other taxa studied through spatial patterning (Sato et al. 2015; Wilbert et al. 2020).

The salivary microbiome is derived from different regions of the buccal cavity and can be considered as a broad representative of the overall general microbiota. The composition of the salivary microbiome is influenced by oral health conditions more than other niches. In an orally healthy person, the microbiome can contain various species prominently across different geographic locations in the world due to external socio-environmental or socio-economic factors. *Streptococcus*, *Neisseria*, *Rothia*, *Prevotella*, *Actinomyces*, *Granulicatella*, *Porphromonas*, *Haemophilus*, and *Porphyromonas* are the prevalent species present. These species are not always found together but instead exist as cohabiting groups with slight variations on a case-by-case basis. The species in the different clusters are *Prevotella histicola*, *Prevotella melaninogenica*, *Veillonella parvula*, *Veillonella atypica*, *Streptococcus salivarius*, and *Streptococcus parasanguinis*; and *Neisseria flavescens*, *Haemophilus parainfluenzae*, *Porphyromonas pasteri*, *Gemella sanguinis*, and *Granulicatella adiacens*. The salivary microbiome is a proven indicator of oral and systemic health (Yamashita and Takeshita 2017). Good biomarkers for the salivary microbiota belong to the *Oribacterium* spp. (Welch et al. 2020). The oral microbiome is ever-changing, and at any point in time, it will have a unique microbial community/fingerprint. Microbes can also disperse through active fluid flow or by chemotaxis. However, such variations do not deviate greatly. The dynamic community stays within a large cloud of possible microbiome populations.

With more than 700 microbial species, it is imperative to understand the composition of the human oral microbiome to understand the disease dynamics. The expanded Human Oral Microbiome Database (eHOMD) has comprehensively curated the bacterial species present in the human aerodigestive tract, which details the site-specific bacterial composition in the oral cavity (http://www.homd.org/). This database is part of the Human Microbiome Project (HMP) and aims to provide complete information on taxonomy, bibliography, and genomics. Several tools are available for data mining and related analyses. The information about the same organism in different niches is given a unique HOT (Human Oral Taxon) number. This is particularly useful as it interlinks the information of bibliographic, genomic, clinical, phylogenetic, and phenotype of each taxon. This freely available online resource can be used to understand the oral microbiota (Chen et al. 2010).

3 Oral Microbiome: Our Coevolved Holobiont

The composition of the microbial community is influenced by its dynamic interaction with the host. Host factors influence the diversity of the microbiome and intricate balance among different species that promotes symbiosis or dysbiosis of oral health. This bidirectional interaction is governed by several intrinsic and extrinsic components (Fig. 2.1) that contribute toward maintaining oral homeostasis (Cornejo et al. 2019). Studies have shown that the host and the oral microbiome have co-evolved together with well-established biochemical and immune networks forming a "holobiont" (Bordenstein and Theis 2015; Youle et al. 2013). Evidence that the presence of the similar microbial composition in placenta and in the oral cavity of the mother strongly supports the commencement of shaping of this co-evolution even before birth (Aagaard et al. 2014). The initial acquisition of the oral microbiota is a dynamic process that is strongly influenced by birth, diet, other exposures, and horizontal transmissions from parents and other peers (Xiao et al. 2020). Gradually, the microbial composition is developed and established, which forms the strong basis for the innate and acquired immune response and overall physiological health (Idris et al. 2017). Such a bidirectional interaction between the host and the oral microbiome is of primary importance in maintaining the overall physiology and health.

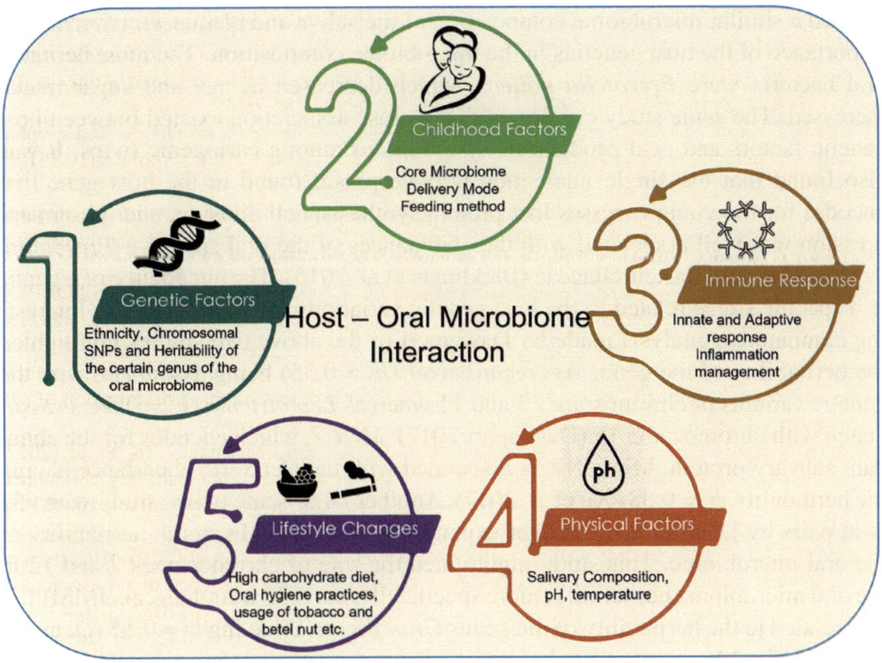

Fig. 2.1 Host–oral microbiome interactions: the intrinsic and extrinsic factors that contribute to the development, establishment, and maintenance of the oral microbiome

Such factors drive the overall balance and imbalance of the oral homeostasis and promote the stable development and establishment of the stable core oral microbiome from infant to early childhood.

The host factors such as genetics, pregnancy period, method of delivery, feeding, and maternal transmission from vaginal and gut influence the oral microbiome transition from infant to early childhood (Xiao et al. 2020). As the core microbiome is established, factors including circadian rhythm, immune response, saliva and salivary proteins, diet, lifestyle, and oral hygiene play an important role (Cornejo et al. 2019).

3.1 Genetic Factors

There is a close association of genetics with all other host factors that contribute to the development and maintenance of the oral microbiome (Blekhman et al. 2015). The salivary components (Lips et al. 2017), diet habits related to taste perception (Kulkarni et al. 2013), hard tissue components (Jeremias et al. 2013), and the host immune response (Moutsopoulos and Konkel 2018) are governed by genetic inheritance. A study on the monozygotic and dizygotic twins in the age group 5–11 confirmed that there is a considerable proportion of the oral microbiome that are heritable (Gomez et al. 2017). It was found that similarity in the human genome showed a similar microbiome composition of the saliva and plaque, reenforcing the importance of the host genetics in the microbiome composition. The most heritable oral bacteria were *Prevotella pallens*, which decreased as age and sugar intake increased. The same study confirmed that a close association existed between host genetic factors and acid production by *S. mutans* among cariogenic twins. It was also found that the single nucleotide polymorphism found in the host gene that encodes for the vital processes like protein synthesis, cell division, and tumor suppression was well associated with the abundances of the oral species—*Prevotella*, *Leptotrichia*, and Pasteurellaceae (Blekhman et al. 2015). The abundance of a genus in a specific site is related to the host genetic variations and vice versa. An interesting comparative analysis made by Davenport of the above two studies highlighted the heritability of the genus *Aggregatibacter* ($h^2 = 0.35$) being associated with the genetic variants of chromosomes 3 and 11 whereas *Leptotrichia* ($h^2 = 0.54$) is associated with chromosome 15 (Davenport 2017). *MUC7*, which encodes for the abundant salivary protein, MUCIN7, is associated with the *Neisseria* abundance having the heritability, $h^2 = 0.38$ (Xu et al. 2017). Another large-scale twins' study with 752 twin pairs by Demmit 2017 further expands the role of SNPs in the heritability of the oral microbiome. This study highlighted the role of chromosomes 7 and 12 in the oral microbiome heritability; more specifically, the SNP near the gene, IMMPL2, was related to the heritability of the genus *Granulicatella* having $h^2 = 0.55$ (Demmitt et al. 2017). More interestingly, studies have shown an interesting relationship between host immune response and acquiring fungal infections. It was shown that the individuals defective in STAT3, which is a gene responsible for the cytokine regulation ((IL-6), IL-21, IL-10, and IL-23), are prone to *C. albicans*-mediated oral infections (Abusleme et al. 2018; Holland et al. 2007). This in turn increased the abundance of *Streptococcus mutans*-mediated dental caries. Ethnicity is also a factor that was proved to play a role in the oral microbiome and its associated homeostasis (Liu et al. 2012). Further in-depth studies are required to clearly establish the relationship between host genetic factors and components of the oral microbiome, such as composition, host immune response, and susceptibility to an infection. The above studies give a strong clue that such factors do play a role in the oral ecosystem and overall health.

3.2 Childhood

Maternal vertical transmission is the first step toward the development of the oral microbiome (Xiao et al. 2020). The acquisition either from the vaginal microbiome or skin-derived microbiome is determined by the mode of delivery (Dominguez-Bello et al. 2010). The infants born through vaginal birth were observed to have improved microbial diversity as compared to the C-section birth. More specifically, in the case of vaginal birth, the microbial profile closely resembles vaginal microbiome—*Lactobacillus*, *Sneathia* spp., *Prevotella*, Bacteroids, and TM7. The cesarean section infants have a microbial profile like the mothers' skin having *Staphylococcus*, *Propionibacterium* spp., *Corynebacterium*, *Veillonella*, etc. (Drell et al. 2017; Lif Holgerson et al. 2011). It was also observed that the acquisition of *S. mutans* was faster in the C-section infants than in the vaginal birth, making them prone to early childhood caries (Li et al. 2005). The importance of the *Lactobacillus* sp. is highlighted in the recent researches in overcoming oral dysbiosis (Caufield et al. 2015). According to the Developmental Origins of Health and Disease (DoHaD) theory, the first 1000 days in the development of healthy children is critical. In this regard, it was shown that the type of feeding—breastfeeding and formula feeding—shapes the oral microbiota. The vital *Lactobacillus* sp. bacteria were found to be predominant in breastfed infants, which were absent in the formula-fed infants (Holgerson et al. 2013; Vestman et al. 2013). A recent long-term study explored the link between the feeding type and adolescent oral microbiota composition. The study considered approximately 11,400 adolescents who were grouped based on the initial 6 months feeding pattern in addition to other factors. Even though there was no significant difference in the oral microbiota, *Veillonella* and *Eubacteria* were found to be higher in the adolescents who were given breast milk only (Eshriqui et al. 2020). The importance of *Veillonella*, which is found in breast milk, is associated with the prevention of dental plaque by metabolizing the acid production by plaque pathogens. Thus, the crucial role of breast milk in the healthy oral microbiota is established. The sustenance of the oral microbiome is further taken care of by the other necessary host factors, which include immune response, lifestyle changes, and microbial interactions with itself and with the host.

3.3 Immune Response

The oral mucosa is home to a wide range of microbial species, and the responsibility of differentiating commensals and pathogens rests on the host immune system. Although surface colonization of species is harmless, under certain conditions innate residents can turn pathogenic and thus induce stress on the host immune system. Hence, the immune system pertaining to the oral cavity is inclined toward a tolerogenic state as assertive immunogenic responses against harmless colonizing bacteria are metabolically redundant and damaging to the host tissue (Feller et al. 2013).

There are several host defense mechanisms in the oral ecosystem that continuously function to provide homeostasis. The coordinated interaction between innate and adaptive immunity helps in maintaining tolerance to the commensals and eliminating pathogens. The innate immune response is imparted by the mucosal surface and enamel along with the physical barriers such as saliva, gingival crevicular fluid (GCF), and transmigrating polymorphonuclear leukocytes (PMNs). They provide the first line of defense against bacterial invasion and pathogenesis (Meyle et al. 2017). Several antimicrobial peptides released by the epithelial cells and neutrophils monitor the oral environment by preventing bacterial buildup (Hans and Veenu 2014). The role of the innate immune system is crucial even though it is short term and nonspecific.

Adaptive immune surveillance is much broader; specifically, the role of inflammation is important in the maintenance of oral homeostasis. Even though the inflammatory response is a protective immune process, the unresolved inflammation leads to several infectious and systemic disorders (Kleinstein et al. 2020). The resident dendritic cells present in the oral mucosa are inclined toward a tolerant state. Dendritic cells are inherently APCs (antigen presenting cells), which release proinflammatory cytokines that activate the adaptive immune system. However, dendritic cells of the oral mucosal secrete anti-inflammatory immune modulators, including interleukin 10 (IL-10), transforming growth factor-beta (TGF-β), and prostaglandin E2, which suppress T-regulatory cell formation. It is also to be noted that pathogen-associated molecular patterns (PAMPs) pertaining to the oral mucosa do not trigger inflammatory responses against commensals, and since the bacterial molecular patterns do not change despite a shift toward pathogenicity, the PAMP immune response continues to remain redundant (Sultan et al. 2018).

Periodontitis and gingivitis are infective oral diseases predominantly driven by prolonged inflammation that has failed to resolve and restore homeostasis. The host inflammatory response plays a significant role by responding to harmful stimuli including pathogens. However, contrary to the intended function, the inflammatory host response favors the progression of gingivitis and periodontitis, inducing bacteria, especially *Porphyromonas gingivalis* and *Fusobacterium nucleatum*. The onset of inflammation results in tissue damage and release of degraded collagen, heme, amino acids, and iron, which in turn serve as a rich source of nutrients for the facilitation of subgingival proteolytic and saccharolytic bacteria (Hajishengallis 2014; Diaz et al. 2016). Nitrate release as a by-product of inflammation also facilitates the growth of Enterobacteriaceae, which take up nitrates as an electron acceptor for anaerobic respiration. Only certain bacteria which can surpass the inflammatory response survive in this niche, and these species are known inflammophilic pathobionts. They tend to take over the inherent microbiome as other species which are unable to survive this environment perish, leading to a state of dysbiosis. Thus, begins a vicious cycle where the dysbiosis biofilm induces host cytokine production leading to inflammatory responses, which further facilitate the growth of dysbiosis pathobionts. The inflammation-dysbiosis dynamics act as a sustained feed forward loop aggravating the onset of periodontitis (Kleinstein et al. 2020). A recent study highlighted the possibility of nitrate as the potential prebiotic where dysbiosis causing genera *Streptococcus*, *Veillonella*, *Porphyromonas*, *Fusobacterium*, *Prevotella*, and *Alloprevotella* were significantly reduced and health associated genera (*Neisseria* and *Rothia*) were enhanced significantly (Rosier et al. 2020). Hence, most therapeutic research directs focus toward anti-inflammatory treatment options, which have proven to be successful in the restoration of the eubiotic microbiome and in tackling periodontitis.

3.4 Physical Factors

Saliva is the most predominant component of the oral ecosystem and plays multiple roles in the homeostasis maintenance of the oral microbiome. Saliva provides the required nutrients and trace elements for the survival of the microorganisms (Pedersen et al. 2018). The nutrients include sugars, amino acids, protein, hormones, and vitamins (Marsh et al. 2016). Studies have shown that *Aggregatibacter actinomycetemcomitans*, *P. gingivalis* (García-Gómez et al. 2013; Kumar 2013), and *Treponema denticola* (Clark and Soory 2006) can utilize the testosterone and cholesterol, respectively, as the nutrient source, which is provided by saliva. In addition to providing lubrication, it also plays a role in the formation of acquired pellicle, which forms the initial step for the adhesion of oral bacteria to a surface. Saliva contains several proteins, such as glycoproteins and other factors such as lactoferrin, statherin, histatins, defensins, and mucins, which play a protective role in inhibiting the adhesion of microbes to the oral surfaces (Lynge et al. 2019). Secretory immunoglobulin A is also present in the saliva, which controls the adhesion and colonization of the pathogenic microbes. Interestingly, the commensals, *S. mitis*, *S. oralis*, and *S. sanguinis* can produce IgA proteases, which can neutralize S-IgA in the saliva (Kilian et al. 1996; Feller et al. 2013), but the host factor and the commensals of the oral microbiome have perfected the balance to co-exist and co-evolve together to maintain ecological balance.

Another important role of saliva is the maintenance of the salivary pH. pH plays a predominant role in the oral microbiome homeostasis. Studies have shown the importance of an acidic environment in the establishment of virulence by the caries-causing pathogens (Tanner et al. 2018). Saliva has a buffering role in maintaining the normal pH of 6.75–7.25 (Samaranayake and Matsubara 2017). Currently, studies are focused on improving the alkalizing capacity of the saliva to prevent the growth and pathogenesis of cariogenic organisms (Bijle et al. 2019). The efficient salivary role can be enhanced by following proper lifestyle habits, particularly dietary habits. Salivary flow and temperature also influence the oral ecosystem. The salivary flow decides the colonization of the specific species in the oral habitat (Marsh et al. 2016).

3.5 Lifestyle

The paradigm shifts in the current oral microbiome composition, when compared to the ancient period, emphasize the importance of lifestyle in the oral microbiome (Adler et al. 2013; Cross et al. 2018). Among the different lifestyle changes, diet and oral hygiene play a crucial role (Chapple et al. 2017). The major causative factor for oral dysbiosis is the diet, specifically, the excessive consumption of fermentable carbohydrates. Such fermentable carbohydrates are converted into organic acids by microbes, which creates an acidic environment. This gives a selective advantage for the acidophilic and acid-tolerating pathogens to bring about an imbalance in the oral ecosystem (Kato et al. 2017). The notable example is *Streptococcus mutans*, a primary driver of dental caries, which is both acidogenic and aciduric (Lemos et al. 2019). In addition, certain micronutrient deficiencies have also led to the development of periodontal diseases. In particular, it was shown that deficiency in vitamin B-complex (Neiva et al. 2005), vitamin C (Yussif et al. 2016), vitamin D (Miley et al. 2009), minerals, carotenoids, and flavanoids leads to dysbiosis in the oral microbiome (Kaur et al. 2019).

One of the well-known, yet often overlooked, factors affecting oral homeostasis is oral hygiene, which includes regular brushing, usage of oral hygiene products, and the usage of tobacco, betel nut, etc. Studies have shown that regular brushing with fluoride-based oral hygiene products is effective in removing dental plaques mechanically, but it should be noted that the oral hygiene products should target only the pathogens and not the commensals in the cleaning process (Figuero et al. 2017; Kumar et al. 2016). The current oral hygiene products compromise these factors by eliminating both the pathogens and the commensals, leading to oral ecosystem dysbiosis.

Human practices such as cigarette smoking (Kato et al. 2016; Wu et al. 2016; Al Kawas et al. 2021; Darwazeh et al. 2010) and betel nut consumption (Hernandez et al. 2017) have an impact on the imbalance in the oral ecosystem. Oral cavity is the first system that encounters the toxicants of cigarettes and irritants of betel nut due to which it suffers dysbiosis. The onset of the periodontitis and the abundance of *Streptococcus* sp. in cigarette smoking individuals as well as betel nut consumers strongly establishes the imbalance created by such practices. While it is established that such practices compromise oral hygiene, they also have a profound effect on the commensals of the oral microbiome (Hernandez et al. 2017). The alterations in the oral microbiome were shown to be associated with an increased acidic environment, depletion of oxygen, and the toxic chemicals impairing the host immunity. In the cigarette smoking individuals, there was a vast difference at the phylum level, where

Firmicutes and Actinobacteria were found in abundance compared to the Proteobacteria (Kato et al. 2016). A notable observation is a lack of or impaired xenobiotic degradation pathways, which explains the accumulation and dissemination of toxic compounds in the oral cavity, causing overall damage to the health. Strikingly Gram-positive anaerobes (*Streptococcus* sp.) were increased as compared to the aerobes *Neisseria* and *Corynebacterium* sp. (Sellappa et al. 2015). The current understanding of the effect of lifestyle in the oral microbiome is limited to the specific set of population, sample processing, time, age, and other factors. Although further studies are required to clearly link the lifestyle changes and oral microbiome maintenance, it can be concluded that lifestyle plays an important role in the overall maintenance of the oral ecosystem.

4 Oral Microbiome: The Link to Health and Diseases

Oral microbiome dysbiosis has often been linked to systemic disorders. While the evidence on this is certainly unclear and most studies demonstrate a correlation rather than a cause-effect relationship, this section will cover what has been reported in the literature. However, the authors of this chapter forewarn the readers that the summation of findings here is not definitive and do not strongly support or deny that oral dysbiosis is the causative factor for these systemic conditions and more research in this area is warranted. The inter-relationships between oral bacteria and susceptibility to systemic diseases were based on the theory of focal infection. The focal infection theory expands the understanding that bacteria or their metabolites, toxins, and inflammatory products from the existing focal point of the host niche disseminate to distant parts of various organ systems (Paster et al. 2001; Shchipkova et al. 2010). The dissemination of the dispersal of biofilm cells and their by-products to non-oral sites is being favored by the vascular supply of the periodontium, breakdown of the epithelial integrity, and simple oral hygiene and dental procedures (Baltch et al. 1988; Carroll and Sebor 1980). Also, the problem exists with the episodes of bacteraemia where the biofilm cells or microcolonies disseminate to the non-oral sites of the immunocompromised individuals, contribute to systemic inflammations, and cause several disorders (Hernichel-Gorbach et al. 1994; Offenbacher et al. 1993). In this regard, oral microbial dysbiosis is probably connected to various systemic disorders, which include arthritis, inflammatory bowel syndrome, Alzheimer's, cardiovascular diseases, oral cancer, and related inflammatory-linked systemic disorders (Fig. 2.2). The oral microbiota dysbiosis leads to the invasion of the pathogens. In most of the systemic diseases, *P. gingivalis*, *F. nucleatum*, and *Neisseria* are the most common pathogens commonly isolated. In addition to the above pathogens, *C. pneumoniae* is strongly associated with respiratory disorders, *T. forsythia*, *T. denticola*, *P. intermedia*, and *Camphylobacter rectus* are associated with cardiovascular disorders, *Rothia aeria* is found commonly in rheumatoid arthritis, and Spirochaetes are found in the brain-related disorders such as Alzheimer's disease. Site-specific pathogens also interact with the disseminated oral pathogens and target the host immune system.

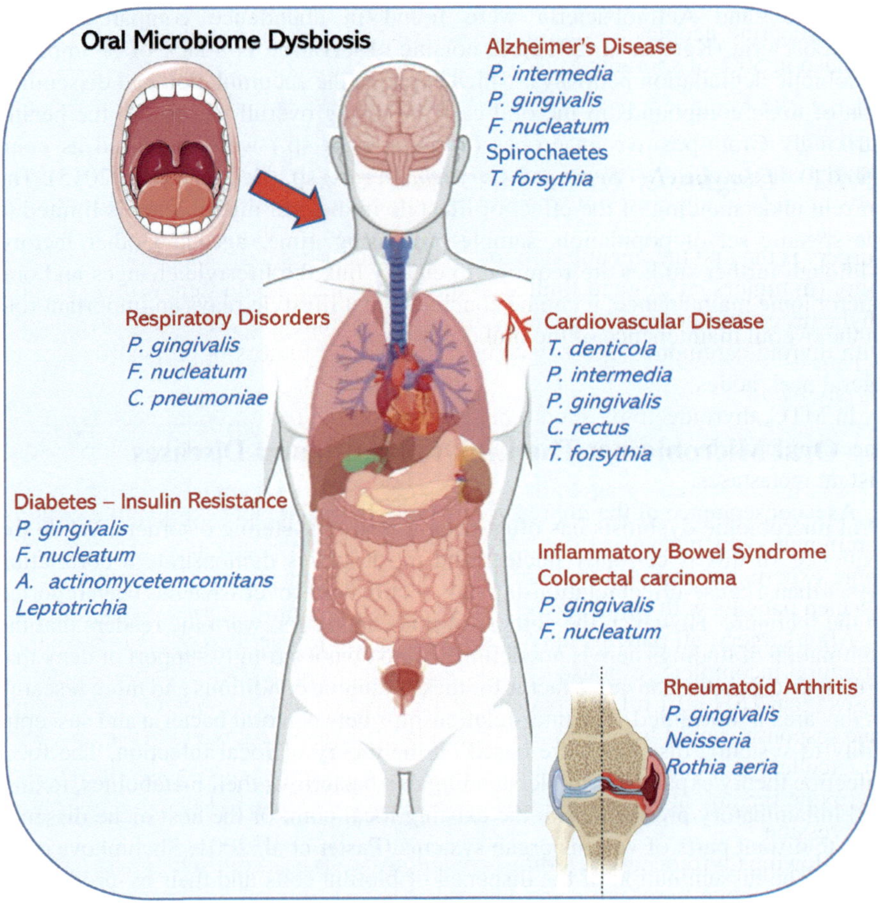

Fig. 2.2 Schematic showing common microbiota that cause oral diseases, which have also been isolated from other immune-inflammatory diseases elsewhere in the body. While the exact link between the systems is unclear, a large body of evidence does correlate oral infectious diseases to increased inflammatory diseases elsewhere in the body

The host adaptive immune system, more specifically the inflammatory network, plays an important role in maintaining oral homeostasis. The problem arises when the inflammation is prolonged, leading to abnormalities in cytokine levels and phagocytosis. One of the classical links between the oral microbiome-induced chronic inflammation and systemic disorder is autoimmune disorder. A recent study has provided an insight into their association with the autoimmune disorders such as osteoarthritis (OA) and rheumatoid arthritis (RA) (Diaz et al. 2016). OA is mediated by the innate immune system with chronic, low-grade inflammation and damages the joints of the hands, knees, hips, and spine in due course of time, whereas RA is an autoimmune disorder where the body's own immune system affects the fluid secretion required for cartilage lubrication (Li et al. 2016). The oral microbiota released into systemic circulation is the root cause for stimulating specific inflammatory processes such as cytokines, which decrease joint cartilage amount (Li et al. 2016). There are reports confirming the links between gut–oral microbiota and disorders like OA and RA, majorly the influence of nutrition and inflammation-reducing dietary factors on the ratio of the intestinal microbiota for improving the arthritis disease state (Sakaguchi et al. 2011; Conti et al. 2015). One of these findings supports the role of environmental factors, in particular, that of microbial presence, in the causal genetic anomaly of ZAP-70 mutation and a polymorphism of the MHC gene as it is genetically susceptible to OA and RA initiation (Sakaguchi et al. 2011). The biomarkers between the arthritis patients and normal healthy subjects were quantified using rRNA gene amplicon sequencing, which determined eight unique signatures (*Prevotella melaninogenica*, *Veillonella dispar*, *Prevotella*, *Neisseria*, *Porphyromonas*, *Veillonella*, *Haemophilus*, *Rothia*, *Streptococcus*, *Actinomyces*, *Granulicatella*, *Leptotrichia*, *Lautropia*, and *Fusobacterium*) in the oral microbiome (Scher et al. 2012). Notably, the microbial diversity in the oral microbiota of OA and RA was higher than in healthy individuals. Recent studies show the direct link of *P. gingivalis* to RA through citrullination, where it induces anti-peptidyl citrulline antibodies reacting to citrullinated human self-proteins to increase the severity of RA (Rutger Persson 2012). Although studies speculate the influence of gut–oral microbiota in OA and RA, a deeper understanding is required to decode the biological complexities of involving several confounding factors that trigger inflammation in the progression of disease pathogenesis. Another important inflammation-related disorder is inflammatory bowel disease (IBD), which is the chronic inflammation of the bowel. Chronic inflammation exists due to the dysbiosis in the gut microbiome due to the involvement of the Gram-negative oral bacteria LPS-induced release of pro-inflammatory cytokines such as TNF-alpha, IL-6, and IL-1 from the macrophages (Hakansson and Molin 2011). Although multiple factors aggravate the pathological condition of IBD, the gut microbiota is a key driver of disease progression, and food supplements with refined sugars or regular use of certain toothpaste was observed to promote the growth of specific bacterial species that modulate the host's immune homeostasis (Becker et al. 2015). A notable observation in IBD disease state is the depletion of anti-inflammatory bacteria, such as Clostridium Cluster IV and XIVa, which produce short-chain fatty acids and an increase in the level of inflammatory members belonging to the Proteobacteriaceae and Pasteurellaceae group (Burman et al. 2016; Sokol et al. 2009). Another important brain disorder, Alzheimer's disease (AD), is a pathological condition where cerebral amyloid-β(Aβ) plaque deposition in the brain results in synaptic disconnection, leading to progressive neuronal death (Aguayo et al. 2018). Several scientific reports show a connection of the periodontal pathogen *P. gingivalis* (Poole et al. 2015) in AD. The endotoxin lipopolysaccharide (LPS) is derived from the brain samples of AD, and this aligns with the previous observation where LPS is directly proportional to the Aβ deposition in AD patients (Poole et al. 2013; Hauss-Wegrzyniak and Wenk 2002; Lee et al. 2008). Importantly, a study showed the ApoE−/− mice orally infected with *P. gingivalis* for 24 weeks leading to the direct infiltration of the pathogen to affect the pyramidal neurons of the hippocampus and cause memory-related responses (Poole et al. 2015). However, the study demonstrated a direct influence of the periodontal pathogen *P. gingivalis* in contrast to an indirect effect via inflammatory mediators like LPS, which promote releasing of peptides that act as porins. These porins may increase the permeability of blood–brain barrier (BBB) to permit bacterial penetration into the brain and cause cognitive decay (Poole et al. 2015). Other than *P. gingivalis*, *Enterococcus faecalis*, known to cause root canal and chronic periodontal infection, migrate into the brain and form abscesses, suggesting a potential link to the pathogenesis of AD (Mylona et al. 2012; Underly et al. 2015).

Finally, one of the most studied and most devastating oral diseases is oral cancer. Combinations of factors including poor oral hygiene and usage of tobacco and alcohol lead to the colonization of inflammatory bacterial members and increase the risk of the most common oral cancer, squamous cell carcinoma (OSCC) (Cankovic et al. 2013; Tanaka and Ishigamori 2011; Sarode et al. 2017). The pathology of the squamous cell carcinoma was observed between the vermilion border of the lips, and the junction of the hard and soft palates or even may extend to the posterior end of the tongue. The colonization of inflammatory bacterial members is likely to modulate the tumor microenvironment via stimulation of chronic inflammation, cell proliferation, inhibition of cellular apoptosis, and promotion of cellular invasion (Al-Hebshi et al. 2019; Kudo et al. 2016). However, an inflammatory response triggered by the colonized microbes or their by-products, toxins, etc., plays a crucial role in all stages of cancer development. It was shown that the reactive oxygen species (ROS) indirectly affect the transcription factor of NF-kß (nuclear factor-kappa β) and cytokine production. These cytokines are formed in a dysregulated manner and play a role in processes like cell growth, inhibition of tumor suppression, immune system, and even survival. Even though not widely accepted, the link between the oral microbiome and oral cancer cannot be disregarded. Several studies associate the involvement of *Streptococcus* sp., *Peptostreptococcus* sp., *Prevotella* sp., *P. gingivalis*, and *Capnocytophaga gingivalis* in causing OSCC (Sasaki et al. 2005; Mager et al. 2005; Katz et al. 2011; Pushalkar et al. 2012; Atanasova and Yilmaz 2014; Galvão-Moreira and da Cruz 2016; Lee et al. 2017). Interestingly, *Capnocytophaga gingivalis*, *Prevotella melaninogenica*, and *Streptococcus mitis* were found to be in higher proportions in the saliva of OSCC patients and scored as significant diagnostic markers (Karpiński 2019). The effect of chemotherapy on oral mucositis was recently reported, which impairs the oral epithelium and thereby brings about oral dysbiosis (Hong et al. 2019), in addition to enhanced tissue destruction by *Candida albicans* (Sobue et al. 2018). Hence, the effect is bidirectional, oral dysbiosis affecting the systemic health and vice versa.

In addition to the above disorders, diabetes, cardiovascular diseases, and respiratory disorders are related to the shift in the oral microbiota dynamics. In the case of diabetes, there are strong reports that prove the enhancement of the pathogenic microorganisms by delaying the inflammatory process, leading to periodontitis (Kuo et al. 2008; Lalla and Papapanou 2011; Pacios et al. 2012). A recent study reported the role of the exaggerated expression of IL17, which induces the overexpression of the cytokines IL6 and RANKL, which in turn prolongs the inflammatory response (Xiao et al. 2017). *P. gingivalis* is also shown to increase insulin resistance (Andriankaja et al. 2012; Kuo et al. 2008). The relationship between type 2 diabetes and oral microbiota establishes that *Betaproteobacteria* are found in abundance in diabetes patients (Almeida-Santos et al. 2021). Like diabetes, cardiovascular diseases are the risk factors for the development of clinical periodontitis. It is well known that atherosclerosis, where cholesterol accumulates, is accompanied by inflammation (Ketelhuth and Hansson 2016). The inflamed arteries serve as the breeding ground for the pathogens that escape from the oral microbiota dysbiosis. The PAVE (Periodontal and Vascular Events) study highlights the reduction in the risk of cardiovascular disease by treating periodontitis. Even though significant results were not obtained, this study paved the way for the researchers to further explore the link between CVD and periodontitis (Beck et al. 2008). Later, several reports were published to insist on the lack of oral hygiene and its associated risk of CVD (Palm et al. 2016). The increased incidence of the species *Anaeroglobus* were found in the atherosclerosis patients (Sen et al. 2018). It was proved that the management of periodontitis reduced subclinical atherosclerosis in a 3-year follow-up study (Fåk et al. 2015). Also, in CVD, *P. gingivalis* is the major player accompanied by *Aggregatibacter actinomycetemcomitans* (Pussinen et al. 2007). The lipopolysaccharide, virulence factors, and the whole bacteria themselves act as mediators for the onset of the prolonged inflammation and are even used as biomarkers for the diagnosis. The unusual increase in the C-reactive protein levels, IL-6, IgG response, and IgA seropositivity are associated with the presence of *P. gingivalis* and *Aggregatibacter actinomycetemcomitans* (Pietiäinen et al. 2018).

The respiratory tract encounters the inflow of oral bacteria exposing the lung to the diverse oral microbiota. The relationship between the oral and lung microbiota is defined by the prominent contradicting theories—topological continuity theory, which states that the oral microbiome and lung microbiome are similar or indistinguishable, and island biogeography theory, which states that each anatomic location is different from each other (Mammen and Sethi 2016). Studies have shown that the benign oral microbial species *Pervotella* sp. regulate IL 17, a pulmonary inflammatory response. A dysbiosis in the oral microbiome is related to poor oral hygiene and leads to severe lung disorders such as pneumonia, cystic fibrosis, and other pulmonary disorders (Pu et al. 2020). The acquisition of the pathogenic microbes in the lung is from oral microbiota through aspiration, airborne translocation, or destruction of salivary pellicles by adhesion and colonization of pathogens. In any of the cases, the useful microbes are replaced with pathogenic strains such as *K. pnuemoniae* and *P. aeruginosa* (Dickson et al. 2015; Munro and Grap 2004). Here, the cytokines released from the pathogens destruct the mucosal layer and neutralize the salivary mucins, which leads to hospital-acquired pneumonia and ventilator-acquired pneumonia (Huffnagle et al. 2017). Thus, oral hygiene is given high priority to prevent nosocomial infections (Price et al. 2014). This evidence supports the evolutionary symbiotic relationship between the host and the microbiome, making the human microbiome a "superorganism."

5 Oral Microbiome: CRISPR Cas: Versatility vs Fatality

The second most diverse microbial community is housed in the mouth, and it harbors over 700 species of microbial species playing a significant role in our physiology and health (Kilian et al. 2016). In this context, research studies provide evidence of the involvement of a natural adaptive immune process known as CRISPR-Cas in their ubiquitous and physiological existence in the oral microbiome (Mojica et al. 2000, 2005; Barrangou et al. 2007). CRISPR (clustered regularly interspaced short palindromic repeats) are short palindromic repeats of base sequences, which are separated by short "spacer DNA." Cas (CRISPR-associated) protein, with helicase and nuclease motifs, utilize the memory sequences called "spacers" to recognize the sequences of phages presented to target and cleave the invading DNA in a sequence-specific manner (Naidu et al. 2014; Pride et al. 2012). However, the recognition of the foreign single-stranded target DNA with the single guide RNA (sgRNA), a known chimera of CRISPR-RNA (crRNA) and trans-activating cRNA (tracrRNA), is regulated by a protospacer motif(PAMs), 5'NGG adjacent to the target DNA (N1-N20), and is critical for the CRISPR/Cas9 system to differentiate self- and non-self-DNA (Moon et al. 2019). Furthermore, the CRISPR-Cas systems are classified into various types: I, II, III, IV, and V, with several subtypes in each category. Most bacterial species possess one or a combination of two CRISPR-Cas types, whereas the oral bacterial species, *Streptococcus*, harbors all types of CRISPR-Cas systems except IV and V. In the oral microbiome perspective, the most widely studied oral microbes are *Streptococcus mutans*, *Enterococcus faecalis*, and *P. gingivalis*. Figure 2.3 depicts the various reported roles of the CRISPR-Cas system on oral microbiome: Broadly, the CRISPR-Cas system is proved to be deleterious to the host system by promoting the pathogenicity and also reducing the interspecies diversity, causing dysbiosis in the oral microbiota. At the same time, such discoveries have led to developing CRISPR-Cas-based antimicrobials.

The major role of this system is to provide the bacteria with an adaptive immune mechanism. This is well-documented in *S. mutans* in the regulation of stress responses associated with DNA repair and virulence factors. The mutations in the genes that involve regulatory factors, stress responses, and virulence are found to affect a differential pattern of expression in the type II Cas protein (Xie et al. 2010; Kajfasz et al. 2010, 2011; Liu et al. 2011). It is also observed that in the oral bacteria, *S. mutans* UA159 CRISPR-Cas systems, there exists a sequence homology between the Cas1 and 2 proteins, and the deletion of their respective genes decreased the survivability of *S. mutans* UA159, on exposure to DNA-damaging agents, mitomycin C (MMC) and ultraviolet (UV), thereby suggesting a role for Cas protein involvement in the protection of bacteria against stress-induced conditions (Serbanescu et al. 2015). The same study highlighted the inter-relationship of a VicR/K two-component regulatory system with Cas gene expression to regulate the environmental stress tolerance. Thus, the role of highly conserved Cas1 protein in DNA repair is crucial (Ka et al. 2016). Furthermore, the involvement of the second messenger cyclic-di-AMP bipolar effect in the expression of CRISPR1 and 2 was

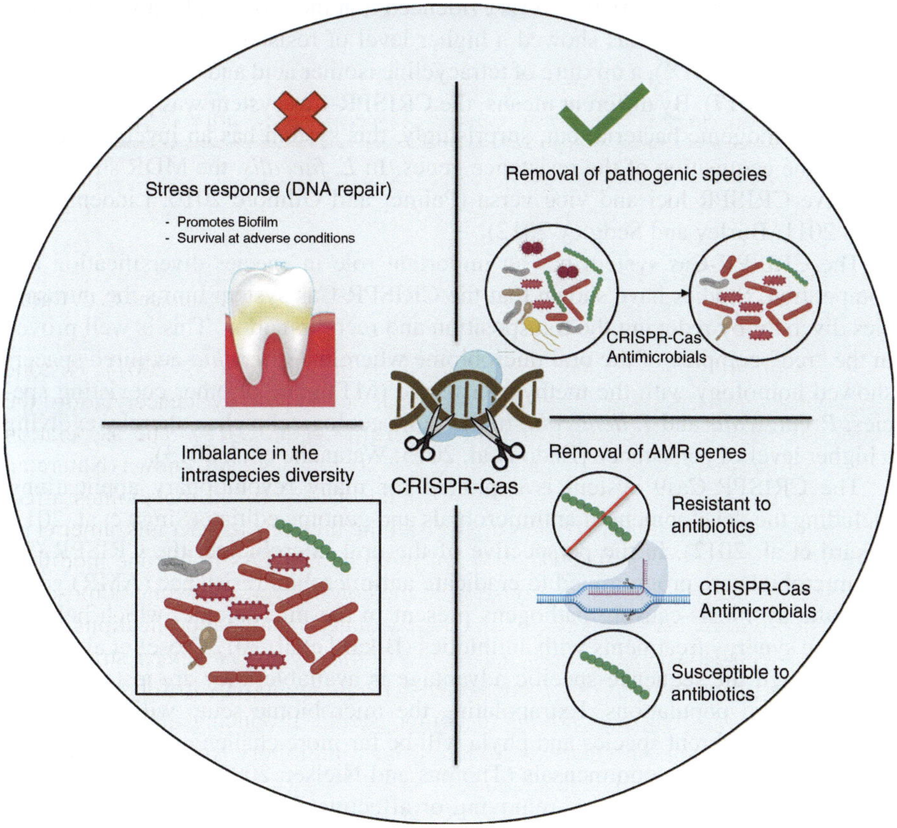

Fig. 2.3 CRISPR-Cas and oral microbiome: the fatality and versatility aspects of the CRISPR-Cas system

found to enhance the adaptive nature of *S. mutans* toward hydrogen peroxide and EPS production (Cheng et al. 2016). The above observation was further confirmed with *S. mutans* Cas1 and 2-deficient mutants in comparison to the wild-type (WT) strains, where the survival fitness of the cells was higher at low acidic pH (5.5) and also susceptible to stress induced by H_2O_2, paraquat, SDS, and high temperatures, suggesting their involvement in acid, oxidative, cell membrane, and temperature stress response (Serbanescu et al. 2015). Clinical isolates of *S. mutans* expressing CRISPR1 and CRISPR2 were shown to enhance the expression of *gtfB* and *gtfC* genes, which are required for strong biofilm formation (Chen et al. 2017). Even though the exact role of the CRISPR-Cas system in promoting biofilm formation is yet to be studied, the preliminary results obtained show that this system might have an important role in biofilm formation (Tong et al. 2017).

Another important aspect of this system is the changes associated with bacterial physiology. A link between CRISPR loci and acquired antibiotic resistance genes in the root canal pathogen *Enterococcus faecalis* was well established (Hullahalli et al.

2017, 2018; Price et al. 2019). It was evidenced that the root canal pathogen *E. faecalis* lacking CRISPR-Cas showed a higher level of resistance to irritants such as chlorohexidine (CHX), a mixture of tetracycline isomer acid and detergent (MTAD) (Tong et al. 2017). By different means, the CRISPR-Cas system was proven to protect the pathogenic bacteria, but, surprisingly, this system has an inverse relationship to the acquisition of the resistance genes. In *E. faecalis*, the MDR strain does not have CRISPR loci and vice versa (Palmer and Gilmore 2010; Lindenstrauss et al. 2011; Burley and Sedgley 2012).

The CRISPR-Cas system has an important role in species diversification and competition. Studies have shown that the CRISPR-Cas system limits the intraspecies diversity by reducing the transposition and recombination. This is well proved in the "red" complex of the oral microbiome wherein *T. forsythia*-acquired spacers showed homology with the methyltransferase (MT) gene of other coexisting species, *P. gingivalis* and *T. denticola*, in the same ecological niches, thereby evolving a higher level of persistence (Endo et al. 2015; Watanabe et al. 2013).

The CRISPR-Cas9 system is exploited for many revolutionary applications, including the development of antimicrobials and genome editing (Singh et al. 2017; Bikard et al. 2012). In the perspective of the oral microbiome, the CRISPR-Cas antimicrobials are programmed to eradicate antimicrobial resistance (AMR) genes from the dysbiosis-causing pathogens present in the microbiome, which helps in effective synergy treatments with antibiotics (Bikard et al. 2012; Yosef et al. 2015). Even though the sequence-specific advantage is available, they are tested only in limited clonal populations. Extrapolating the microbiome setup with billions of microbes of different species and phyla will be far-more challenging and may lead to the lethality of the commensals (Thomas and Nielsen 2005). First, it may have unwanted knock-on effects of removing or affecting the growth/metabolism of a strain in a population that may allow the outgrowth of clinically problematic pathogens, and such shifts in species in the oral microbiome structure may affect oral health (Theriot et al. 2014; Jorth et al. 2014). Second, over the course of time, the resistance mutation will be mapped in the CRISPR-Cas loci or will inactivate the CRISPR to deactivate AMR genes by selecting anti-CRISPR loci that are critical for cleaving or encoding factors (Bikard et al. 2014; Jiang et al. 2013; Vercoe et al. 2013). However, the resistance through mutation may be overcome by using the multiplexing approach of targeting multiple sequences to decrease the chance of recurrence of resistance against CRISPR-Cas antimicrobials (Bikard and Barrangou 2017). The hurdles of using CRISPR-Cas antimicrobials in natural microbial communities require a better optimization process for a safe delivery and active engagement with communities and highly adhere to clear guidelines and score as a next-generation drug class (Pursey et al. 2018).

6 Conclusion

The co-existence and co-evolution of the host and the microbiome have led to a deeper understanding of the role of the oral microbiome in health and disease. The diversification in the oral microbiota is naturally set to protect and be resilient to the perturbations caused. With everyday technological advancements, the purpose of each microorganism in the oral microbiome is understood. The varied landscape of the mouth, host, and microbiome-derived factors decides the core microbiome. The different causative factors for dysbiosis in the oral ecosystem are mapped. The triangular relationship between host factors—mainly the immune inflammatory network—oral microbiome, and systemic disorders are well established. With such advanced and ever-growing knowledge of the oral microbiome, it is imperative to take necessary measures to maintain a healthy oral microbiome and in turn healthy life. The core microbiome is established from childbirth; thus, the awareness and importance of the healthy oral microbiome should be educated. Preventative measures such as healthy lifestyle choices and proper oral hygiene practices, which maintain the correct balance of the healthy microbiome, should be given priority. The treatment measures should be directed toward safeguarding or promoting the beneficial microbiota rather than destroying it with indiscriminate use of antimicrobials. More focus should be given on improving the host immune system, right from the beginning. There should be a paradigm shift from the current antimicrobials and mechanical based management to ecological based proactive management of the oral ecosystem. The repositories like HOMD have helped researchers to understand the diversification of the oral microbiota and its importance in health and diseases. Such deeper understanding will help researchers to understand the specific markers of the systemic diseases, which can be exploited for the diagnosis and drug development purpose focusing on personalized medicines.

Acknowledgments This work was supported by the MHRD, India Scheme for Promotion of Academic and Research Collaboration—SPARC Cell, SPARC/2018-2019/P27. SV wishes to express her gratitude to DST-INSPIRE (IF170369) for the financial support.

References

Aagaard K, Ma J, Antony KM, Ganu R, Petrosino J, Versalovic J (2014) The placenta harbors a unique microbiome. Sci Transl Med 6(237):237ra65. https://doi.org/10.1126/scitranslmed.3008599

Abusleme L, Diaz PI, Freeman AF, Greenwell-Wild T, Brenchley L, Desai JV et al (2018) Human defects in STAT3 promote oral mucosal fungal and bacterial dysbiosis. JCI Insight 3(17):e122061. https://doi.org/10.1172/jci.insight.122061

Adler CJ, Dobney K, Weyrich LS, Kaidonis J, Walker AW, Haak W et al (2013) Sequencing ancient calcified dental plaque shows changes in oral microbiota with dietary shifts of the Neolithic and industrial revolutions. Nat Genet 45(4):450–455., 455e1. https://doi.org/10.1038/ng.2536

Aguayo S, Schuh CMAP, Vicente B, Aguayo LG (2018) Association between Alzheimer's disease and oral and gut microbiota. Are pore forming proteins the missing link? J Alzheimer's Dis: JAD 65(1):29–46. https://doi.org/10.3233/JAD-180319

Al Kawas S, Al-Marzooq F, Rahman B, Shearston JA, Saad H, Benzina D, Weitzman M (2021) The impact of smoking different tobacco types on the subgingival microbiome and periodontal health. A pilot study. Sci Rep 11(1):1113. https://doi.org/10.1038/s41598-020-80937-3

Al-Hebshi NN, Borgnakke WS, Johnson NW (2019) The microbiome of oral squamous cell carcinomas. A functional perspective. Curr Oral Health Rep 6(2):145–160. https://doi.org/10.1007/s40496-019-0215-5

Almeida-Santos A, Martins-Mendes D, Gayà-Vidal M, Pérez-Pardal L, Beja-Pereira A (2021) Characterization of the oral microbiome of medicated Type-2 diabetes patients. Front Microbiol 12:610370. https://doi.org/10.3389/fmicb.2021.610370

Andriankaja OM, Galicia J, Dong G, Xiao W, Alawi F, Graves DT (2012) Gene expression dynamics during diabetic periodontitis. J Dent Res 91(12):1160–1165. https://doi.org/10.1177/0022034512465292

Atanasova KR, Yilmaz O (2014) Looking in the Porphyromonas gingivalis cabinet of curiosities. The microbium, the host and cancer association. Mol Oral Microbiol 29(2):55–66. https://doi.org/10.1111/omi.12047

Baltch AL, Pressman HL, Schaffer C, Smith RP, Hammer MC, Shayegani M, Michelsen P (1988) Bacteremia in patients undergoing oral procedures. Study following parenteral antimicrobial prophylaxis as recommended by the American Heart Association, 1977. Arch Intern Med 148(5):1084–1088. https://doi.org/10.1001/archinte.148.5.1084

Barrangou R, Fremaux C, Deveau H, Richards M, Boyaval P, Moineau S et al (2007) CRISPR provides acquired resistance against viruses in prokaryotes. Science (New York, NY) 315(5819):1709–1712. https://doi.org/10.1126/science.1138140

Beck JD, Couper DJ, Falkner KL, Graham SP, Grossi SG, Gunsolley JC et al (2008) The periodontitis and vascular events (PAVE) pilot study. Adverse events. J Periodontol 79(1):90–96. https://doi.org/10.1902/jop.2008.070223

Becker C, Neurath MF, Wirtz S (2015) The intestinal microbiota in inflammatory bowel disease. ILAR J 56(2):192–204. https://doi.org/10.1093/ilar/ilv030

Bernardi S, Continenza MA, Al-Ahmad A, Karygianni L, Follo M, Filippi A, Macchiarelli G (2019) Streptococcus spp. and fusobacterium nucleatum in tongue dorsum biofilm from halitosis patients. A fluorescence in situ hybridization (FISH) and confocal laser scanning microscopy (CLSM) study. New Microbiol 42(2):108–113

Bijle MNA, Ekambaram M, Lo ECM, Yiu CKY (2019) The combined antimicrobial effect of arginine and fluoride toothpaste. Sci Rep 9(1):8405. https://doi.org/10.1038/s41598-019-44612-6

Bikard D, Barrangou R (2017) Using CRISPR-Cas systems as antimicrobials. Curr Opin Microbiol 37:155–160. https://doi.org/10.1016/j.mib.2017.08.005

Bikard D, Euler CW, Jiang W, Nussenzweig PM, Goldberg GW, Duportet X et al (2014) Exploiting CRISPR-Cas nucleases to produce sequence-specific antimicrobials. Nat Biotechnol 32(11):1146–1150. https://doi.org/10.1038/nbt.3043

Bikard D, Hatoum-Aslan A, Mucida D, Marraffini LA (2012) CRISPR interference can prevent natural transformation and virulence acquisition during in vivo bacterial infection. Cell Host Microbe 12(2):177–186. https://doi.org/10.1016/j.chom.2012.06.003

Blekhman R, Goodrich JK, Huang K, Sun Q, Bukowski R, Bell JT et al (2015) Host genetic variation impacts microbiome composition across human body sites. Genome Biol 16:191. https://doi.org/10.1186/s13059-015-0759-1

Bordenstein SR, Theis KR (2015) Host biology in light of the microbiome. Ten principles of Holobionts and Hologenomes. In. PLoS Biol 13(8):e1002226. https://doi.org/10.1371/journal.pbio.1002226

Burley KM, Sedgley CM (2012) CRISPR-Cas, a prokaryotic adaptive immune system, in endodontic, oral, and multidrug-resistant hospital-acquired enterococcus faecalis. J Endod 38(11):1511–1515. https://doi.org/10.1016/j.joen.2012.07.004

Burman S, Hoedt EC, Pottenger S, Mohd-Najman N-S, Ó'Cuív P, Morrison M (2016) An (anti)-inflammatory microbiota. Defining the role in inflammatory bowel disease? Digestive Diseases (Basel, Switzerland) 34(1–2):64–71. https://doi.org/10.1159/000443759

Cankovic M, Bokor-Bratic M, Loncar J, Marinoski J, Ilic M (2013) Bacterial flora on the surface of oral squamous cell carcinoma. Arch Oncol 21(2):62–64. https://doi.org/10.2298/AOO1302062C

Carroll GC, Sebor RJ (1980) Dental flossing and its relationship to transient bacteremia. J Periodontol 51(12):691–692. https://doi.org/10.1902/jop.1980.51.12.691

Caufield PW, Schön CN, Saraithong P, Li Y, Argimón S (2015) Oral lactobacilli and dental caries. A model for niche adaptation in humans. J Dent Res 94(9 Suppl):110S–118S. https://doi.org/10.1177/0022034515576052

Chapple ILC, Bouchard P, Cagetti MG, Campus G, Carra M-C, Cocco F et al (2017) Interaction of lifestyle, behaviour or systemic diseases with dental caries and periodontal diseases. Consensus report of group 2 of the joint EFP/ORCA workshop on the boundaries between caries and periodontal diseases. J Clin Periodontol 44(Suppl 18):S39–S51. https://doi.org/10.1111/jcpe.12685

Chen J, Li T, Zhou X, Cheng L, Huo Y, Zou J, Li Y (2017) Characterization of the clustered regularly interspaced short palindromic repeats sites in Streptococcus mutans isolated from early childhood caries patients. Arch Oral Biol 83:174–180. https://doi.org/10.1016/j.archoralbio.2017.07.023

Chen T, Yu W-H, Izard J, Baranova OV, Lakshmanan A, Dewhirst FE (2010) The human oral microbiome database. A web accessible resource for investigating oral microbe taxonomic and genomic information. Database (Oxford) 2010:baq013. https://doi.org/10.1093/database/baq013

Cheng X, Zheng X, Zhou X, Zeng J, Ren Z, Xu X et al (2016) Regulation of oxidative response and extracellular polysaccharide synthesis by a diadenylate cyclase in Streptococcus mutans. Environ Microbiol 18(3):904–922. https://doi.org/10.1111/1462-2920.13123

Clark DT, Soory M (2006) The metabolism of cholesterol and certain hormonal steroids by Treponema denticola. Steroids 71(5):352–363. https://doi.org/10.1016/j.steroids.2005.11.006

Conti V, Leone MC, Casato M, Nicoli M, Granata G, Carlesimo M (2015) High prevalence of gluten sensitivity in a cohort of patients with undifferentiated connective tissue disease. Eur Ann Allergy Clin Immunol 47(2):54–57

Cornejo UP, van der Veen MH, Krom BP (2019) Review. Modulation of the oral microbiome by the host to promote ecological balance. In. Odontology 107(4):437–448. https://doi.org/10.1007/s10266-019-00413-x

Cross KL, Chirania P, Xiong W, Beall CJ, Elkins JG, Giannone RJ et al (2018) Insights into the evolution of host association through the isolation and characterization of a novel human periodontal Pathobiont, Desulfobulbus oralis. MBio 9(2). https://doi.org/10.1128/mBio.02061-17

Darwazeh AM-G, Al-Dwairi ZN, Al-Zwairi AA-W (2010) The relationship between tobacco smoking and oral colonization with Candida species. J Contemp Dent Pract 11(3):17–24

Davenport ER (2017) Tooth be told, genetics influences oral microbiome. Cell Host Microbe 22(3):251–253. https://doi.org/10.1016/j.chom.2017.08.018

Demmitt BA, Corley RP, Huibregtse BM, Keller MC, Hewitt JK, McQueen MB et al (2017) Genetic influences on the human oral microbiome. BMC Genomics 18(1):659. https://doi.org/10.1186/s12864-017-4008-8

Dewhirst FE, Chen T, Izard J, Paster BJ, Tanner ACR, Yu W-H et al (2010) The human oral microbiome. J Bacteriol 192(19):5002–5017. https://doi.org/10.1128/JB.00542-10

Diaz PI, Hoare A, Hong B-Y (2016) Subgingival microbiome shifts and community dynamics in periodontal diseases. J Calif Dent Assoc 44(7):421–435

Dickson RP, Erb-Downward JR, Freeman CM, McCloskey L, Beck JM, Huffnagle GB, Curtis JL (2015) Spatial variation in the healthy human lung microbiome and the adapted island model of lung biogeography. Ann Am Thorac Soc 12(6):821–830. https://doi.org/10.1513/AnnalsATS.201501-029OC

Dominguez-Bello MG, Costello EK, Contreras M, Magris M, Hidalgo G, Fierer N, Knight R (2010) Delivery mode shapes the acquisition and structure of the initial microbiota across multiple body habitats in newborns. Proc Natl Acad Sci U S A 107(26):11971–11975. https://doi.org/10.1073/pnas.1002601107

Drell T, Štšepetova J, Simm J, Rull K, Aleksejeva A, Antson A et al (2017) The influence of different maternal microbial communities on the development of infant gut and oral microbiota. Sci Rep 7(1):9940. https://doi.org/10.1038/s41598-017-09278-y

Endo A, Watanabe T, Ogata N, Nozawa T, Aikawa C, Arakawa S et al (2015) Comparative genome analysis and identification of competitive and cooperative interactions in a polymicrobial disease. ISME J 9(3):629–642. https://doi.org/10.1038/ismej.2014.155

Eshriqui I, Viljakainen HT, Ferreira SRG, Raju SC, Weiderpass E, Figueiredo RAO (2020) Breastfeeding may have a long-term effect on oral microbiota. Results from the fin-HIT cohort. Int Breastfeed J 15(1):42. https://doi.org/10.1186/s13006-020-00285-w

Fåk F, Tremaroli V, Bergström G, Bäckhed F (2015) Oral microbiota in patients with atherosclerosis. Therosclerosis 243(2):573–578. https://doi.org/10.1016/j.atherosclerosis.2015.10.097

Feller L, Altini M, Khammissa RAG, Chandran R, Bouckaert M, Lemmer J (2013) Oral mucosal immunity. Oral Surg Oral Med Oral Pathol Oral Radiol 116(5):576–583. https://doi.org/10.1016/j.oooo.2013.07.013

Figuero E, Nóbrega DF, García-Gargallo M, Tenuta LMA, Herrera D, Carvalho JC (2017) Mechanical and chemical plaque control in the simultaneous management of gingivitis and caries. A systematic review. J Clin Periodontol 44(Suppl 18):S116–S134. https://doi.org/10.1111/jcpe.12674

Filoche S, Wong L, Sissons CH (2010) Oral biofilms. Emerging concepts in microbial ecology. J Dent Res 89(1):8–18. https://doi.org/10.1177/0022034509351812

Galvão-Moreira LV, da Cruz MCFN (2016) Oral microbiome, periodontitis and risk of head and neck cancer. Oral Oncol 53:17–19. https://doi.org/10.1016/j.oraloncology.2015.11.013

García-Gómez E, González-Pedrajo B, Camacho-Arroyo I (2013) Role of sex steroid hormones in bacterial-host interactions. Biomed Res Int 2013:928290. https://doi.org/10.1155/2013/928290

Gomez A, Espinoza JL, Harkins DM, Leong P, Saffery R, Bockmann M et al (2017) Host genetic control of the oral microbiome in health and disease. Cell Host Microbe 22(3):269–278.e3. https://doi.org/10.1016/j.chom.2017.08.013

Hajishengallis G (2014) The inflammophilic character of the periodontitis-associated microbiota. Mol Oral Microbiol 29(6):248–257. https://doi.org/10.1111/omi.12065

Hakansson A, Molin G (2011) Gut microbiota and inflammation. Nutrients 3(6):637–682. https://doi.org/10.3390/nu3060637

Hannig C, Hannig M, Kensche A, Carpenter G (2017) The mucosal pellicle–an underestimated factor in oral physiology. Arch Oral Biol 80:144–152. https://doi.org/10.1016/j.archoralbio.2017.04.001

Hans M, Veenu MH (2014) Epithelial antimicrobial peptides. Guardian of the oral cavity. Int J Pept 2014:370297. https://doi.org/10.1155/2014/370297

Hauss-Wegrzyniak B, Wenk GL (2002) Beta-amyloid deposition in the brains of rats chronically infused with thiorphan or lipopolysaccharide. The role of ascorbic acid in the vehicle. Neurosci Lett 322(2):75–78. https://doi.org/10.1016/s0304-3940(02)00087-3

Hernandez BY, Zhu X, Goodman MT, Gatewood R, Mendiola P, Quinata K, Paulino YC (2017) Betel nut chewing, oral premalignant lesions, and the oral microbiome. PLoS One 12(2):e0172196. https://doi.org/10.1371/journal.pone.0172196

Hernichel-Gorbach E, Kornman KS, Holt SC, Nichols F, Meador H, Kung JT, Thomas CA (1994) Host responses in patients with generalized refractory periodontitis. J Periodontol 65(1):8–16. https://doi.org/10.1902/jop.1994.65.1.8

Holgerson PL, Vestman NR, Claesson R, Ohman C, Domellöf M, Tanner ACR et al (2013) Oral microbial profile discriminates breast-fed from formula-fed infants. J Pediatr Gastroenterol Nutr 56(2):127–136. https://doi.org/10.1097/MPG.0b013e31826f2bc6

Holland SM, DeLeo FR, Elloumi HZ, Hsu AP, Uzel G, Brodsky N et al (2007) STAT3 mutations in the hyper-IgE syndrome. N Engl J Med 357(16):1608–1619. https://doi.org/10.1056/NEJMoa073687

Hong B-Y, Sobue T, Choquette L, Dupuy AK, Thompson A, Burleson JA et al (2019) Chemotherapy-induced oral mucositis is associated with detrimental bacterial dysbiosis. Microbiome 7(1):66. https://doi.org/10.1186/s40168-019-0679-5

Huffnagle GB, Dickson RP, Lukacs NW (2017) The respiratory tract microbiome and lung inflammation. A two-way street. Mucosal Immunol 10(2):299–306. https://doi.org/10.1038/mi.2016.108

Hullahalli K, Rodrigues M, Nguyen UT, Palmer K (2018) An attenuated CRISPR-Cas system in enterococcus faecalis permits DNA acquisition. MBio 9(3). https://doi.org/10.1128/mBio.00414-18

Hullahalli K, Rodrigues M, Palmer KL (2017) Exploiting CRISPR-Cas to manipulate enterococcus faecalis populations. elife 6. https://doi.org/10.7554/eLife.26664

Idris A, Hasnain SZ, Huat LZ, Koh D (2017) Human diseases, immunity and the oral microbiota–insights gained from metagenomic studies. Oral Sci Int 14(2):27–32. https://doi.org/10.1016/S1348-8643(16)30024-6

Jeremias F, Koruyucu M, Küchler EC, Bayram M, Tuna EB, Deeley K et al (2013) Genes expressed in dental enamel development are associated with molar-incisor hypomineralization. Arch Oral Biol 58(10):1434–1442. https://doi.org/10.1016/j.archoralbio.2013.05.005

Jiang W, Maniv I, Arain F, Wang Y, Levin BR, Marraffini LA (2013) Dealing with the evolutionary downside of CRISPR immunity. Bacteria and beneficial plasmids. PLoS Genet 9(9):e1003844. https://doi.org/10.1371/journal.pgen.1003844

Jorth P, Turner KH, Gumus P, Nizam N, Buduneli N, Whiteley M (2014) Metatranscriptomics of the human oral microbiome during health and disease. MBio 5(2):e01012–e01014. https://doi.org/10.1128/mBio.01012-14

Ka D, Lee H, Jung Y-D, Kim K, Seok C, Suh N, Bae E (2016) Crystal structure of streptococcus pyogenes Cas1 and its interaction with Csn2 in the type II CRISPR-Cas system. Structure (London, England: 1993) 24(1):70–79. https://doi.org/10.1016/j.str.2015.10.019

Kajfasz JK, Abranches J, Lemos JA (2011) Transcriptome analysis reveals that ClpXP proteolysis controls key virulence properties of Streptococcus mutans. Microbiology 157(Pt 10):2880–2890. https://doi.org/10.1099/mic.0.052407-0

Kajfasz JK, Rivera-Ramos I, Abranches J, Martinez AR, Rosalen PL, Derr AM et al (2010) Two Spx proteins modulate stress tolerance, survival, and virulence in Streptococcus mutans. J Bacteriol 192(10):2546–2556. https://doi.org/10.1128/JB.00028-10

Karpiński TM (2019) Role of oral microbiota in cancer development. Microorganisms 7(1). https://doi.org/10.3390/microorganisms7010020

Kato I, Vasquez A, Moyerbrailean G, Land S, Djuric Z, Sun J et al (2017) Nutritional correlates of human oral microbiome. J Am Coll Nutr 36(2):88–98. https://doi.org/10.1080/07315724.2016.1185386

Kato I, Vasquez AA, Moyerbrailean G, Land S, Sun J, Lin H-S, Ram JL (2016) Oral microbiome and history of smoking and colorectal cancer. J Epidemiol Res 2(2):92–101. https://doi.org/10.5430/jer.v2n2p92

Katz J, Onate MD, Pauley KM, Bhattacharyya I, Cha S (2011) Presence of Porphyromonas gingivalis in gingival squamous cell carcinoma. Int J Oral Sci 3(4):209–215. https://doi.org/10.4248/IJOS11075

Kaur K, Sculley D, Wallace J, Turner A, Ferraris C, Veysey M et al (2019) Micronutrients and bioactive compounds in oral inflammatory diseases. J Nutr Intermediary Metabolism 18(7):100105. https://doi.org/10.1016/j.jnim.2019.100105

Ketelhuth DFJ, Hansson GK (2016) Adaptive response of T and B cells in atherosclerosis. Circ Res 118(4):668–678. https://doi.org/10.1161/CIRCRESAHA.115.306427

Kilian M, Chapple ILC, Hannig M, Marsh PD, Meuric V, Pedersen AML et al (2016) The oral microbiome–an update for oral healthcare professionals. Br Dent J 221(10):657–666. https://doi.org/10.1038/sj.bdj.2016.865

Kilian M, Reinholdt J, Lomholt H, Poulsen K, Frandsen EV (1996) Biological significance of IgA1 proteases in bacterial colonization and pathogenesis. Critical evaluation of experimental evidence. APMIS 104(5):321–338. https://doi.org/10.1111/j.1699-0463.1996.tb00724.x

Kleinstein SE, Nelson KE, Freire M (2020) Inflammatory networks linking Oral microbiome with systemic health and disease. J Dent Res 99(10):1131–1139. https://doi.org/10.1177/0022034520926126

Kudo Y, Tada H, Fujiwara N, Tada Y, Tsunematsu T, Miyake Y, Ishimaru N (2016) Oral environment and cancer. Genes and Environment 38:13. https://doi.org/10.1186/s41021-016-0042-z

Kulkarni GV, Chng T, Eny KM, Nielsen D, Wessman C, El-Sohemy A (2013) Association of GLUT2 and TAS1R2 genotypes with risk for dental caries. Caries Res 47(3):219–225. https://doi.org/10.1159/000345652

Kumar PS (2013) Sex and the subgingival microbiome. Do female sex steroids affect periodontal bacteria? Periodontol 61(1):103–124. https://doi.org/10.1111/j.1600-0757.2011.00398.x

Kumar S, Tadakamadla J, Johnson NW (2016) Effect of Toothbrushing frequency on incidence and increment of dental caries. A systematic review and meta-analysis. J Dent Res 95(11):1230–1236. https://doi.org/10.1177/0022034516655315

Kuo L-C, Polson AM, Kang T (2008) Associations between periodontal diseases and systemic diseases. A review of the inter-relationships and interactions with diabetes, respiratory diseases, cardiovascular diseases and osteoporosis. Public Health 122(4):417–433. https://doi.org/10.1016/j.puhe.2007.07.004

Lalla E, Papapanou PN (2011) Diabetes mellitus and periodontitis. A tale of two common interrelated diseases. Nat Rev Endocrinol 7(12):738–748. https://doi.org/10.1038/nrendo.2011.106

Lee W-H, Chen H-M, Yang S-F, Liang C, Peng C-Y, Lin F-M et al (2017) Bacterial alterations in salivary microbiota and their association in oral cancer. Sci Rep 7(1):16540. https://doi.org/10.1038/s41598-017-16418-x

Lee JW, Lee YK, Yuk DY, Choi DY, Ban SB, Oh KW, Hong JT (2008) Neuro-inflammation induced by lipopolysaccharide causes cognitive impairment through enhancement of beta-amyloid generation. J Neuroinflammation 5:37. https://doi.org/10.1186/1742-2094-5-37

Lemos JA, Palmer SR, Zeng L, Wen ZT, Kajfasz JK, Freires IA et al (2019) The biology of Streptococcus mutans. Microbiol Spectrum 7(1). https://doi.org/10.1128/microbiolspec.GPP3-0051-2018

Li Y, Caufield PW, Dasanayake AP, Wiener HW, Vermund SH (2005) Mode of delivery and other maternal factors influence the acquisition of Streptococcus mutans in infants. J Dent Res 84(9):806–811. https://doi.org/10.1177/154405910508400905

Li Y, Luo W, Deng Z, Lei G (2016) Diet-intestinal microbiota Axis in osteoarthritis. A possible role. Mediat Inflamm 2016:3495173. https://doi.org/10.1155/2016/3495173

Lif Holgerson P, Harnevik L, Hernell O, Tanner ACR, Johansson I (2011) Mode of birth delivery affects oral microbiota in infants. J Dent Res 90(10):1183–1188. https://doi.org/10.1177/0022034511418973

Lindenstrauss AG, Pavlovic M, Bringmann A, Behr J, Ehrmann MA, Vogel RF (2011) Comparison of genotypic and phenotypic cluster analyses of virulence determinants and possible role of CRISPR elements towards their incidence in enterococcus faecalis and enterococcus faecium. Syst Appl Microbiol 34(8):553–560. https://doi.org/10.1016/j.syapm.2011.05.002

Lips A, Antunes LS, Antunes LA, Abreu JGB d, Barreiros D, Oliveira DSB d et al (2017) Genetic polymorphisms in DEFB1 and miRNA202 are involved in salivary human β-Defensin 1 levels and caries experience in children. Caries Res 51(3):209–215. https://doi.org/10.1159/000458537

Liu B, Faller LL, Klitgord N, Mazumdar V, Ghodsi M, Sommer DD et al (2012) Deep sequencing of the oral microbiome reveals signatures of periodontal disease. PLoS One 7(6):e37919. https://doi.org/10.1371/journal.pone.0037919

Liu J, Wu C, Huang I-H, Merritt J, Qi F (2011) Differential response of Streptococcus mutans towards friend and foe in mixed-species cultures. Microbiology 157(Pt 9):2433–2444. https://doi.org/10.1099/mic.0.048314-0

Lynge P, Marie A, Belstrøm D (2019) The role of natural salivary defences in maintaining a healthy oral microbiota. J Dent 80(Suppl 1):S3–S12. https://doi.org/10.1016/j.jdent.2018.08.010

Mager DL, Haffajee AD, Devlin PM, Norris CM, Posner MR, Goodson JM (2005) The salivary microbiota as a diagnostic indicator of oral cancer. A descriptive, non-randomized study of cancer-free and oral squamous cell carcinoma subjects. J Transl Med 3:27. https://doi.org/10.1186/1479-5876-3-27

Mammen MJ, Sethi S (2016) COPD and the microbiome. Respirol (Carlton, Vic) 21(4):590–599. https://doi.org/10.1111/resp.12732

Marsh PD, Do T, Beighton D, Devine DA (2016) Influence of saliva on the oral microbiota. Periodontol 70(1):80–92. https://doi.org/10.1111/prd.12098

Meyle J, Dommisch H, Groeger S, Giacaman RA, Costalonga M, Herzberg M (2017) The innate host response in caries and periodontitis. J Clin Periodontol 44(12):1215–1225. https://doi.org/10.1111/jcpe.12781

Miley DD, Garcia MN, Hildebolt CF, Shannon WD, Couture RA, Spearie A, Catherine L et al (2009) Cross-sectional study of vitamin D and calcium supplementation effects on chronic periodontitis. J Periodontol 80(9):1433–1439. https://doi.org/10.1902/jop.2009.090077

Mojica FJM, Díez-Villaseñor C, García-Martínez J, Soria E (2005) Intervening sequences of regularly spaced prokaryotic repeats derive from foreign genetic elements. J Mol Evol 60(2):174–182. https://doi.org/10.1007/s00239-004-0046-3

Mojica FJ, Díez-Villaseñor C, Soria E, Juez G (2000) Biological significance of a family of regularly spaced repeats in the genomes of archaea, bacteria and mitochondria. Mol Microbiol 36(1):244–246. https://doi.org/10.1046/j.1365-2958.2000.01838.x

Moon SB, Kim DY, Ko J-H, Kim J-S, Kim Y-S (2019) Improving CRISPR genome editing by engineering guide RNAs. Trends Biotechnol 37(8):870–881. https://doi.org/10.1016/j.tibtech.2019.01.009

Moutsopoulos NM, Konkel JE (2018) Tissue-specific immunity at the Oral mucosal barrier. Trends Immunol 39(4):276–287. https://doi.org/10.1016/j.it.2017.08.005

Munro CL, Grap MJ (2004) Oral health and care in the intensive care unit. State of the science. In. Am J Crit Care 13(1):25–33. Discussion 34

Mylona E, Vadala C, Papastamopoulos V, Skoutelis A (2012) Brain abscess caused by enterococcus faecalis following a dental procedure in a patient with hereditary hemorrhagic telangiectasia. J Clin Microbiol 50(5):1807–1809. https://doi.org/10.1128/JCM.06658-11

Naidu M, Robles-Sikisaka R, Abeles SR, Boehm TK, Pride DT (2014) Characterization of bacteriophage communities and CRISPR profiles from dental plaque. BMC Microbiol 14:175. https://doi.org/10.1186/1471-2180-14-175

Neiva RF, Al-Shammari K, Nociti FH, Soehren S, Wang H-L (2005) Effects of vitamin-B complex supplementation on periodontal wound healing. J Periodontol 76(7):1084–1091. https://doi.org/10.1902/jop.2005.76.7.1084

Offenbacher S, Heasman PA, Collins JG (1993) Modulation of host PGE2 secretion as a determinant of periodontal disease expression. J Periodontol 64(Suppl 5S):432–444. https://doi.org/10.1902/jop.1993.64.5s.432

Pacios S, Kang J, Galicia J, Gluck K, Patel H, Ovaydi-Mandel A et al (2012) Diabetes aggravates periodontitis by limiting repair through enhanced inflammation. FASEB J 26(4):1423–1430. https://doi.org/10.1096/fj.11-196279

Palm F, Pussinen PJ, Aigner A, Becher H, Buggle F, Bauer MF et al (2016) Association between infectious burden, socioeconomic status, and ischemic stroke. Atherosclerosis 254:117–123. https://doi.org/10.1016/j.atherosclerosis.2016.10.008

Palmer KL, Gilmore MS (2010) Multidrug-resistant enterococci lack CRISPR-cas. MBio 1(4). https://doi.org/10.1128/mBio.00227-10

Paster BJ, Boches SK, Galvin JL, Ericson RE, Lau CN, Levanos VA et al (2001) Bacterial diversity in human subgingival plaque. J Bacteriol 183(12):3770–3783. https://doi.org/10.1128/JB.183.12.3770-3783.2001

Pedersen AML, Sørensen CE, Proctor GB, Carpenter GH, Ekström J (2018) Salivary secretion in health and disease. J Oral Rehabil 45(9):730–746. https://doi.org/10.1111/joor.12664

Pietiäinen M, Liljestrand JM, Kopra E, Pussinen PJ (2018) Mediators between oral dysbiosis and cardiovascular diseases. Eur J Oral Sci 126(Suppl 1):26–36. https://doi.org/10.1111/eos.12423

Poole S, Singhrao SK, Chukkapalli S, Rivera M, Velsko I, Kesavalu L, Crean SJ (2015) Active invasion of Porphyromonas gingivalis and infection-induced complement activation in ApoE−/− mice brains. J Alzheimer's Dis: JAD 43(1):67–80. https://doi.org/10.3233/JAD-140315

Poole S, Singhrao SK, Kesavalu L, Curtis MA, Crean SJ (2013) Determining the presence of periodontopathic virulence factors in short-term postmortem Alzheimer's disease brain tissue. J Alzheimer's Dis: JAD 36(4):665–677. https://doi.org/10.3233/JAD-121918

Price R, MacLennan G, Glen J (2014) Selective digestive or oropharyngeal decontamination and topical oropharyngeal chlorhexidine for prevention of death in general intensive care. Systematic review and network meta-analysis. BMJ 348:g2197. https://doi.org/10.1136/bmj.g2197

Price VJ, McBride SW, Hullahalli K, Chatterjee A, Duerkop BA, Palmer KL (2019) Enterococcus faecalis CRISPR-Cas is a robust barrier to conjugative antibiotic resistance dissemination in the murine intestine. mSphere 4(4). https://doi.org/10.1128/mSphere.00464-19

Pride DT, Salzman J, Relman DA (2012) Comparisons of clustered regularly interspaced short palindromic repeats and viromes in human saliva reveal bacterial adaptations to salivary viruses. Environ Microbiol 14(9):2564–2576. https://doi.org/10.1111/j.1462-2920.2012.02775.x

Proctor DM, Shelef KM, Gonzalez A, Davis CL, Dethlefsen L, Burns AR et al (2020) Microbial biogeography and ecology of the mouth and implications for periodontal diseases. Periodontol 82(1):26–41. https://doi.org/10.1111/prd.12268

Pu CY, Seshadri M, Manuballa S, Yendamuri S (2020) The Oral microbiome and lung diseases. Curr Oral Health Rep 7(1):79–86. https://doi.org/10.1007/s40496-020-00259-1

Pursey E, Sünderhauf D, Gaze WH, Westra ER, van Houte S (2018) CRISPR-Cas antimicrobials. Challenges and future prospects. PLoS Pathog 14(6):e1006990. https://doi.org/10.1371/journal.ppat.1006990

Pushalkar S, Ji X, Li Y, Estilo C, Yegnanarayana R, Singh B et al (2012) Comparison of oral microbiota in tumor and non-tumor tissues of patients with oral squamous cell carcinoma. BMC Microbiol 12:144. https://doi.org/10.1186/1471-2180-12-144

Pussinen PJ, Tuomisto K, Jousilahti P, Havulinna AS, Sundvall J, Salomaa V (2007) Endotoxemia, immune response to periodontal pathogens, and systemic inflammation associate with incident cardiovascular disease events. Arterioscler Thromb Vasc Biol 27(6):1433–1439. https://doi.org/10.1161/ATVBAHA.106.138743

Rosier BT, Buetas E, Moya-Gonzalvez EM, Artacho A, Mira A (2020) Nitrate as a potential prebiotic for the oral microbiome. Sci Rep 10(1):12895. https://doi.org/10.1038/s41598-020-69931-x

Ruby J, Goldner M (2007) Nature of symbiosis in oral disease. J Dent Res 86(1):8–11. https://doi.org/10.1177/154405910708600102

Rutger Persson G (2012) Rheumatoid arthritis and periodontitis–inflammatory and infectious connections. Review of the literature. J Oral Microbiol 4. https://doi.org/10.3402/jom.v4i0.11829

Sakaguchi S, Tanaka S, Tanaka A, Ito Y, Maeda S, Sakaguchi N, Hashimoto M (2011) Thymus, innate immunity and autoimmune arthritis. Interplay of gene and environment. FEBS Lett 585(23):3633–3639. https://doi.org/10.1016/j.febslet.2011.10.026

Samaranayake L, Matsubara VH (2017) Normal oral flora and the oral ecosystem. Dent Clin N Am 61(2):199–215. https://doi.org/10.1016/j.cden.2016.11.002

Sarode G, Sarode SC, Tupkari J, Patil S (2017) Is oral squamous cell carcinoma unique in terms of intra- and inter-tumoral heterogeneity? In. Translational Res Oral Oncol 2:2057178X1770357. https://doi.org/10.1177/2057178X17703578

Sasaki M, Yamaura C, Ohara-Nemoto Y, Tajika S, Kodama Y, Ohya T et al (2005) Streptococcus anginosus infection in oral cancer and its infection route. Oral Dis 11(3):151–156. https://doi.org/10.1111/j.1601-0825.2005.01051.x

Sato Y, Yamagishi J, Yamashita R, Shinozaki N, Ye B, Yamada T et al (2015) Inter-individual differences in the oral Bacteriome are greater than intra-day fluctuations in individuals. PLoS One 10(6):e0131607. https://doi.org/10.1371/journal.pone.0131607

Scher JU, Ubeda C, Equinda M, Khanin R, Buischi Y, Viale A et al (2012) Periodontal disease and the oral microbiota in new-onset rheumatoid arthritis. Arthritis Rheum 64(10):3083–3094. https://doi.org/10.1002/art.34539

Sellappa S, Rajamanickam U, Selvaraj V, RafiqKhan M, Vijayakumar S (2015) Association of bacterial growth in the oral cavity between tobacco smokers and tobacco chewers. Int J Toxicol Pharmacol Res 7(4):228–231

Sen S, Giamberardino LD, Moss K, Morelli T, Rosamond WD, Gottesman RF et al (2018) Periodontal disease, regular dental care use, and incident ischemic stroke. Stroke 49(2):355–362. https://doi.org/10.1161/STROKEAHA.117.018990

Serbanescu MA, Cordova M, Krastel K, Flick R, Beloglazova N, Latos A et al (2015) Role of the Streptococcus mutans CRISPR-Cas systems in immunity and cell physiology. J Bacteriol 197(4):749–761. https://doi.org/10.1128/Jb.02333-14

Shchipkova AY, Nagaraja HN, Kumar PS (2010) Subgingival microbial profiles of smokers with periodontitis. J Dent Res 89(11):1247–1253. https://doi.org/10.1177/0022034510377203

Singh V, Braddick D, Dhar PK (2017) Exploring the potential of genome editing CRISPR-Cas9 technology. Gene 599:1–18. https://doi.org/10.1016/j.gene.2016.11.008

Sobue T, Bertolini M, Thompson A, Peterson DE, Diaz PI, Dongari-Bagtzoglou A (2018) Chemotherapy-induced oral mucositis and associated infections in a novel organotypic model. Mol Oral Microbiol 33(3):212–223. https://doi.org/10.1111/omi.12214

Sokol H, Seksik P, Furet JP, Firmesse O, Nion-Larmurier I, Beaugerie L et al (2009) Low counts of Faecalibacterium prausnitzii in colitis microbiota. Inflamm Bowel Dis 15(8):1183–1189. https://doi.org/10.1002/ibd.20903

Sonnenburg JL, Fischbach MA (2011) Community health care. Therapeutic opportunities in the human microbiome. Sci Transl Med 3(78):78ps12. https://doi.org/10.1126/scitranslmed.3001626

Sultan AS, Kong EF, Rizk AM, Jabra-Rizk MA (2018) The oral microbiome. A lesson in coexistence. PLoS Pathog 14(1):e1006719. https://doi.org/10.1371/journal.ppat.1006719

Tanaka T, Ishigamori R (2011) Understanding carcinogenesis for fighting oral cancer. J Oncol 2011:603740. https://doi.org/10.1155/2011/603740

Tanner ACR, Kressirer CA, Rothmiller S, Johansson I, Chalmers NI (2018) The caries microbiome. Implications for reversing Dysbiosis. Adv Dent Res 29(1):78–85. https://doi.org/10.1177/0022034517736496

Theriot CM, Koenigsknecht MJ, Carlson PE, Hatton GE, Nelson AM, Li B et al (2014) Antibiotic-induced shifts in the mouse gut microbiome and metabolome increase susceptibility to Clostridium difficile infection. Nat Commun 5:3114. https://doi.org/10.1038/ncomms4114

Thomas CM, Nielsen KM (2005) Mechanisms of, and barriers to, horizontal gene transfer between bacteria. Nat Rev Microbiol 3(9):711–721. https://doi.org/10.1038/nrmicro1234

Tong Z, Du Y, Ling J, Huang L, Ma J (2017) Relevance of the clustered regularly interspaced short palindromic repeats of enterococcus faecalis strains isolated from retreatment root canals on periapical lesions, resistance to irrigants and biofilms. Exp Ther Med 14(6):5491–5496. https://doi.org/10.3892/etm.2017.5205

Turnbaugh PJ, Ley RE, Hamady M, Fraser-Liggett CM, Knight R, Gordon JI (2007) The human microbiome project. Nature 449(7164):804–810. https://doi.org/10.1038/nature06244

Underly R, Song M-S, Dunbar GL, Weaver CL (2015) Expression of Alzheimer-type neurofibrillary epitopes in primary rat cortical neurons following infection with enterococcus faecalis. Front Aging Neurosci 7:259. https://doi.org/10.3389/fnagi.2015.00259

Vercoe RB, Chang JT, Dy RL, Taylor C, Gristwood T, Clulow JS et al (2013) Cytotoxic chromosomal targeting by CRISPR/Cas systems can reshape bacterial genomes and expel or remodel pathogenicity islands. PLoS Genet 9(4):e1003454. https://doi.org/10.1371/journal.pgen.1003454

Vestman NR, Timby N, Holgerson PL, Kressirer CA, Claesson R, Domellöf M et al (2013) Characterization and in vitro properties of oral lactobacilli in breastfed infants. BMC Microbiol 13:193. https://doi.org/10.1186/1471-2180-13-193

Watanabe T, Nozawa T, Aikawa C, Amano A, Maruyama F, Nakagawa I (2013) CRISPR regulation of intraspecies diversification by limiting IS transposition and intercellular recombination. Genome Biol Evol 5(6):1099–1114. https://doi.org/10.1093/gbe/evt075

Welch M, Jessica L, Dewhirst FE, Borisy GG (2019) Biogeography of the oral microbiome. The site-specialist hypothesis. Annu Rev Microbiol 73:335–358. https://doi.org/10.1146/annurev-micro-090817-062503

Welch M, Jessica L, Ramírez-Puebla S, Tabita; Borisy, Gary G. (2020) Oral microbiome geography. Micron-scale habitat and niche. Cell Host Microbe 28(2):160–168. https://doi.org/10.1016/j.chom.2020.07.009

Welch M, Jessica L, Rossetti BJ, Rieken CW, Dewhirst FE, Borisy GG (2016) Biogeography of a human oral microbiome at the micron scale. Proc Natl Acad Sci U S A 113(6):E791–E800. https://doi.org/10.1073/pnas.1522149113

Welch M, Jessica L, Utter DR, Rossetti BJ, Welch M, David B, Eren A, Murat; Borisy, Gary G. (2014) Dynamics of tongue microbial communities with single-nucleotide resolution using oligotyping. Front Microbiol 5:568. https://doi.org/10.3389/fmicb.2014.00568

Wilbert SA, Welch M, Jessica L, Borisy GG (2020) Spatial ecology of the human tongue dorsum microbiome. Cell Rep 30(12):4003–4015.e3. https://doi.org/10.1016/j.celrep.2020.02.097

Wu J, Peters BA, Dominianni C, Zhang Y, Pei Z, Yang L et al (2016) Cigarette smoking and the oral microbiome in a large study of American adults. ISME J 10(10):2435–2446. https://doi.org/10.1038/ismej.2016.37

Xiao J, Fiscella KA, Gill SR (2020) Oral microbiome. Possible harbinger for children's health. Int J Oral Sci 12(1):12. https://doi.org/10.1038/s41368-020-0082-x

Xiao E, Mattos M, Vieira GHA, Chen S, Corrêa JD, Wu Y et al (2017) Diabetes enhances IL-17 expression and alters the Oral microbiome to increase its pathogenicity. Cell Host Microbe 22(1):120–128.e4. https://doi.org/10.1016/j.chom.2017.06.014

Xie Z, Okinaga T, Niu G, Qi F, Merritt J (2010) Identification of a novel bacteriocin regulatory system in Streptococcus mutans. Mol Microbiol 78(6):1431–1447. https://doi.org/10.1111/j.1365-2958.2010.07417.x

Xu D, Pavlidis P, Taskent RO, Alachiotis N, Flanagan C, DeGiorgio M et al (2017) Archaic hominin introgression in Africa contributes to functional salivary MUC7 genetic variation. Mol Biol Evol 34(10):2704–2715. https://doi.org/10.1093/molbev/msx206

Yamashita Y, Takeshita T (2017) The oral microbiome and human health. J Oral Sci 59(2):201–206. https://doi.org/10.2334/josnusd.16-0856

Yosef I, Manor M, Kiro R, Qimron U (2015) Temperate and lytic bacteriophages programmed to sensitize and kill antibiotic-resistant bacteria. Proc Natl Acad Sci U S A 112(23):7267–7272. https://doi.org/10.1073/pnas.1500107112

Youle M, Knowlton N, Rohwer F, Gordon J, Relman DA (2013) Superorganisms and Holobionts. Microbe Magazine 8(4):152–153. https://doi.org/10.1128/microbe.8.152.1

Yussif NM, Aziz A, Manar A, Rahman A, Ahmed R (2016) Evaluation of the anti-inflammatory effect of locally delivered vitamin C in the treatment of persistent gingival inflammation. Clinical and histopathological study. J Nutr Metabolism 2016:2978741. https://doi.org/10.1155/2016/2978741

Zijnge V, van Leeuwen M, Barbara M, Degener JE, Abbas F, Thurnheer T, Gmür R, Harmsen HJM (2010) Oral biofilm architecture on natural teeth. PLoS One 5(2):e9321. https://doi.org/10.1371/journal.pone.0009321

Chapter 3
Emerging Role of Gut Microbiota in Functional Gastrointestinal Disorders

Philip Augustine, Uday Chand Ghoshal, Rizwan Ahamed Zulfikar, and Cyriac Abby Philips

1 Introduction

The brain–gut interrelation plays a central role in linking psychological factors and gut dysfunction that clinically present with gastrointestinal (GI) symptoms and disease. The major clinical domains that involve brain–gut axis function include organic (structural pathology at macro-and micro-level), motility (measurable organ dysfunction), and functional GI disorders (FIGD). The latter is specifically defined in the presence of "illness experiences," symptoms rather than signs, strongly linked to psychosocial impact and diagnosed by specific subjective (Rome) criteria. The biopsychosocial concept of disorders of the GI system linked genetics, culture, and environmental factors to stress, personality traits, psychology, coping, cognition, and social functions further to central (CNS) and enteric (ENS) nervous system influences that formed the pathophysiological basis of FGID. In this regard, current research has demonstrated that the motility, sensation, immune function, and mucosal physiology of the gut influenced by food and dietary habits have been linked to alterations of the intestinal microbiota and its functional metabolism. This liaison between the gut microbiota, the local (enteric), and central nervous systems have been shown to influence symptoms, severity, and behavior among patients with

P. Augustine (✉) · R. A. Zulfikar
Gastroenterology and Advanced GI Endoscopy, Center of Excellence in Gastrointestinal Sciences, Rajagiri Hospital, Aluva, Kerala, India
e-mail: abbyphilips@theliverinst.in

U. C. Ghoshal
Department of Gastroenterology, Sanjay Gandhi Post Graduate Institute of Medical Sciences, Lucknow, Uttar Pradesh, India

C. A. Philips
Clinical and Translational Hepatology, The Liver Institute, Center of Excellence in Gastrointestinal Sciences, Rajagiri Hospital, Aluva, Kerala, India

S. Thomas (ed.), *Human Microbiome*,
https://doi.org/10.1007/978-981-16-7672-7_3

FGID. Thus, FGID is a syndrome of "clustered" GI symptoms, related to GI functioning, associated with perturbed gut–brain interaction and gut microbiota, associated with visceral hypersensitivity, motility disturbance, and altered mucosal and immune function in the presence of disturbed CNS processing, diagnosed by the Rome Criteria. The Rome IV Criteria classified FGIDs into 33 adult and 20 pediatric variants, primarily based on symptoms into anatomic regions (esophageal, gastroduodenal, bowel, biliary, and anorectal and centrally mediated disorders of GI pain) for easy utilization in clinical practice (Drossman 2016; Schmulson and Drossman 2017).

The role of the luminal microenvironment, especially the microbiota and associated functional metabolism and its relationship with the enteric neuromuscular apparatus and its central connections through the gut–brain axis, was initially noticed in patients who developed and sustained FGID-type symptoms after enteric infections. Gut–brain axis communications are dependent on several complex signaling pathways that involve the sympathetic (splanchnic) and parasympathetic (vagal) nerves, the ENS, hypothalamus–pituitary axis, and CNS that are in turn affected by intestinal microbiota (microbiota–gut–brain axis) and psychosocial factors—a bidirectional interaction. Our knowledge on gut microbiota-associated changes and strong links to FGIDs stems from translational research studies encompassing direct and indirect intestinal microbiota modulation predominantly in patients with functional dyspepsia (FD) and irritable bowel syndrome (IBS) and its subtypes. The interaction between psychosocial and dietary factors (food/food components) trigger morphological changes to the gut epithelium and alters the mucosal endocrine signaling, leading to perturbation in local and systemic immune and inflammatory responses that culminate in FGID—whether the role of intestinal microbiota is a cause or an effect to this ultimate event is a matter of further research (Fig. 3.1). Nonetheless, specific gut microbiota changes have been shown to directly and indirectly (through functional metabolism) promote symptoms, affect the severity, and engage treatment responses in patients with FGIDs (Barbara et al. 2016).

2 Gut Microbiota Associations in FGID

The microbiota–gut–brain axis activity in FGID was demonstrated initially in small animal experiments. When male rat pups were stressed by separating them from their mothers in the immediate postnatal period, increases in plasma cortisol and alterations in fecal microbiota were noticeable compared with an unseparated control group. Similarly, when germ-free mice underwent fecal microbiota transplantation (FMT) with stool derived from severe depressive patients, they demonstrated anxiety-depression behavior compared to a control group undergoing FMT from "normal" human controls. These studies represent the bidirectionality of the role of gut microbiota in FGIDs (Luo et al. 2018). The introduction of pathogenic bacteria or short or repeated antibiotic feeding courses in healthy mice was associated with

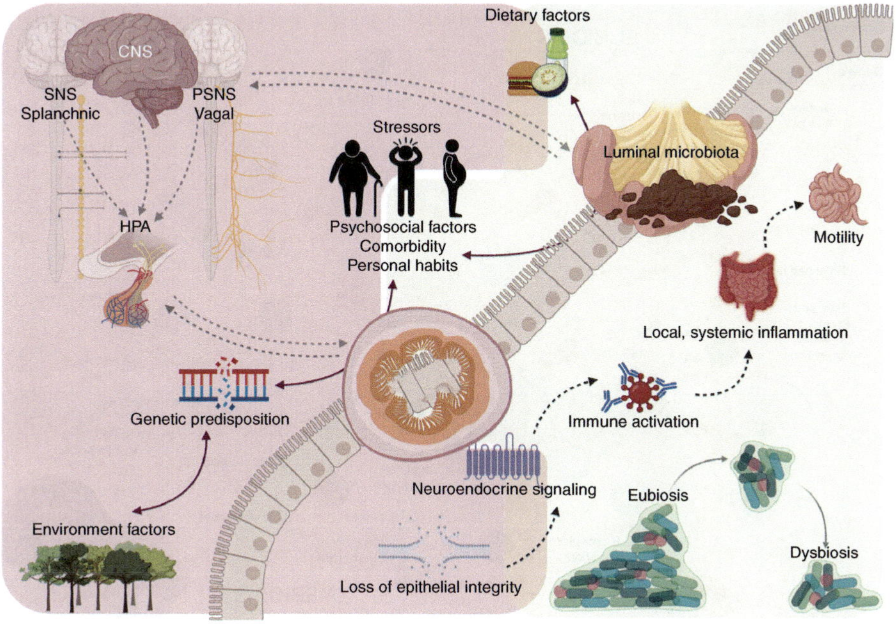

Fig. 3.1 The bidirectional, multifactorial pathophysiology of functional gastrointestinal disorders development. *CNS*, central nervous system; *SNS*, sympathetic nervous system; *PSNS*, parasympathetic nervous system; *HPA*, hypothalamus–pituitary axis

mood changes, anxiety-like behavior, and cognitive decline in the short and long term that occurred in tandem with a modulation of the intestinal microbiota. Germ-free mice demonstrated developmental changes and perturbed gut mucosal immunity, which was partially reversible through recolonization using stool transfer from healthy mice. Correspondingly, mice receiving feces from diarrhea-predominant irritable bowel syndrome (IBS) patients exhibited faster gastrointestinal transit, gut barrier dysfunction, innate immune activation, and anxiety-like responses, demonstrating a strong association between intestinal dysbiosis and intestinal and behavioral manifestations in FGID (De Palma et al. 2017; Ceylani et al. 2018; Kwon et al. 2020). Much of our understanding of microbiota (bacterial taxa) in FGID emanates from quantitative polymerase chain reaction (qPCR), PCR-denaturing gradient gel electrophoresis (DGGE), fluorescent in situ hybridization, pyrosequencing, and the recent, next-generation sequencing (NGS) studies conducted in patients with functional dyspepsia (FD) and IBS. Multiple studies have looked at gut microbial diversities within the adult and pediatric groups, between genders, and at various sites of the GI tract resulting in heterogeneous findings across populations and regions.

Furthermore, striking differences between the luminal microbiota and the mucosa-associated microbiota (MAM), the latter, considered more stable and reflective of host–disease interaction, also portend differences across studies in

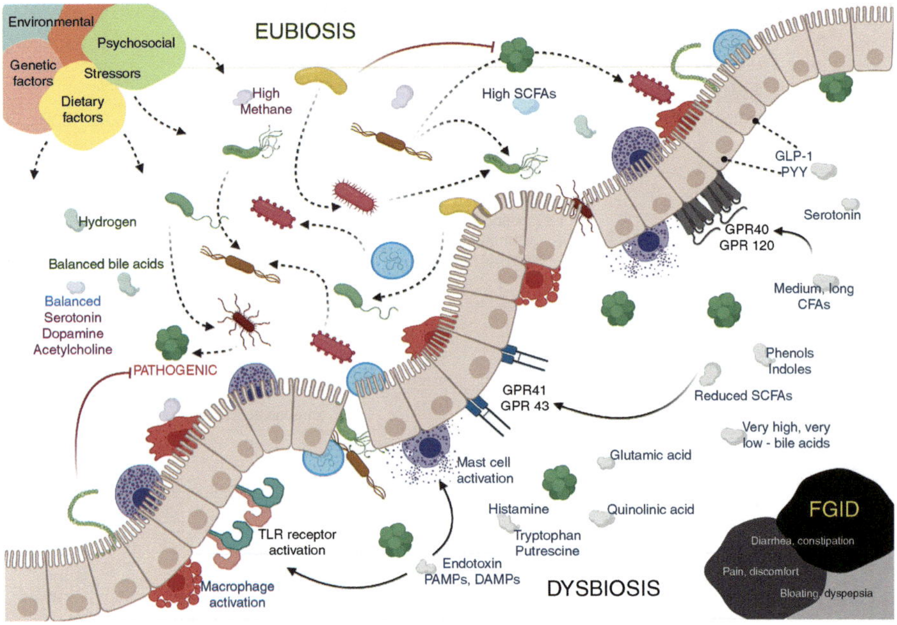

Fig. 3.2 Schematic representation of gut dysbiosis and microbial and related functional changes in patients with functional gastrointestinal disorders (FGIDs). The unbroken red lines with perpendicular dashed ends represent inhibition. *SCFA*, short-chain fatty acid; *TLR*, toll-like receptor; *PAMP*, pathogen-associated molecular pattern; *DAMP*, damage-associated molecular pattern; *GPR*, G-protein coupled receptor; *GLP*, glucagon-like peptide; *PYY*, peptide YY

FGID. Nonetheless, taken together, from a bird's eye view, all these studies provide insights into common associations between specific bacterial taxa and the type of FGID considered (Mottawea et al. 2019) (Fig. 3.2).

2.1 Gut Microbiota in Functional Dyspepsia

In patients with FD [post-prandial distress syndrome (PDS), epigastric pain syndrome (EPS), or PDS–EPS overlap], higher levels of *Prevotella* were notable in gastric fluid aspirate, and those with PDS, an inverse correlation between *Prevotella* abundance and disease severity was noted (Nakae et al. 2016). Similarly, in the gastric fluid, at the phylum level, higher *Bacteroides* compared to *Proteobacteria* and absence of *Acidobacteria* were remarkable in FD patients compared to healthy controls (Igarashi et al. 2017). In studies that looked at MAM (gastric and small intestinal mucosa biopsies), an inverse relationship between *Streptococcus* and *Prevotella* (Zhong et al. 2017), negative correlation between the abundance of *Veillonella* and gastric emptying time (Shanahan et al. 2018), and higher levels of *Firmicutes*, especially *Streptococcus*, positively correlated with symptoms (Fukui et al. 2020) in patients with FD. Studies have also demonstrated important interactions between *Helicobacter pylori* and gut microbiota in patients with nonulcer and ulcer dyspepsia. *H. pylori*-negative biopsy-proven gastritis was found to be associated with greater enrichment of *Firmicutes*, *Fusobacteria*, *Bacteroidetes*, and *Actinobacteria*. At the same time, in those patients positive for *H. pylori*, the fecal samples were enriched with *Proteobacteria*. Interestingly, it was seen that bacterial species and richness diversity were higher among persons living in less industrialized nonmodern regions in whom *H. pylori* incidence was also very low. Patients with nonulcerative dyspepsia had a greater abundance of *Cutibacterium acnes* at the species level (Gantuya et al. 2019; Chua et al. 2019).

2.2 Gut Microbiota in Irritable Bowel Syndrome

In patients with IBS (diarrhea or constipation-predominant or mixed type and unclassified), deep molecular analysis of microbiota has revealed specific bacterial taxa changes associated with symptoms and severity compared to healthy controls. The first such study to utilize state-of-the-art techniques was performed in 2007 in which authors identified changes in *Coprococcus*, *Collinsella*, and *Coprobacillus* abundances in patients with IBS (Kassinen et al. 2007). In general, a "healthy" microbiota is characterized by a higher prevalence of *Firmicutes* and *Bacteroidetes* and a lack of *Proteobacteria*. The most crucial aspect of gut microbiota related to IBS stems from observations and studies in small animal models and patients with post-infection (following acute gastroenteritis) IBS. It was noted that approximately 10% to 14% of patients within 3 to 12 months after an acute GI infection developed IBS symptoms driven by bacterial and host factors, local and systemic immune activation, and enteric neuronal changes that ultimately led to changes in intestinal motility and development of symptoms. In small animals infected with *Campylobacter jejuni*, it was shown that post-infection, alteration in stool form, increase in rectal lymphocytes, reduction in interstitial cells of Cajal, and bacterial growth predominated along with the production of cytolethal distending toxin resulting in subsequent autoimmunity to enterocyte adhesion protein vinculin impressing the fact that IBS development and progression had strong links to bacteria-driven local as well as systemic immune-related and neuroendocrine changes (Pimentel et al. 2015).

The parasite *Giardia duodenalis* has been shown to reduce thickness and disrupt extracellular matrix compositions and structural integrity of the mucosal microbiota biofilms leading to over-representation of *Clostridiales* and a decreased amount of *Phascolarctobacterium* species in experimental models of post-infection IBS, which was also clearly demonstrated among IBS patients from Italy and Norway (Beatty et al. 2017). From a dysbiosis point of view, at the phylum level, a twofold increase in *Firmicutes* to *Bacteroidetes* ratio, increase in *Actinobacteria*, and reduction in *Bifidobacterium* correlated with symptoms and severity in patients with IBS (Jeffery et al. 2012; Pimentel and Lembo 2020). A large body of evidence from gut microbiota studies in IBS emphasizes the relative richness of pro-inflammatory bacterial species (Enterobacteriaceae) associated with a parallel decline in beneficial species *Bifidobacterium* and *Lactobacillus*. The differential dysbiosis in IBS has been demonstrated between patients with and without abdominal bloating, further classified into different subtypes based on bowel habits and between patients from various regions. *Subdoligranulum* and *Anaerovorax* (belonging to the families Ruminococcaceae and Eubacteriaceae, respectively) were found to increase in those without bloating. In patients with constipation-predominant IBS, *Collinsella* was increased, while among those with predominantly diarrhea, members of the *Firmicutes* phyla *(Oscillibacter*, *Anaerovorax*, *Streptococcus*, and *Eubacteriaceae)* were significantly decreased (Ringel-Kulka et al. 2016; Zhuang et al. 2017; Ringel et al. 2018). Similarly, in a study on fecal and mucosal microbiota, researchers found that IBS symptom severity was associated negatively with microbial diversity or richness, exhaled methane levels, presence of methanogens, and reduced prevalence of Methanobacteriales or *Prevotella* species. Only two previous studies have shown the predominant role of *Pseudomonas aeruginosa* among patients with IBS (Kerckhoffs et al. 2011; Shukla et al. 2015; Ghoshal et al. 2018a, b).

Contrary to study findings on bacterial communities, in a study published from Korea, authors found that halophilic archaea such as *Halorubrum* and *Halococcus* species predominated. This was possibly due to high-salt food intake notable among Korean populations, implying the role of dietary factors on qualitative and quantitative gut microbial aspects, which is not yet fully weighed into studies on IBS and its subtypes (Nam et al. 2008). Metagenomic analysis on fecal samples from patients with constipation-predominant IBS revealed predominant microbiota directed anti-inflammatory activity when transferred to conventional mice due to increased *Akkermansia muciniphila* even with a decrease in the relative abundance of *Bacteroides*, *Roseburia*, and *Eubacterium rectale* and an increase in pathogens belonging to *Enterobacteriaceae* and *Desulfovibrio* species, demonstrating the importance of cross-talk between microbial taxa and host intestinal homeostasis (Gobert et al. 2016). A recent study analyzed the fecal and mucosa-associated bacterial composition along the GI tract in patients with IBS. The authors found that feces' bacterial profiles and the sigmoid colon mucosa, but not duodenum, differed between IBS patients. The IBS-specific bacterial profiles were linked to the colonic antibacterial gene expression. Furthermore, the fecal bacterial profile differed between IBS subtypes, while the mucosa-associated bacterial profile was significantly associated with IBS symptom severity (Sundin et al. 2020). Large-scale studies on specific microbial changes and interactions in patients with IBS subsets seem incomplete in current literature. Nonetheless, a recent study showed that diversity richness was reduced, and levels of *Faecalibacterium* and *Dorea* were lower and higher, respectively, in patients with diarrhea-predominant IBS (Maharshak et al. 2018). Another study demonstrated an increased abundance of *Prevotella* and association with a high risk of diarrhea-predominant IBS in the Chinese population (Su et al. 2018). In a systematic review and meta-analysis that included differential expression of intestinal microbiota in patients with IBS versus healthy controls and subgroup analysis, authors found lower levels of *Lactobacillus*, *Bifidobacterium*, and *Faecalibacterium prausnitzii* in patients with diarrhea-predominant IBS (Liu et al. 2017). A recent systematic review involving 777 patients and 461 healthy controls demonstrated that, for most studies, those with IBS had lower α-diversity in both fecal and mucosal samples. Relatively consistent findings on intestinal microbiota analyses included increased *Firmicutes*, decreased *Bacteroidetes*, and increased *Firmicutes:Bacteroidetes* ratio at the phylum level and increased *Clostridia* as well as decreased *Bacteroides* (Duan et al. 2019). A more recent meta-analysis showed that the family Enterobacteriaceae (phylum Proteobacteria), the family Lactobacillaceae, and the genus *Bacteroides* were increased. In contrast, uncultured Clostridiales I, *Faecalibacterium*, and *Bifidobacterium* were decreased in patients with IBS (Pittayanon et al. 2019). A study comparing fecal and mucosal gut microbial signatures among patients with inflammatory bowel disease (IBD), IBS, and healthy controls showed that *Erysipelotrichi* was a potential biomarker of IBS. In contrast, *Enterococcus* was significantly identified in patients with IBD (Lo Presti et al. 2019). Several authors have described changes associated with intestinal bacterial communities at the luminal and mucosal level in patients with IBS since the original description more than a decade ago. A summary of gut microbial (bacterial) interactions in patients with the FGIDs, FD, and IBS and its subsets are illustrated in Fig. 3.3.

3 Gut Microbiota and Functional Metabolites in FGID

Metabolomics, the exhaustive study and profile generation of small-molecule metabolic products of cells, tissues, and organisms at a specific point in time, is a novel approach to analyzing complex interactions between gut microbiota functions. Mass spectrometry and magnetic resonance spectroscopy are powerful tools to identify, quantify, and apply biostatics and mathematical models and discern biologically significant metabolites from large data sets (Liu and Locasale 2017). Apart from this, analysis and identification of functional metabolism based on metagenomic data and biomarker discovery, in the absence of quantification, can be achieved through bioinformatic pipelines such as Phylogenetic Investigation of Communities by Reconstruction of Unobserved States (PICRUSt), which allows inference of the functional profile (significant pathways of metabolism) of a microbial community using operational taxonomic units-based marker gene sequence and survey (Douglas et al. 2020). Fermentation of polysaccharides and generation of short-chain fatty acids (SCFAs; acetate, propionate, butyrate) by intestinal bacteria lead to hydrogen, methane, and other by-products that affect gut mucosal barrier, gut permeability, and bowel motility. Intestinal bacteria also play a central role in gut–brain interactions through the production or degradation of various locally acting neurenteric and systemic neuroactive substances such as the anti-inflammatory S-adenosylmethionine, neurotoxin quinolinic acid, glutamic acid, hydroxybutyric acid, dopamine, acetylcholine, kynurenine, histamine, and serotonin (Valles-Colomer et al. 2019). The bacteria also harbor hormonal receptors that mediate

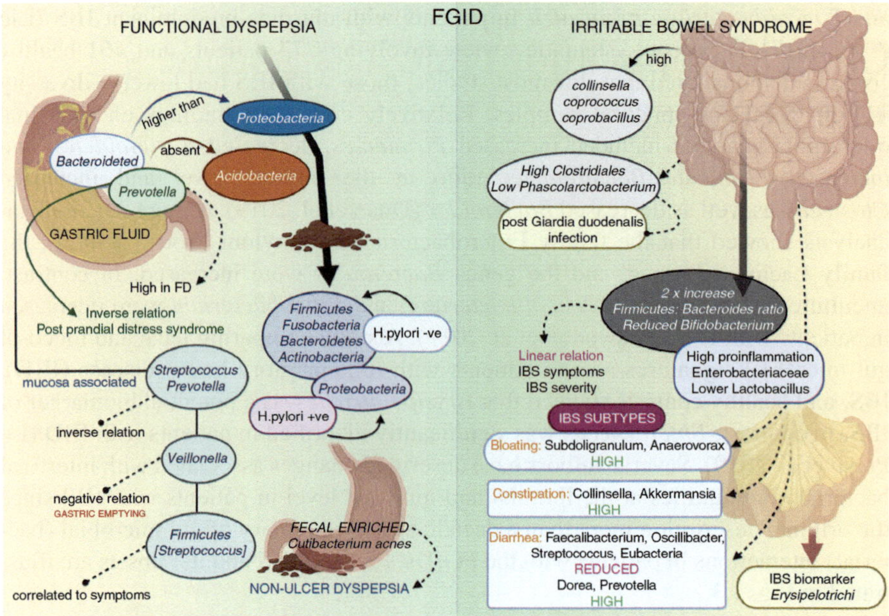

Fig. 3.3 Summary of bacterial interactions and associated relationships with symptoms in patients with functional dyspepsia and irritable bowel syndrome (IBS). FGID, functional gastrointestinal disorder; *H. pylori*, *Helicobacter pylori*

cross-talk between the host and microbiota. SCFAs are an important energy source for colonocytes but are also chemical messengers or signaling molecules for various host cells. The G-protein coupled receptors (GPR41, GPR43), also known as the free fatty acid receptor types 3 and 2 (FFAR), have been identified as receptors for SCFAs and expressed in a variety of cells, including colonic endocrine L cells, adipose tissue, neutrophils, monocytes, and mucosal mast cells. SCFAs upregulate the secretion of glucagon-like peptide 1 (GLP-1) and peptide tyrosine-tyrosine (PYY) that take part in "ileal-brake," a primary inhibitory feedback mechanism to control the transit of a meal through the GI tract to optimize nutrient digestion and absorption. Similarly, medium—and long-chain fatty acids produced by bacterial activity on dietary substrates within the host act on GPR40 and GPR120 receptors on the enterocytes and promote cholecystokinin's secretion and regulatory activity and glucose-dependent insulinotropic polypeptide/gastric inhibitory polypeptide. Similarly, reducing certain gut microbiota metabolites and subsequent increase in certain others (due to host, environment, or associated comorbidity) lead to mucosal barrier dysfunction leading to leakage of pathogens into the lamina propria of intestinal mucosa triggering the mucosal immune system. This results in pro-inflammatory cytokine production and stimulation through direct bacterial or indirect bacterial products-related activity on the toll-like receptors on the enterocytes, mast cell, and macrophage activation. This local and subsequent systemic immune activation and pro-inflammatory profile have been considered to play a central role in neurovisceral sensitivity and associated symptoms in patients with FGID, especially IBS (Fukui et al. 2018). Thus, it is evident that secondary bacterial metabolites play a central role in intestinal function and systemic neuroendocrine and immune regulation in humans. In patients with FGID, especially those with IBS, it was found that perturbed metabolite profiles in fecal samples, such as an increase in gaseous hydrogen, phenols, and indoles, were associated with symptoms and disease severity. Colonic spore-forming bacteria belonging to the *Clostridiales* order, enriched in *Ruminococcaceae* and *Lachnospiraceae*, were associated with biosynthesis and release serotonin from intestinal enterochromaffin cells and modulate intestinal motility, a serotonergic dysfunction notable in patients with IBS (Yano et al. 2015; Labus et al. 2019). Metabolism of polysaccharides in the gut lumen leads to the production of hydrogen and methane by-products. Colonic bacteria also produce short-chain fatty acids such as acetate, propionate, and butyrate that affect intestinal permeability and motility. Methane gas production, specifically by methanogens in the colon, slows intestinal transit and augments small intestinal contractility. It was shown that reduction in butyrate production and butyrate-producing taxa was found among patients with diarrhea and mixed (diarrhea, constipation)-type IBS. Similarly, lower methane production in the intestinal lumen was notable in patients with diarrhea-predominant IBS, while higher levels were noted in patients with constipation-predominant IBS. The symptoms of flatulence in FGID patients have been linked to reduced hydrogen gas removal from the colon due to decreased sulfate-reducing bacterial taxa (Pozuelo et al. 2015; Tap et al. 2016; Chong et al. 2019). The role of metabolite production and its effects on gut motility, mucosal immunity, local immune regulation, and symptom development in IBS and FD patients have been identified through prebiotic interventional studies and subsequent observations. It was shown that inulin-type fructans and arabinoxylan oligosaccharides fermentation capacity by Bifidobacteria strains depended on bacterial cooperation, and the metabolites rich in short-chain fatty acids thus produced acted on metabolite-sensing G-protein-coupled receptors to regulate inflammatory responses and motility. An essential metabolite of the human colon, butyrate, the central energy source for the colon epithelial cells, maintains gut mucosal integrity and promotes immunomodulatory and anti-inflammatory properties within the intestinal milieu. *Bifidobacteria* and other butyrate-producing bacteria, such as *Faecalibacterium prausnitzii*, *Anaerostipes*, *Eubacterium*, and *Roseburia,* interact with each other through cross-feeding pathways that generate metabolites beneficial for the host (Rivière et al. 2016, 2018; Chong et al. 2019). Bile acid metabolites are produced from cholesterol in hepatocytes with cholic and chenodeoxycholic acids. Intestinal bacteria deconjugate bile acids to form secondary bile acids such as lithocholic acid and deoxycholic acid. These stimulate enterocyte secretion through their actions on sodium and chloride channels. Excessive bile acid secretion is negatively controlled by fibroblast growth factor 19 (FGF-19). This inhibitory molecule and its regulation are affected by gut microbiota functions and metabolite production. The levels of bile acids within the intestinal lumen decide motility functions—high levels lead to diarrhea-like symptoms. In contrast, very low levels result in constipation in the host. Thus, gut microbial metabolite generation affects the host, which depends on the quality and diversity of intestinal microbiota (Raskov et al. 2016). In a China study, authors identified gut microbiota and metabolite signature in patients with IBS using gas chromatography coupled to time-of-flight mass spectrometry (GC-TOFMS) and 16S rDNA amplicon sequencing. They found that metabolites ornithine, putrescine, N-acetyl tryptophan, and L-tryptophan were associated with abdominal pain and discomfort and stool characteristics. In contrast, eicosatrienoic acid, oxoadipic acid, L-phenylalanine, L-valine, and gamma-aminobutyric acid were associated with the duration of symptoms in patients with IBS (Zhu et al. 2019). Thus, it is clear that the quality and type of bacterial taxa, their interactions, and beneficial cooperation lead to favorable metabolite generation that acts at the local and systemic levels to promote or improve intestinal function. These findings led to the use of healthy donor FMT or prebiotic use, which promotes advantageous metabolite generation for the treatment of FGID, especially in patients with IBS.

4 Fecal Microbiota Transplantation in FGID

Fecal microbiota transplantation or FMT is the infusion of screened fresh or stored (frozen or encapsulated) feces from a healthy donor into the GI tract of a patient with a specific disease addressable to intestinal dysbiosis, intending to restore microbial homeostasis and advantageous functionality toward the host. The FMT procedure gained interest with its extremely beneficial therapeutic role in patients with a mild and severe recurrent form of *Clostridium difficle* infection, a condition well known to be associated with intestinal dysbiosis (Cammarota et al. 2017; Cheng et al. 2020). The use of FMT includes stepwise, scrutinized, protocol-based donor screening followed by different methods for feces infusion that is dependent on the treating unit's expertise. In brief, donor screening must include a thorough clinical history, including history of chronic as well as recent drug and medications such as antibiotics and proton pump inhibitors; GI symptoms, history of travel within 3 months; neuropsychiatric disorders, the latter, in the donor as well as first degree relatives; and physical examination and blood investigations to rule out acute as well as chronic infections, metabolic disorders, and possible transmissible diseases. A minimum of 30 g of freshly donated or frozen stool material (stored at -80 °C with added glycerol to a final concentration of 10%) homogenized with normal saline (three to five times larger volume of solvent) through blending and gauze filtering or manual/device straining can be infused into the recipient through a colonoscope into the lower GI tract or through a fluoroscopy-guided, nasally placed tube or gastroduodenoscopy-directed introduction into the upper GI tract (Wang et al. 2019; Kim and Gluck 2019; Cammarota et al. 2019). Researchers from China extracted and analyzed microbiota in feces from constipated donors who had undergone effective therapy with FMT and transplanted the extracted microbiota into pseudo-germ-free mice while measuring parameters of intestinal motility. They found that the treated mice developed lower pellet frequency and stool water percentage, smaller pellet size, delayed gastrointestinal transit time, and weaker spontaneous contractions of colonic smooth muscle. To identify the mechanism underlying delayed gut motility in detail, the authors evaluated microbial metabolites. They found that SCFAs and secondary bile acids were decreased in mice receiving microbiota from constipated donors. They also demonstrated that the compositional changes of gut microbiota in constipated patients (taxa and the species richness and alpha diversity) were greater than healthy volunteers (Ge et al. 2017). The effect of allogenic and autologous FMT on IBS symptoms, visceral sensitivity, and compositional changes in fecal and mucosa-adherent microbiota was studied by researchers from Finland and Sweden in a randomized controlled study. They showed that single FMT via colonoscopy might have beneficial effects in patients with IBS. Still, allogenic fecal material was not superior to autologous feces, suggesting that prior bowel cleansing may contribute to symptoms and gut microbiota changes in IBS. This study sheds light on discovering standardized practices that minimize inadvertent microbiota modulation in patients treated with FMT (Holster et al. 2019). A single-arm open-labeled study included patients with IBS who underwent colonoscopy-directed FMT with a change in Bristol stool form scale (to types 3 or 4) at 4 weeks post-treatment as the primary endpoint. The authors noted that among responders to FMT, stool bacterial diversity increased with improved psychological status (measured using the Hamilton Rating Scale), especially in donor feces enriched in *Bifidobacterium* (Mizuno et al. 2017). Improvement in depression and anxiety symptoms after FMT in a group of patients with IBS, functional diarrhea, or functional constipation associated with microbial alpha diversity improvement was demonstrated in an open-label observational study from Japan (Kurokawa et al. 2018). A randomized placebo-controlled, double-blind study from Denmark on FMT in patients with moderate to severe IBS demonstrated significant improvement in symptoms gauged by amelioration in the IBS-Severity Scoring System (IBS-SSS) in patients receiving placebo compared to those on fecal capsules. Even though FMT improved bacterial richness and diversity compared to placebo, clinical improvements per predefined primary endpoints were notably absent in the FMT group (Halkjær et al. 2018).

In another double-blind, randomized, placebo-controlled, parallel-group, single-center trial from Norway, 90 participants with moderate to severe IBS were randomly assigned to receive either freshly processed feces (50 to 80 g stool in 200 mL saline and 50 mL of 85% glycerol) or patients' own feces as placebo along with loperamide for retention benefit. On modified intention-to-treat analysis (55 in the active treatment group and 28 in the placebo group), 65% of participants receiving active treatment versus 43% receiving placebo showed a response in the form of graded symptom improvements at 3 months (Johnsen et al. 2018). The previous Danish and the current Norwegian studies were contrasting, probably because of the higher dosing and better route of FMT utilized in the latter, which beckons standardization of the FMT procedure in specific subsets of patients with FGID. This was also confirmed in a recent study in IBS patients wherein authors repeated the FMT procedure by infusing 60 g of freshly prepared faces into the duodenum through a gastroscope in patients not responding to the initial 30 g volume FMT. It was shown that repeated and higher dosed FMT improved responses and alleviated symptoms in patients who did not initially respond to the treatment (El-Salhy et al. 2019).

A more recent study showed that in patients with diarrhea-predominant IBS, several intestinal microbiota taxa and SCFAs, which were significantly different in the patients at baseline compared to their donors, normalized by the third week following FMT in parallel with significant improvement in symptoms and quality of life that was maintained up to 28 weeks post-treatment (Mazzawi et al. 2019). In a meta-analysis of eight single arm (SATs) and five randomized controlled trials ($N = 105$ patients on FMT and 105 controls), the authors found that 59.5% of IBS patients had significant improvement in the former of symptoms. In contrast, there were no differences between FMT and control treatment in IBS symptom, severity, or quality of life in the latter. This meant that randomized controlled trial results were dependent on and affected by the placebo effect; dosing, route, and FMT source were confounding factors. The effectiveness of FMT was dependent on the IBS subtype (Myneedu et al. 2019).

In a meta-analysis, the authors examined the efficacy of FMT in 267 IBS patients. They found that, for all individuals, there was no improvement in IBS symptoms as compared to placebo and concluded that the dose and method of delivery might have influenced response, and that fresh or frozen donor stool delivered by colonoscopy or nasojejunal tube may be associated with the better response, which needs further validation through larger more rigorously conducted trials (Ianiro et al. 2019). In a more recent systematic review and meta-analysis on 742 citations with ultimately 254 eligible participants with IBS undergoing FMT, the authors noted no significant difference in the global improvement of IBS symptoms at 12 weeks in those receiving FMT to placebo. The heterogeneity among studies was significant, and subgroup analyses revealed benefits of single-dose FMT using colonoscopy and nasojejunal tubes in comparison with autologous FMT for placebo treatment (number needed to treat = 5, RR = 1.59) and a reduction in the likelihood of improvement of multiple-dose capsule FMT RCTs (number needed to harm = 3, RR = 0.54). The authors also found that the placebo response was 33.7% in non-oral FMT RCTs and 67.8% in capsule FMT RCTs. Thus, current evidence from RCTs does not suggest a benefit of FMT for global IBS symptoms (Xu et al. 2019) (Fig. 3.4).

5 Conclusions and Future Directions

Current evidence sheds light on the influential role of intestinal dysbiosis in patients with FGIDs. However, our knowledge regarding specific taxa and their functions in different subsets of FGID remains limited to patients with IBS and, to some extent, FD. Even so, there remain wide variations in observed changes in the species and genera of these patients dependent on the region (Asian versus European), methodology (16 s RNA sequencing versus shotgun sequencing), and sites (fecal versus mucosal; duodenum versus colon) studied. With improved standardization of study methods, our comprehension of the precise role of qualitative and quantitative gut

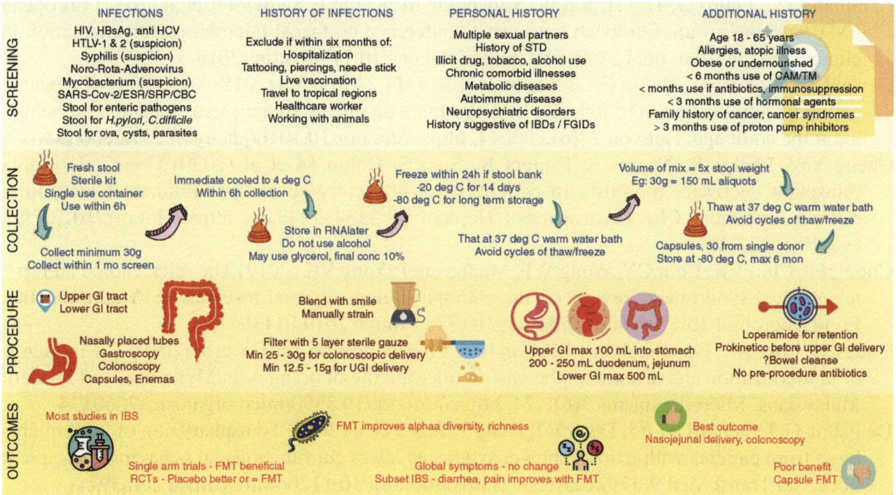

Fig. 3.4 Brief summary of FMT in IBS patients. *HIV*, human immunodeficiency virus; *HTLV*, human T-lymphotropic virus; *SARS-Cov-2*, novel coronavirus; *ESR*, erythrocyte sedimentation rate; *CRP*, C-reactive protein; *CBC*, omplete blood counts; *STD*, sexually transmitted disease; *IBD*, inflammatory bowel disease; *FGID*, functional gastrointestinal disorder; *CAM*, complementary and alternative medicine; *TM*, traditional medicine; *mo*, month(s); deg. *C*, degree centigrade; *GI*, gastrointestinal; *RCT*, randomized controlled trial

microbial functions in patients with FGID and subgroups has been steadily improving. In similar lines, FMT's use as a therapeutic option in patients with FGID, especially those with IBS, has not yielded favorable results due to differences in dosing, route, and duration of therapy utilized across studies. In the future, understanding and identifying specific groups of patients with FGID in whom intestinal dysbiosis plays a central role in the pathogenesis of the disease, independent of other factors, who would benefit from gut microbial modulation, through large population-based observational and randomized controlled interventions needs effectuation.

References

Barbara G, Feinle-Bisset C, Ghoshal UC, Quigley EM, Santos J, Vanner S et al (2016) The intestinal microenvironment and functional gastrointestinal disorders. Gastroenterology S0016-5085(16):00219–00215. https://doi.org/10.1053/j.gastro.2016.02.028

Beatty JK, Akierman SV, Motta JP, Muise S, Workentine ML, Harrison JJ et al (2017) Giardia duodenalis induces pathogenic dysbiosis of human intestinal microbiota biofilms. Int J Parasitol 47(6):311–326. https://doi.org/10.1016/j.ijpara.2016.11.010

Cammarota G, Ianiro G, Kelly CR, Mullish BH, Allegretti JR, Kassam Z et al (2019) International consensus conference on stool banking for faecal microbiota transplantation in clinical practice. Gut 68(12):2111–2121. https://doi.org/10.1136/gutjnl-2019-319548

Cammarota G, Ianiro G, Tilg H, Rajilić-Stojanović M, Kump P, Satokari R et al (2017) European FMT working group. European consensus conference on faecal microbiota transplantation in clinical practice. Gut 66(4):569–580. https://doi.org/10.1136/gutjnl-2016-313017

Ceylani T, Jakubowska-Doğru E, Gurbanov R, Teker HT, Gozen AG (2018) The effects of repeated antibiotic administration to juvenile BALB/c mice on the microbiota status and animal behavior at the adult age. Heliyon 24(6):e00644. https://doi.org/10.1016/j.heliyon.2018.e00644

Cheng YW, Phelps E, Nemes S, Rogers N, Sagi S, Bohm M et al (2020) Fecal microbiota transplant decreases mortality in patients with refractory severe or fulminant Clostridium difficile infection. Clin Gastroenterol Hepatol 10:2234–2243.e1. https://doi.org/10.1016/j.cgh.2019.12.029

Chong PP, Chin VK, Looi CY, Wong WF, Madhavan P, Yong VC (2019) The microbiome and irritable bowel syndrome–a review on the pathophysiology, current research and future therapy. Front Microbiol 10:1136. https://doi.org/10.3389/fmicb.2019.01136

Chua EG, Loke MF, Gunaletchumy SP, Gan HM, Thevakumar K, Tay CY et al (2019) The influence of modernization and disease on the gastric microbiome of orang Asli, Myanmars and Modern Malaysians. Microorganisms 7(6):174. https://doi.org/10.3390/microorganisms7060174

De Palma G, Lynch MD, Lu J, Dang VT, Deng Y, Jury J et al (2017) Transplantation of fecal microbiota from patients with irritable bowel syndrome alters gut function and behavior in recipient mice. Sci Transl Med 9(379):eaaf6397. https://doi.org/10.1126/scitranslmed.aaf6397

Douglas GM, Maffei VJ, Zaneveld JR, Yurgel SN, Brown JR, Taylor CM et al (2020) MGI. PICRUSt2 for prediction of metagenome functions. Nat Biotechnol 38(6):685–688. https://doi.org/10.1038/s41587-020-0548-6

Drossman DA (2016) Functional gastrointestinal disorders: history, pathophysiology, Clinical Features and Rome IV. Gastroenterology S0016-5085(16):00223–00227. https://doi.org/10.1053/j.gastro.2016.02.032

Duan R, Zhu S, Wang B, Duan L (2019) Alterations of gut microbiota in patients with irritable bowel syndrome based on 16S rRNA-targeted sequencing: a systematic review. Clin Transl Gastroenterol 10(2):e00012. https://doi.org/10.14309/ctg.0000000000000012

El-Salhy M, Hausken T, Hatlebakk JG (2019) Increasing the dose and/or repeating Faecal microbiota transplantation (FMT) increases the response in patients with irritable bowel syndrome (IBS). Nutrients 11(6):1415. https://doi.org/10.3390/nu11061415

Fukui A, Takagi T, Naito Y, Inoue R, Kashiwagi S, Mizushima K et al (2020) Higher levels of streptococcus in upper gastrointestinal mucosa associated with symptoms in patients with functional dyspepsia. Digestion 101(1):38–45. https://doi.org/10.1159/000504090

Fukui H, Xu X, Miwa H (2018) Role of gut microbiota-gut hormone Axis in the pathophysiology of functional gastrointestinal disorders. J Neurogastroenterol Motil 24(3):367–386. https://doi.org/10.5056/jnm18071

Gantuya B, El-Serag HB, Matsumoto T, Ajami NJ, Oyuntsetseg K, Azzaya D et al (2019) Gastric microbiota in helicobacter pylori-negative and–positive gastritis among high incidence of gastric cancer area. Cancers (Basel) 11(4):504. https://doi.org/10.3390/cancers11040504

Ge X, Zhao W, Ding C, Tian H, Xu L, Wang H et al (2017) Potential role of fecal microbiota from patients with slow transit constipation in the regulation of gastrointestinal motility. Sci Rep 7(1):441. https://doi.org/10.1038/s41598-017-00612-y

Ghoshal UC, Sachdeva S, Pratap N, Verma A, Karyampudi A, Misra A et al (2018b) Indian consensus on chronic constipation in adults: a joint position statement of the Indian motility and functional diseases association and the Indian society of gastroenterology. Indian J Gastroenterol 37(6):526–544. https://doi.org/10.1007/s12664-018-0894-1

Ghoshal UC, Srivastava D, Misra A (2018a) A randomized double-blind placebo-controlled trial showing rifaximin to improve constipation by reducing methane production and accelerating colon transit: a pilot study. Indian J Gastroenterol 37(5):416–423. https://doi.org/10.1007/s12664-018-0901-6

Gobert AP, Sagrestani G, Delmas E, Wilson KT, Verriere TG, Dapoigny M et al (2016) The human intestinal microbiota of constipated-predominant irritable bowel syndrome patients exhibits anti-inflammatory properties. Sci Rep 6:39399. https://doi.org/10.1038/srep39399

Halkjær SI, Christensen AH, Lo BZS, Browne PD, Günther S, Hansen LH et al (2018) Faecal microbiota transplantation alters gut microbiota in patients with irritable bowel syndrome: results from a randomised, double-blind placebo-controlled study. Gut 67(12):2107–2115. https://doi.org/10.1136/gutjnl-2018-316434

Holster S, Lindqvist CM, Repsilber D, Salonen A, de Vos WM, König J et al (2019) The effect of allogenic versus autologous fecal microbiota transfer on symptoms, visceral perception and fecal and mucosal microbiota in irritable bowel syndrome: a randomized controlled study. Clin Transl Gastroenterol 10(4):e00034. https://doi.org/10.14309/ctg.0000000000000034

Ianiro G, Eusebi LH, Black CJ, Gasbarrini A, Cammarota G, Ford AC (2019) Systematic review with meta-analysis: efficacy of faecal microbiota transplantation for the treatment of irritable bowel syndrome. Aliment Pharmacol Ther 50(3):240–248. https://doi.org/10.1111/apt.15330

Igarashi M, Nakae H, Matsuoka T, Takahashi S, Hisada T, Tomita J et al (2017) Alteration in the gastric microbiota and its restoration by probiotics in patients with functional dyspepsia. BMJ Open Gastroenterol 4(1):e000144. https://doi.org/10.1136/bmjgast-2017-000144

Jeffery IB, O'Toole PW, Öhman L, Claesson MJ, Deane J, Quigley EM et al (2012) An irritable bowel syndrome subtype defined by species-specific alterations in faecal microbiota. Gut 61(7):997–1006. https://doi.org/10.1136/gutjnl-2011-301501

Johnsen PH, Hilpüsch F, Cavanagh JP, Leikanger IS, Kolstad C, Valle PC (2018) Faecal microbiota transplantation versus placebo for moderate-to-severe irritable bowel syndrome: a double-blind, randomised, placebo-controlled, parallel-group, single-Centre trial. Lancet Gastroenterol Hepatol 3(1):17–24. https://doi.org/10.1016/S2468-1253(17)30338-2

Kassinen A, Krogius-Kurikka L, Mäkivuokko H, Rinttilä T, Paulin L, Corander J et al (2007) The fecal microbiota of irritable bowel syndrome patients differs significantly from that of healthy subjects. Gastroenterology 133(1):24–33. https://doi.org/10.1053/j.gastro.2007.04.005

Kerckhoffs APM, Ben-Amor K, Samsom M, van der Rest ME, de Vogel J, Knol J et al (2011) Molecular analysis of faecal and duodenal samples reveals significantly higher prevalence and numbers of *Pseudomonas aeruginosa* in irritable bowel syndrome. J Med Microbiol 60(Pt 2):236–245. https://doi.org/10.1099/jmm.0.022848-0

Kim KO, Gluck M (2019) Fecal microbiota transplantation: an update on clinical practice. Clin Endosc 52(2):137–143. https://doi.org/10.5946/ce.2019.009

Kurokawa S, Kishimoto T, Mizuno S, Masaoka T, Naganuma M, Liang KC et al (2018) The effect of fecal microbiota transplantation on psychiatric symptoms among patients with irritable bowel syndrome, functional diarrhea and functional constipation: an open-label observational study. J Affect Disord 235:506–512. https://doi.org/10.1016/j.jad.2018.04.038

Kwon HJ, Mohammed AE, Eltom KH, Albrahim JS, Alburae NA (2020) Evaluation of antibiotic-induced behavioral changes in mice. Physiol Behav 223:113015. https://doi.org/10.1016/j.physbeh.2020.113015

Labus JS, Osadchiy V, Hsiao EY, Tap J, Derrien M, Gupta A et al (2019) Evidence for an association of gut microbial clostridia with brain functional connectivity and gastrointestinal sensorimotor function in patients with irritable bowel syndrome, based on tripartite network analysis. Microbiome 7(1):45. https://doi.org/10.1186/s40168-019-0656-z

Liu X, Locasale JW (2017) Metabolomics: a primer. Trends Biochem Sci 42(4):274–284. https://doi.org/10.1016/j.tibs.2017.01.004

Liu HN, Wu H, Chen YZ, Chen YJ, Shen XZ, Liu TT (2017) Altered molecular signature of intestinal microbiota in irritable bowel syndrome patients compared with healthy controls: a systematic review and meta-analysis. Dig Liver Dis 49(4):331–337. https://doi.org/10.1016/j.dld.2017.01.142

Lo Presti A, Zorzi F, Del Chierico F, Altomare A, Cocca S, Avola A et al (2019) Fecal and mucosal microbiota profiling in irritable bowel syndrome and inflammatory bowel disease. Front Microbiol 10:1655. https://doi.org/10.3389/fmicb.2019.01655

Luo Y, Zeng B, Zeng L, Du X, Li B, Huo R et al (2018) Gut microbiota regulates mouse behaviors through glucocorticoid receptor pathway genes in the hippocampus. Transl Psychiatry 8(1):187. https://doi.org/10.1038/s41398-018-0240-5

Maharshak N, Ringel Y, Katibian D, Lundqvist A, Sartor RB, Carroll IM et al (2018) Fecal and Mucosa-Associated Intestinal Microbiota in Patients with Diarrhea-Predominant Irritable Bowel Syndrome. Dig Dis Sci 63(7):1890–1899. https://doi.org/10.1007/s10620-018-5086-4

Mazzawi T, Hausken T, Hov JR, Valeur J, Sangnes DA, El-Salhy M (2019) Clinical response to fecal microbiota transplantation in patients with diarrhea-predominant irritable bowel syndrome is associated with normalization of fecal microbiota composition and short-chain fatty acid levels. Scand J Gastroenterol 54(6):690–699. https://doi.org/10.1080/00365521.2019.1624815

Mizuno S, Masaoka T, Naganuma M, Kishimoto T, Kitazawa M, Kurokawa S (2017) Bifidobacterium-rich fecal donor may be a positive predictor for successful fecal microbiota transplantation in patients with irritable bowel syndrome. Digestion 96(1):29–38. https://doi.org/10.1159/000471919

Mottawea W, Butcher J, Li J, Abujamel T, Manoogian J, Mack D et al (2019) The mucosal-luminal interface: an ideal sample to study the mucosa-associated microbiota and the intestinal microbial biogeography. Pediatr Res 85(6):895–903. https://doi.org/10.1038/s41390-019-0326-7

Myneedu K, Deoker A, Schmulson MJ, Bashashati M (2019) Fecal microbiota transplantation in irritable bowel syndrome: a systematic review and meta-analysis. United European Gastroenterol J 7(8):1033–1041. https://doi.org/10.1177/2050640619866990

Nakae H, Tsuda A, Matsuoka T, Mine T, Koga Y (2016) Gastric microbiota in the functional dyspepsia patients treated with probiotic yogurt. BMJ Open Gastroenterol 3(1):e000109. https://doi.org/10.1136/bmjgast-2016-000109

Nam YD, Chang HW, Kim KH, Roh SW, Kim MS, Jung MJ et al (2008) Bacterial, archaeal, and eukaryal diversity in the intestines of Korean people. J Microbiol 46(5):491–501. https://doi.org/10.1007/s12275-008-0199-7

Pimentel M, Lembo A (2020) Microbiome and its role in irritable bowel syndrome. Dig Dis Sci 65(3):829–839. https://doi.org/10.1007/s10620-020-06109-5

Pimentel M, Morales W, Pokkunuri V, Brikos C, Kim SM, Kim SE et al (2015) Autoimmunity links vinculin to the pathophysiology of chronic functional bowel changes following campylobacter jejuni infection in a rat model. Dig Dis Sci 60(5):1195–1205. https://doi.org/10.1007/s10620-014-3435-5

Pittayanon R, Lau JT, Yuan Y, Leontiadis GI, Tse F, Surette M (2019) Gut microbiota in patients with irritable bowel syndrome-a systematic review. Gastroenterology 157(1):97–108. https://doi.org/10.1053/j.gastro.2019.03.049

Pozuelo M, Panda S, Santiago A, Mendez S, Accarino A, Santos J (2015) Reduction of butyrate– and methane-producing microorganisms in patients with irritable bowel syndrome. Sci Rep 5:12693. https://doi.org/10.1038/srep12693

Raskov H, Burcharth J, Pommergaard HC, Rosenberg J (2016) Irritable bowel syndrome, the microbiota and the gut-brain axis. Gut Microbes 7(5):365–383. https://doi.org/10.1080/19490976.2016.1218585

Ringel Y, Katibian D, Lundqvist A, Sartor RB, Carroll IM, Ringel-Kulka T (2018) Fecal and mucosa-associated intestinal microbiota in patients with diarrhea-predominant irritable bowel syndrome. Dig Dis Sci 63(7):1890–1899. https://doi.org/10.1007/s10620-018-5086-4

Ringel-Kulka T, Benson AK, Carroll IM, Kim J, Legge RM, Ringel Y (2016) Molecular characterization of the intestinal microbiota in patients with and without abdominal bloating. Am J Physiol Gastrointest Liver Physiol 310(6):G417–G426. https://doi.org/10.1152/ajpgi.00044.2015

Rivière A, Selak M, Geirnaert A, Van den Abbeele P, De Vuyst L (2018) Complementary mechanisms for degradation of inulin-type Fructans and Arabinoxylan oligosaccharides among Bifidobacterial strains suggest bacterial cooperation. Appl Environ Microbiol 84(9):e02893–e02817. https://doi.org/10.1128/AEM.02893-17

Rivière A, Selak M, Lantin D, Leroy F, De Vuyst L (2016) Bifidobacteria and butyrate-producing colon bacteria: importance and strategies for their stimulation in the human gut. Front Microbiol 7:979. https://doi.org/10.3389/fmicb.2016.00979

Schmulson MJ, Drossman DA (2017) What is new in Rome IV. J Neurogastroenterol Motil 23(2):151–163. https://doi.org/10.5056/jnm16214

Shanahan ER, Shah A, Do A, Fairlie T, Ghasemi P, Hansen TJ et al (2018) Duodenal mucosa-associated microbiota (MAM) and gastric emptying: veillonella in the duodenal MAM linked to slow gastric emptying. Gastroenterology 154:S40. https://doi.org/10.1016/S0016-5085(18)30604-8

Shukla R, Ghoshal U, Dhole TN, Ghoshal UC (2015) Fecal microbiota in patients with irritable bowel syndrome compared with healthy controls using real-time polymerase chain reaction: an evidence of dysbiosis. Dig Dis Sci 60(10):2953–2962. https://doi.org/10.1007/s10620-015-3607-y

Su T, Liu R, Lee A, Long Y, Du L, Lai S (2018) Altered intestinal microbiota with increased abundance of prevotella is associated with high risk of diarrhea-predominant irritable bowel syndrome. Gastroenterol Res Pract 2018:6961783. https://doi.org/10.1155/2018/6961783

Sundin J, Aziz I, Nordlander S, Polster A, Hu YOO, Hugerth LW et al (2020) Evidence of altered mucosa-associated and fecal microbiota composition in patients with irritable bowel syndrome. Sci Rep 10(1):593. https://doi.org/10.1038/s41598-020-57468-y

Tap J, Derrien M, Törnblom H, Brazeilles R, Cools-Portier S, Doré J et al (2016) Identification of an intestinal microbiota signature associated with severity of irritable bowel syndrome. Gastroenterology 152(1):111–123.e8. https://doi.org/10.1053/j.gastro.2016.09.049

Valles-Colomer M, Falony G, Darzi Y, Tigchelaar EF, Wang J, Tito RY et al (2019) The neuroactive potential of the human gut microbiota in quality of life and depression. Nat Microbiol 4(4):623–632. https://doi.org/10.1038/s41564-018-0337-x

Wang JW, Kuo CH, Kuo FC, Wang YK, Hsu WH, Yu FJ et al (2019) Fecal microbiota transplantation: review and update. J Formos Med Assoc 118(Suppl 1):S23–S31. https://doi.org/10.1016/j.jfma.2018.08.011

Xu D, Chen VL, Steiner CA, Berinstein JA, Eswaran S, Waljee AK et al (2019) Efficacy of fecal microbiota transplantation in irritable bowel syndrome: a systematic review and meta-analysis. Am J Gastroenterol 114(7):1043–1050. https://doi.org/10.14309/ajg.0000000000000198

Yano JM, Yu K, Donaldson GP, Shastri GG, Ann P, Ma L (2015) Indigenous bacteria from the gut microbiota regulate host serotonin biosynthesis. Cell 161(2):264–276. https://doi.org/10.1016/j.cell.2015.02.047

Zhong L, Shanahan ER, Raj A, Koloski NA, Fletcher L, Morrison M et al (2017) Dyspepsia and the microbiome: time to focus on the small intestine. Gut 66(6):1168–1169. https://doi.org/10.1136/gutjnl-2016-312574

Zhu S, Liu S, Li H, Zhang Z, Zhang Q, Chen L et al (2019) Identification of gut microbiota and metabolites signature in patients with irritable bowel syndrome. Front Cell Infect Microbiol 9:346. https://doi.org/10.3389/fcimb.2019.00346

Zhuang X, Xiong L, Li L, Li M, Chen M (2017) Alterations of gut microbiota in patients with irritable bowel syndrome: a systematic review and meta-analysis. J Gastroenterol Hepatol 32(1):28–38. https://doi.org/10.1111/jgh.13471

Chapter 4
The Human Gut Microbiota and Gastrointestinal Cancer: Current Status and Therapeutic Perspectives

Goutam Chowdhury and Dharanidharan Ramamurthy

1 Introduction

The human gastrointestinal tract (GIT) is one of the most complex networks in the body and is colonized by trillions of microorganisms, including bacteria, viruses, archaea, parasites, and fungi (Faith et al. 2013). Bacteria are considered to be the foremost domain of microbiota colonizing in the GIT. Bacterial populations occupying the gut differ significantly between individuals, depending on host specificities such as environmental conditions, lifestyle, age, smoking habit, genetic factors, antibiotic therapy, and contact with pathogenic organisms (Gill et al. 2006). Microorganisms participate in host health via synthesis of essential amino acids, short-chain fatty acids (SCFAs), and vitamins (Marchesi et al. 2016). The gut microbiota has emerged as a critical player in the maintenance of human health, influencing not only the GI tract but also distal organs such as the brain, liver, and pancreas (Flint et al. 2012). Various studies have shown that microbial alterations, and a relative decrease in levels of beneficial bacteria, relate to the development of gastrointestinal and extra-gastrointestinal cancers (Panebianco et al. 2018a, b). Several lines of evidence suggest that the gut microbiota is related to a variety of cancers, which may educate potential development of cancer therapies targeted at the gut microbiome (Meng et al. 2018). This chapter provides a complete review of the studies on the human gut microbiota and GIT cancers, specifically gastric, esophageal, colorectal, and pancreatic cancers.

G. Chowdhury (✉)
Division of Bacteriology, ICMR-National Institute of Cholera and Enteric Diseases, Kolkata, India

D. Ramamurthy
Medical Biotechnology & Immunotherapy Research Unit, Institute of Infectious Disease and Molecular Medicine, Faculty of Health Sciences, University of Cape Town, Cape Town, South Africa

S. Thomas (ed.), *Human Microbiome*,
https://doi.org/10.1007/978-981-16-7672-7_4

Cancer is considered a main reason for mortality and morbidity in the world. According to the latest report released by the World Health Organization (WHO), the global cancer burden is estimated to have risen to 18.1 million new cases and 9.6 million deaths in 2018 (WHO 2018). Currently, cancers affecting the GIT system are well known as a major health problem, and according to the Global Cancer Statistics 2018, GIT cancers have high incidence and mortality rates (Bray et al. 2018). The conventional and traditional treatments of cancer are surgery, radiotherapy, and chemotherapy (Sohda and Kuwano 2017). Recent novel therapeutic approaches include stem-cell therapy, hormone-based therapy, and immunotherapy. Both conventional and modern therapy strategies however encounter some limitations and side effects such as specific toxicity toward normal body cells (Axelrad et al. 2016). The surgical procedure for the excision of tumors is a limited and insufficient approach on its own since it needs to be followed up by chemotherapy and radiotherapy (Sohda and Kuwano 2017). Besides, as a conventional and most used type of cancer treatment, chemotherapy could lead to the development of multidrug-resistant (MDR) cells in some patients (Arruebo et al. 2011). Therefore, there is a need for novel therapeutic agents with fewer complications. Several bacterial species have been used in live, attenuated, or genetically modified and bacterial products (including bacterial peptides, bacteriocins, enzymes, and toxins) that could multiply selectively in tumors and inhibit their growth (Elsalem et al. 2020). In addition, there is substantial data indicating the role of GIT microbiota in modulating tumor response to anticancer drugs such as conventional chemotherapy and molecular-targeted therapeutics (Cheng et al. 2020). Bacterial microbiota are therefore attractive targets for modulation, prevention, or treatment of GIT cancers.

2 Gut Microbiota and GIT Cancer

During the last few decades, the relationship between the gut microbiota and carcinogenesis was documented promptly (Elsalem et al. 2020). In the gut, the microbial dysbiosis or individual bacteria can promote carcinoma or induce cancer process through a variety of mechanisms such as damaging DNA, production of carcinogenic metabolites, stimulation of chronic inflammation, activating oncogenic pathways, inhibition of antitumor immunity, deleterious alterations in physiological host processes like inflammation, antigen-driven lymph proliferation, damaging host DNA, and induction of hormones that increase epithelial cell proliferation (Schwabe and Jobin 2013; Francescone et al. 2014; Fulbright et al. 2017). Several bacteria may also possess and promote cancer through direct effects on cell transformation or through the production of toxic, carcinogenic metabolites that activate the β-catenin signal pathway involved in carcinogenesis (Kamada et al. 2013).

Gut dysbiosis or alterations in gut microbiota composition have been implicated in the initiation and development of various pathological conditions within the gut microbiota, ranging from bowel inflammation, to diabetes, obesity, neurodegenerative diseases, and development of various cancers, including gastric cancer, colorectal cancer, hepatocellular carcinoma (HCC), and pancreatic cancer (Garrett 2015; Virtue et al. 2019). In gut dysbiosis, certain bacterial pathogens can negatively disturb the host's metabolism and immune system functionalities, and many bacteria can damage DNA by releasing specific metabolites, thereby triggering tumor growth and promoting GIT cancer (Coker et al. 2018). Epidemiological evidence supports the opinion that long-term antibiotic exposures change the composition by decreasing the diversity of gut microbiota and increasing the risk of GIT cancers (Wong et al. 2017). Recent studies also suggest that due to dysbiosis or microbial commensal imbalance, 20% of gut microbial pathogens have been driven to GIT tumorigenesis or malignancies (Sethi et al. 2018), and long-term antibiotic use was highly correlated with increased colorectal tumor progression. Remarkably, there is convincing evidence linking bacterial dysbiosis to cancer like *Helicobacter pylori* with gastric cancer (Coker et al. 2018), *Porphyromonas gingivalis* and *Aggregatibacter actinomycetemcomitans* with pancreatic cancer (Fan et al. 2018), *Streptococcus mitis/anginosus* with esophageal cancer (Narikiyo et al. 2004), and *Streptococcus bovis/gallolytucis* with colorectal cancer (CRC) (Rezasoltani et al. 2018).

Gut microbial dysbiosis also has a key role in therapeutic responses toward anticancer treatment. These depend on the microbial ability to metabolize drugs and to influence inflammation as well as immune responses within the tumor microenvironment (Panebianco et al. 2018a, b). However, the association between microbiota and responses to anticancer therapies has been defined as a bidirectional relationship, where both factors can have an extensive effect on each other (Li et al. 2019). Recently, the approaches of using next-generation sequencing, mass spectrometry, and pharmacomicrobiomics have emerged as a new field for investigating and evaluating the microbiota structure and investigating the metabolic, functional, and genetic action of the microbiota (Fessler et al. 2019). Therefore, the role of probiotics and antibiotics either alone or in combination with anticancer drugs has been explored to manipulate the microbiota, which in turn might have positive outcomes for GIT cancer prevention and treatment (Javanmard et al. 2018). The gut microbiota interventions in GIT cancers are explained in Table 4.1.

3 Gut Microbiota and Esophageal Cancer

Esophageal cancer is the eighth most commonly diagnosed cancer and the sixth leading cause of cancer death worldwide (D'Journo and Thomas 2014). Esophageal cancer is divided histologically into two major groups: esophageal squamous cell carcinoma (ESCC) arising from the epithelial cells, with different geographical

Table 4.1 Microbiota interventions in GIT cancers

Cancer	Microbes associated	Virulence or risk factor	Mechanisms	References
Gastric cancer	*H. pylori*	*Cag*-PAI, *vacA*, mucosal alterations, inflammation, chronic superficial gastritis, intestinal metaplasia, and finally to dysplasia and adenocarcinoma	Infection with *CagA* or *vacA H. pylori* promotes the secretion of gastrin, which may induce the proliferation of mucosal cells and affect the epithelial cell barrier, create vacuoles in the host cells, and inhibit the T-cell mediated immune response and inflammation. *CagA* promotes GC through multiple cellular signaling pathways, such as the mitogen-activated protein kinase (MAPK) cascade, NF-κB expression, PI3K/Akt signaling pathways, EMT through the oncogenic associated protein (YAP) pathway, and induction of gene mutation of p53. *VacA* encourages the GC through affecting the epithelial cell barrier and inhibiting the T-cell-mediated immune response.	Palframan et al. (2012); Yang et al. (2015)
Pancreatic cancer	*H. pylori*, *Streptococcus mitis*, *Pseudomonas aeruginosa*	*CagA*, LPS Taste receptor 2 member 38 (T2R38)	*H. Pylori* infection activates STAT3, NF-jB, and AP-1 to mediate pancreatic cancer progression via increasing the level of anti-apoptotic and proproliferative proteins such as B-cell lymphoma-extralarge (Bcl-xL) and myeloid cell leukemia-1 (MCL-1). *P. aeruginosa* infection activates T2R38, induces multidrug resistance protein 1 (ABCB1), and gets involved in cancer invasion and metastasis	Manes et al. (1998); Gaida et al. (2016)

Table 4.1 (continued)

Cancer	Microbes associated	Virulence or risk factor	Mechanisms	References
Colorectal cancer	*S. gallolytucis* Enterotoxigenic *B. fragilis (ETBF)* *Fusobacterium nucleatum*	*S. gallolytucis* toxin *B. fragilis* toxin, fragilysin *FadA*	Increased the expression of proliferation markers and polyamines and production of interleukin IL-1 and IL-8 in the colonic mucosa, which promotes the neoplastic process for CRC ETBF producing metalloprotease fragilisyn activates STAT3 and stimulates the IL-17 production, consequently promoting NF-κB and Wnt pathway activation leading to tumor formation. FadA interacts with E-cadherin on the endothelium, activates β-catenin signaling pathway, resulting in an increased expression of transcription factors and inflammatory and oncogenic responses.	Boleij et al. (2011); Kim et al. (2006); Rubinstein et al. (2019)
Esophageal cancer	Bacteroidetes, Proteobacteria, Fusobacteria, *Veillonella*, *Prevotella*, *Neisseria*, and *Fusobacterium*	LPS Esophageal squamous cell carcinoma (ESCC) and esophageal adenocarcinoma (EAC)	Activating innate immune responses that lead to NF-jB activation, promoting the release of inflammation-associated mediators including IL1b, IL6, IL8, and TNFa, increasing the risk of reflux through relaxing lower esophageal sphincter.	Yang et al. (2009)

distribution, and esophageal adenocarcinoma (EAC) arising from the glandular cells of the distal esophagus (Sohda and Kuwano 2017). The known risk factors of cancer esophageal cancer are genetics, gastroesophageal reflux disease (GERD), alcohol and tobacco consumption, low fiber intake, and obesity (Wang et al. 2016). Bacterial and viral infections also contribute to the formation of esophageal malignant neoplasms. Results from several studies reported that Bacteroidetes, Proteobacteria, Fusobacteria, *Veillonella*, *Prevotella*, *Neisseria*, and *Fusobacterium* were prevalent in patients with esophageal cancer but were not detected in controls (Narikiyo et al. 2004). Gram-negative bacteria participate in the oncogenic process of esophageal cancer through multiple mechanisms, which include activating innate immune responses that lead to NF-kB activation, promoting the release of inflammation-associated mediators including IL1b, IL6, IL8, and TNF-α, increasing the risk of reflux through relaxing the lower esophageal sphincter, and delaying gastric emptying (Runge et al. 2015; Zaidi et al. 2016).

Currently, limited evidence is available regarding the role of microbiomes in esophageal cancer treatment or prevention. A study showed that disturbance of the commensal microbiota using antibiotics reduced the sensitivity of xenograft tumors in animal models (Sawada et al. 2016). Therefore, more studies should be directed toward understanding the pathogenesis of esophageal cancer and the role of microbiomes, which might have diagnostic and therapeutic implications (Snider et al. 2016).

4 Gut Microbiota and Gastric Cancer

Gastric cancer (GC) is ranked as fourth for incidence and one of the leading causes of cancer-related deaths worldwide (Siegel et al. 2017). GC commonly develops through a multifactorial process, depending on genetic, environmental, dietary, and host-related factors, such as age, sex, diet, smoking, and alcohol consumption (Fig. 4.1)

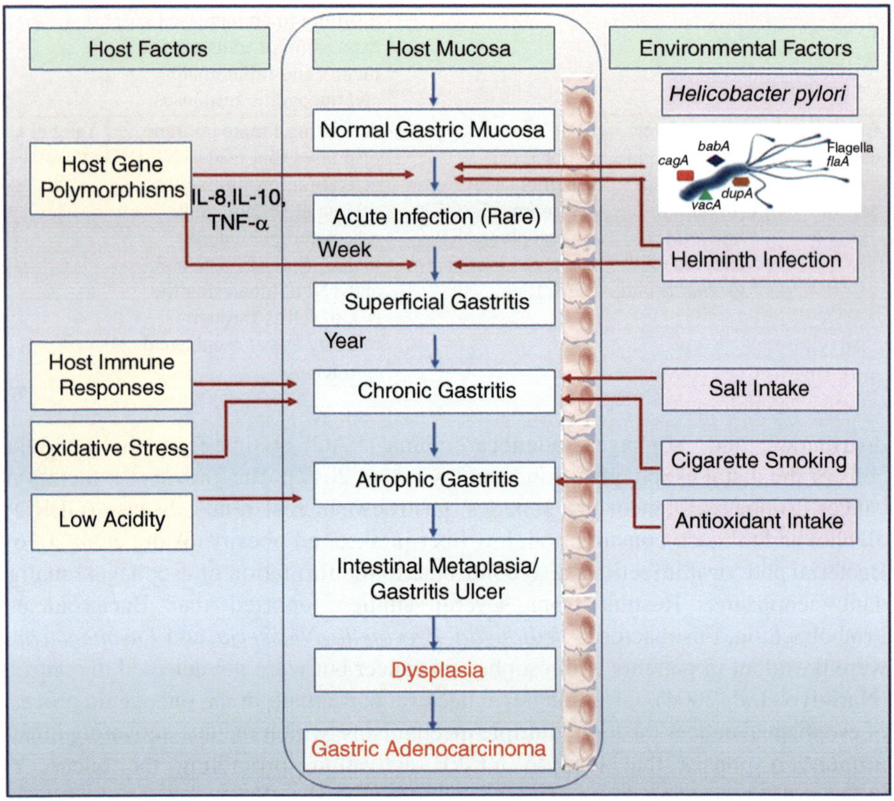

Fig. 4.1 Multifactorial pathway leading to gastric carcinoma. Host, *H. pylori*, and environmental factors act in combination to contribute to the precancerous cascade leading to the development of gastric cancer

(Zabaleta 2012; Ishaq and Nunn 2015). The main risk factor for GC is chronic infection by *H. pylori*, a Gram-negative bacterium living in the gastric mucosa of 50% of the human population (Doorakkers et al. 2016). According to the World Health Organization (WHO 2018), *H. pylori* was described as a class I carcinogen since it has a crucial role in the initiation of GC. *H. pylori* infection may result in intestinal-type GC through different mechanisms such as the indirect processes from inflammation mediation or through direct pathological role caused by bacterial virulence factors (Doorakkers et al. 2016). Chronic inflammation or GC caused by *H. pylori* infection first initiates early pre-neoplastic lesions such as atrophic gastritis and enhances the progression to advanced lesions, including metaplasia, dysplasia, and ultimately the development of gastric adenocarcinomas. The entire inflammatory procedure is multifaceted and incidental, involving the interplay between *H. pylori*, immune cells, acidic environment, reactive oxygen, and nitrogen species, collectively, leading to increased oxidative stress, DNA damage, and the expression of pro-inflammatory mediators (Wroblewski et al. 2010; Guevara and Cogdill 2000; Sukri et al. 2020).

Cytotoxin-associated gene A (*cagA*) and vacuolating cytotoxin A (*vacA*) are among the most frequently investigated critical virulence factors of *H. pylori* (Khatoon et al. 2016). *CagA* is translocated into the host cell by the type IV secretion system and acts as a classic oncogene. *CagA* has been suggested to potentiate the inflammatory reactions, which in turn facilitate the development of gastritis to GC (Odenbreit et al. 2000). Cag pathogenicity island (cag PAI) and *CagA* proteins are delivered into gastric epithelial cells and activate several pathways associated with carcinogenesis (Kwok et al. 2007). Another virulence factor, *VacA*, was found to affect the epithelial cell barrier, create vacuoles in the host cells, and inhibit the T-cell mediated immune response, which results in a favorable environment for *H. pylori* for promoting GI cancer (Yahiro et al. 2015).

H. pylori-positive individuals in GC patients are characterized by an increase in the counts of different microbial community like Acidobacteria, Proteobacteria, and Spirochaetes as well as a decrease in the counts of Bacteroidetes, Actinobacteria, and Firmicutes. Conversely, *H. pylori*-negative individuals carry more abundant phyla of Firmicutes, Bacteroidetes, and Actinobacteria (Maldonado-Contreras et al. 2011). Another study has shown that gastric cancer patients have a much-extended composition of microbiota, demonstrated by the reduction of *Porphyromonas*, *Neisseria*, *Prevotella pallens*, and *Streptococcus sinensis* and simultaneous enrichment of *Lactobacillus coleohominis*, *Klebsiella pneumoniae*, *Acinetobacter baumannii*, and Lachnospiraceae (Dias-Jácome et al. 2016). Other members of Helicobacteraceae also express pathogenic components such as the outer membrane proteins, phospholipase protein, and nickel-binding proteins, assist microbes with colonization in the mucosal layer of the gastric tract, and then promote the process of gastritis, ultimately causing tumorigenesis in the stomach (De Witte et al. 2016).

Eradication of *H. pylori* is the key consideration for the prevention of gastric carcinoma. Many studies have examined the effects of *H. pylori* eradication and prevention of gastric carcinoma (Ajani et al. 2017). According to the WHO and United Nations, probiotics are live microorganisms that, directed in suitable amounts, confer a health benefit for the host (Pormohammad et al. 2019). Results on the application of probiotics in gastric cancer are mainly directed toward the eradication of *H. pylori* infection since it is a major risk factor. The use of probiotics showed inhibitory effects on *H. pylori* infection using animal models (Ford et al. 2020). In addition, findings from a recent meta-analysis on clinical trials investigating the use of probiotics as a supplement with antibiotic therapy reported positive effects, which include a reduction in side effects, better patient compliance, and enhanced eradication (Tang et al. 2020).

5 Gut Microbiota and Pancreatic Cancer

Pancreatic cancer (PC) is one of the most common and deadly cancers worldwide, with more than 432,242 deaths annually (Rawla et al. 2019). There are several risk factors for PC like obesity, alcohol, smoking, chronic pancreatitis, familiarity, and type 2 diabetes (McGuigan et al. 2018). Several studies demonstrated that gut microbiota might influence pancreatic carcinogenesis by promoting inflammation, activating the immune response, and perpetuating cancer-associated inflammation (Farrell et al. 2012). Studies also revealed more than a twofold increase in the risk of pancreatic cancer in patients with high levels of bacterial species like *Streptococcus*, *Actinomyces*, *Prevotella*, *Porphyromonas gingivalis*, and Bacteroidetes (Michaud et al. 2013). Inflammation due to an immunological response to these bacteria and their toxins has been shown to play an important role in PC (Zhang et al. 2010). Another study showed that *Neisseria elongata* and *Streptococcus mitis* were found to achieve the highest discriminatory power between PC patients compared with healthy controls, but *Granulicatella adiacens* and *S. mitis* were significantly altered in patients with PC (Farrell et al. 2012). Microbiota as a target for pancreatic cancer treatment was recently evaluated. *Lactobacillus* is a commensal bacterium that reduces gingival inflammation and cariogenic periodontal pathogenic bacteria (Pushalkar et al. 2018). Olah et al. showed that administration of *Lactobacillus plantarum* to patients with acute pancreatitis reduced the development of pancreatic sepsis compared with control patients (Oláh et al. 2007).

6 Gut Microbiota and Colorectal Cancer

Colorectal cancer (CRC) is the third most commonly diagnosed cancer and cancer-related mortality (Siegel et al. 2020). CRC is the most common histopathological subtype; however, the specific mechanism of the intestinal flora in causing CRC is unclear. Many environmental factors such as smoking, diet, heavy alcohol consumption, and lifestyle strongly impact the pathogenesis risk for CRC (Fleming et al. 2012; Tan and Chen 2016). The human intestine is a perfect habitat for different species of bacteria, with the highest concentration found in the colon. CRC is a development from the colon and is associated with specific bacteria (Jahani-Sherafat et al. 2018; Garrett 2019). Recently, the metagenomic analysis showed that the samples from patients with CRC identified CRC-enriched bacteria, including *Streptococcus gallolyticus*, *Enterococcus faecalis*, enterotoxigenic *B. fragilis* (ETBF), *Escherichia coli*, *Fusobacterium nucleatum*, *Porphyromonas asaccharolytica*, *Parvimonas micra*, *Prevotella intermedia*, *Alistipes finegoldii*, and *Thermanaerovibrio acidaminovorans* (Mira-Pascual et al. 2015; Toprak et al. 2006; Arthur et al. 2012). Results from recent studies showed that the role of *S. gallolytucis* is associated with a carcinoma of the CRC. In addition, patients with CRC showed higher fecal carriage of *S. gallolytucis* in comparison with control subjects (Abdulamir et al. 2011). Animal studies also showed that *S. gallolytucis* increased the expression of proliferation markers and polyamines (Gagnière et al. 2016). *Bacteroides fragilis* is considered normal flora in the colon, with 80% of children and adults being colonized by the bacterium. However, recent studies showed that enterotoxigenic *B. fragilis* (ETBF) was detected significantly higher in the stools of CRC patients compared with controls (Toprak et al. 2006). Virulence of ETBF is essentially due to the *B. fragilis* toxin (BFT), which cleaves the E-cadherin tumor suppressor protein, resulting in enhanced nuclear Wnt/β-catenin signaling that yields increased colonic carcinoma cell proliferation (Purcell et al. 2017). *E. coli* is part of the normal flora in the human colon, but several human studies reported an association between *E. coli* and CRC (Maddocks et al. 2009). A study reported that *E. coli* colonized 71% of mucosa samples from patients with CRC compared with 42% in controls (Martin et al. 2004). *E. coli* harbors cytotoxic necrotizing factor (Cnf) and cytolethal distending toxin (Cdt), which are significantly associated with CRC (Buc et al. 2013). Animal studies showed that infection with *E. coli* caused the formation of sporadic CRC in infected mice (Swidsinski et al. 1998). *F. nucleatum* provided the strongest evidence of the role of carcinogenesis in the colon (Amitay et al. 2017). Metagenomic analysis showed the enrichment of *F. nucleatum* at higher levels in CRC tissues and fecal samples of CRC patients compared to healthy controls. FadA and Fap2 proteins are the virulence factors of *F. nucleatum* that were described to mediate adhesion to E-cadherin, activate β-catenin signaling, and enhance subsequent inflammatory and oncogenic responses (Rubinstein et al. 2013; Yang et al. 2017). *Enterococcus faecalis* is recognized as a human pathogen, and significantly high levels were observed in fecal specimens of CRC patients compared to healthy controls (Balamurugan et al. 2008). The role of *E. faecalis* acting with the release of reactive oxygen and nitrogen species that cause subsequent DNA breaks, point mutations, and chromosomal instability has been suggested as the main mechanism for oncogenic activity in CRC (Pillar and Gilmore 2004).

The gut microbiome-based therapies, such as fecal microbiota transplant (FMT), probiotics, and symbiotic, might be considered attractive targets for the treatment of CRC (Faïs et al. 2016). However, limited clinical evidence is available regarding the role of bacterial eradication in the treatment of CRC (Rossen et al. 2015). FMT is known to be highly effective in bacterial infection and promising for ulcerative colitis, and it significantly increased insulin sensitivity in male patients with metabolic syndrome. FMT could restore CRC-associated dysbiosis and reduce microbiota-induced inflammatory, proliferative, and pro-carcinogenic reactions and genotoxicity (Filip et al. 2018). Manipulation of gut microbiota using probiotics might be considered a novel therapeutic modality for the prevention of CRC development or reduction of chemotherapy-induced adverse effects (Jahani-Sherafat et al. 2018). The probiotic *Lactobacillus casei* was also described to reduce the expression and activity of the drug-metabolizing enzymes, cytochromes P450, which are known to be associated with CRC carcinogenesis (Rafter et al. 2007).

7 Role of Bacteria in the Treatment of Cancer (Bacteriotherapy)

Although some bacterial infections with immunogenic nature are considered to increase the risk of developing cancer or are carcinogenic, other bacteria have conversely demonstrated great potential in the treatment of cancer (Soleimanpour et al. 2020). Different bacterial species have recently shown unexpected potential in invasion and colonization of solid tumors, resulting in tumor suppression (Wei et al. 2007). In addition, different bacterial species release numerous substances with anticancer activity, including bacteriocins, peptides, toxins, enzymes, and spores (Laliani et al. 2020). The bacterial species also can enhance the host immune responses toward malignancy via several mechanisms such as their use as a vehicle for delivering anticancer drugs, decreasing the nutrients required for the cancer cell metabolism, ability to grow in the necrotic and hypoxic regions of the tumor, releasing substances that counteract the microenvironment that promotes tumorigenesis, and formation of biofilms (Yaghoubi et al. 2020). The use of bacteria and their products as anticancer agents are explained in Table 4.2.

7.1 Bacteriocins

Bacteriocins are ribosomal proteins or peptides produced by bacteria. The physiological functions of bacteriocins are aimed toward inhibiting the growth or killing of the competing microorganisms that are related or nonrelated bacterial strains in a particular biological niche (Simons et al. 2020). Several bacteriocins are used as narrow-spectrum antibiotics and food preservatives. Bacteriocins are divided according to their molecular weight or size of microcins (less than 20 kDa), colicin (20 to 90 kDa), and tailocins (high molecular weight) (Lee and Kim 2011). The classification organizes Gram-positive bacteriocins into four major groups based on the physiochemical properties and their structures. Class I contains small thermostable bacteriocins with the size of less than 5 kDa, known as lantibiotics. Some examples of this class include nisin, lacticin, mersacidin lanthionine, methyllanthionine, and dehydroalanine. Class II bacteriocins contained thermostable peptides

Table 4.2 Summaries of studies on anticancer bacterial metabolites

Name of product	Bacterium	Origin	Molecular weight	Human gastrointestinal cancer cells/cell lines	Anti-cancer mechanism	References
Anticancer bacteriocins	Bovicin HC5	*Streptococcus bovis* HC5	2.4 kDa	Liver, hepatocellular carcinoma (HepG2)	Pore in the cell membrane to prompt the potassium efflux in the target cells	Mantovani et al. (2002)
	Nisin A	*Lactococcus lactis*	3.5 kDa	Colon cancer (LS180, SW48, HT29, and Caco2), liver hepatocellular carcinoma (HepG2)	Inhibiting tumor cell growth by compromising the integrity of the cellular membrane and forming the short-lived pores and changes the potential of the membrane	Joo et al. (2012); Norouzi et al. (2018)
	Pediocin	*P. acidilactici*	5–17 kDa	Colon adenocarcinoma (HT29)	Induction of apoptosis by DNA fragmentation	Villarante et al. (2011)
	Fermenticin HV6b	*Lactobacillus fermentum*	6.6 KDa	Hepatocellular carcinoma (HepG2), cervical adenocarcinoma (HeLa)	Anticancer activity induces apoptosis in the vascular endothelial cells, cell contraction, and DNA fragmentation	Kaur et al. (2013)
	Colicin	*Escherichia coli*	40–80 *kDa*	Colon cancer (HCT116)	Hit the target cells through pore development, nonspecific DNAse activity, inhibition of protein biosynthesis by cleaving 16S rRNA or tRNAs, by inhibiting the synthesis of murein	Tomita et al. (2000); Lakey and Slatin (2001)
	Pyocin S2	*Pseudomonas aeruginosa*	73.8 kDa	Hepatocellular carcinoma (HepG2)	Induce cell death by cytotoxic activity on the tumor cells	Abdi-Ali et al. (2004)

(continued)

Table 4.2 (continued)

Name of product	Bacterium	Origin	Molecular weight	Human gastrointestinal cancer cells/cell lines	Anti-cancer mechanism	References
Anticancer peptides	Mixirins A–C	*Bacillus* sp.	1 kDa	Human colon tumor (HCT-116)	Inhibits the proliferation of human colon tumor cell line	Zhang et al. (2004)
	Arenamides A, B	*Salinispora arenicola*		Colon cell line (HCT116)	Inhibits cysteine and cytotoxic effect	Asolkar et al. (2009)
	Azurin	*Pseudomonas aeruginosa*	16 kDa	Liver cell line (HEPG2), colon cell line (HCT116)	Increased intracellular in the receptor tyrosine signaling process, preventing through reducing VEGFR-activity, forming complexes with tumor protein p53, interfering in the receptor tyrosine kinase EphB2-mediated signaling process, preventing angiogenesis through reducing VEGFR-2 tyrosine kinase activity and inhibition of the growth of cancer cells	Yamada et al. (2009); Jia et al. (2011)
	Entap	*Enterococus* sp. strains	6.2 kDa	Gastric adenocarcinoma cells (AGS), colorectal adenocarcinoma (HT-29)	Cancer cell arrest in G1 and induction of autophagous apoptosis, deregulation of proliferation autophagous apoptosis	Karpiński (2012); Karpiński et al. (2013)
	Pep27anal2	*Streptococcus pneumoniae*	3.6 kDa	Gastric cancer cells (SNU-601)	Membrane inducing caspase and cytochrome c independent apoptosis, cellular permeabilization, proliferation in gastric cancer cell	Lee et al. (2005); Sung et al. (2007)
	Proximicins	*Verrucosispora* sp. MG-37	293–246 kDa	Human gastric adenocarcinoma (AGS), human hepatocellular carcinoma (HepG2)	Upregulating the intracellular levels of p53, and the cyclin kinase inhibitor p21	Fiedler et al. (2008)

Anticancer toxins	Diphtheria toxin	*Corynebacterium diphtheria*	60 kDa	Adrenocortical carcinoma (H295R), colon cancer (SW480, SW620, HCT116, CaCo-2, and HT-29)	Inhibiting the tumor proliferation, decreasing the angiogenesis, inducing the apoptosis, acting as an immunological adjuvant, and inhibiting the heparin-binding epidermal growth factor	Vallera et al. (2002); Martarelli et al. (2009)
	Verotoxin	*E. coli*	70 kDa	Human colorectal cancer cell lines (HCT116)	Anticancer activities by blocking the protein synthesis, prevent cell growth, arrest the cell cycle at the S phase, and target the receptor of the membrane	Bhattacharjee et al. (2005)
	Exotoxin A	*Pseudomonas aeruginosa*	66 kDa	Pancreatic cancer (PaCa-2)	Targeting with tumor-related antigens and induction of cytotoxic pathways	Shapira et al. (2011)
	CPE enterotoxin	*Clostridium perfringens*	35-kDa	Human colorectal cancer cell lines (SW480, SW620, HCT116, CaCo-2, and HT-29), human gastric cancer cell lines (SGC7901, MKN45, AGS, MGC803, BGC823, and HGC27)	Anticancer activity by inhibiting the proliferation of different human cancer cells, binding to claudin-3 and cells, pore-forming in cells, inducing apoptosis	Black et al. (2015); Pahle et al. (2017)
	TcdB	*Clostridium difficile*	270 kDa	Colorectal cancer (CT26)	Inhibit proliferation, induce apoptosis	Huang et al. (2014)
Anticancer enzyme	Arginine deiminase	*Mycoplasma hominis*, *M. arginine*	46 3 kDa	Hepatocellular carcinoma (HCC)	Inducing the caspase-independent apoptosis, decreasing the cell proliferation, and inducing autophagy	Fiedler et al. (2015)
Anticancer activity of bacteria	*Salmonella typhimurium*			Human pancreatic cancer (ASPC-1), colon carcinoma (WiDr, CT26)	Used as a target delivery vehicle that encodes the therapeutic proteins and decreases tumor growth	Pawelek et al. (1997)
	Listeria monocytogenes Lm-LLO-E7 and ADXS31-142			Human colon cancer (Colo205)	Activation of NADP+ oxidase, increasing the intracellular level of Ca2+, production of high ROS levels, inducing strong CTL responses, and depletion of TCD8+	Shahabi et al. (2011)
	Lactobacillus spp.			Colon cancer (Caco-2, BGC-823, HT-29)	Producing the antioxidants; decreasing the activity of carcinogenic compounds; increasing TNFα; inducing caspase-3 activity; inactivating the Wnt/β-catenin signaling	Sakatani et al. (2016); Jacouton et al. (2017)

with a molecular weight of 30 kDa with an amphiphilic helix in structure. Some examples of class II bacteriocins are pediocin PA-1, sakacin A, lactacin F, lactococcin G, gassericin A, circularin A, and carnocyclin A. Class III contains large heat-labile bacteriocins with more than 30 kDa molecular weight. Megacins, klebicin, helveticin I, and enterolysin are members of class III bacteriocins. Class IV bacteriocins are characterized as complex proteins containing lipid or carbohydrate moieties (Sahl and Bierbaum 1998; Kaur and Kaur 2015; Baindara et al. 2018). Bacteriocins act through a "membrane-active" mechanism, i.e., by interacting with the negatively charged cell membrane. The electrostatic interaction between cationic bacteriocins and cancer cells plays an important role in their cytotoxic effect (Norouzi et al. 2018). Several studies have shown that bacteriocins with strange toxicities for the treatment of cancer are nonimmunogenic and effective in the treatment of infectious diseases. Bacteriocins can terminate the membrane integrity and initiate the apoptosis of the cancer cells (Kaur and Kaur 2015; Baindara et al. 2017). Bacteriocins are able to suppress the tumor through other mechanisms such as cell cycle alterations, cell shrinkage, induction of apoptosis, preventing biosynthesis of proteins through 16SrRNA or tRNAs breakdown, necrosis, caspase activation, forming pores in the plasma membrane, and inhibiting the cell proliferation and angiogenesis (Yaghoubi et al. 2020). Bovicin belongs to class I bacteriocins that are produced by *Streptococcus bovis* HC5 with a molecular weight of 2.4 kDa. Bovicin exhibits a broad-spectrum antimicrobial activity against various other Gram-positive and Gram-negative bacteria (Paiva et al. 2012). The antimicrobial activity is shown by binding to the cell membrane and disrupting cell membrane integrity by pore formation (Mantovani et al. 2002). Nisin is a low-molecular-weight bacteriocin that belongs to class I bacteriocins, which is also called lantibiotic. Nisin is a polycyclic antibacterial peptide with 34 amino acids and is produced by a Gram-positive bacterium, i.e., *Lactococcus lactis* subsp. Nisin is a heat-stable bacteriocin and has been used as a food preservative for a long time (Begde et al. 2011). Recent studies indicate that nisin exhibits anticancer activity by inhibiting tumor cell growth, the integrity of the cellular membrane, forming the short-lived pores, and changing the potential of the membrane (Joo et al. 2012). Nisin can also inhibit the invasion of tumor cells and metastasis of different human cancer cells and recurrence of colon cancer cell lines including LS180, SW48, HT29, and Caco2 and different human gastrointestinal cancer cells (Norouzi et al. 2018). Pediocins are small peptides (>5 kDa) and belong to class IIa bacteriocins (Papagianni and Anastasiadou 2009). Pediocins exhibit anticancer activity against several cancer cell lines as well as gastrointestinal cancer, such as the mouse spleen lymphoblast cell line (Sp2/O-Ag14) hepatocarcinoma (HepG2) (Balgir et al. 2010). Pediocin has also shown cytotoxic activity that inhibits the proliferation of tumor cells, such as the human colon adenocarcinoma cells (HT29), MCF7, and HeLa cell lines (Kumar et al. 2012; Villarante et al. 2011). Fermenticin HV6b is a class IIa bacteriocin with a molecular mass of 6.6 kDa. Fermenticin HV6b is produced by *Lactobacillus fermentum* HV6b MTCC 10770 isolated from the human vaginal ecosystem (Kaur et al. 2012). In vitro studies show that fermenticin HV6b exhibits anticancer activity through different mechanisms including encouraging apoptosis in the vascular endothelial cells, cell contraction, and DNA fragmentation in various cancerous cells (Kaur et al. 2013). Colicins are plasmid-encoded antibacterial bacteriocins with high molecular mass ranging from 40 to 80 kDa that is secreted by *E. coli* and other related Enterobacteriaceae. Colicins are active against *E. coli* strains and other closely related bacteria, such as Shigella and Salmonella (Braun et al. 1994). Thirty different types of colicins have been identified and differentiated according to their killing activity and the mode of action (Smarda and Smajs 1998). Colicin can hit the target cells through different mechanisms such as pore formation (colicins A, B, E1, Ia, Ib, K, L, N, U, 5, and 10), nonspecific DNAse activity (colicins E2, E7, E8, and E9), or inhibition of protein biosynthesis by cleaving 16S rRNA or tRNAs (colicins E3, E4, E6, E5, and D) or by inhibiting the synthesis of murein (colicins M and pesticin) (Tomita et al. 2000; Lakey and Slatin 2001). Colicins have anticancer activities against a variety of human tumor cell lines in vitro, such as colon cancer, bone cancer, and uteri cell line HeLa. Colicin E1 has cytotoxic activity against HT29 that is a human colon cancer cell line (Chumchalová and Smarda 2003). Pyocins are produced by *Pseudomonas aeruginosa* strains with a molecular mass of 73.8 kDa. In 1954, Jacob first isolated pyocins from the *P. aeruginosa* 10 strain (Jacob 1954). Three different types of pyocins have been identified, namely (i) R-type pyocins resembling nonflexible and contractile tails of bacteriophages, (ii) F-type pyocins resembling phage tails, but with a flexible and noncontractile rod-like structure, and (iii) S-type pyocins that are colicin-like, protease-sensitive bacteriocins. Pyocin S2 is isolated from *P. aeruginosa* 42A, has cytotoxic activity on the tumor cells, and is able to inhibit the growth of different cancer cells such as HepG2, which is a hepatocellular carcinoma cell line (Abdi-Ali et al. 2004).

7.2 Bacterial Peptides

In addition to bacteriocins, several studies suggest that bacterial peptides also have anticancer activity. The peptides have been characterized by special chemical structures including N-terminally attached fatty acids, N formulated residues, D-amino acids, heterocyclic elements, N- and C-methylated residues and glycosylated amino acids, and the phosphorylated residues. Such peptides of bacterial origin, in addition to antimicrobial activities, exhibit a great potential activity toward different types of cancer. The anticancer effects of the bacterial peptides on different types of cancer are described in Table 4.2. Mixirins are cyclic acylpeptides derived from the marine bacterium *Bacillus* species, with a molecular weight of about 1 kDa. Mixirins have three major members called mixirins A, B, and C. Evidence suggests that three cyclic acyl-peptides have anticancer potential to inhibit the propagation of human colon tumor cell line HCT-116 (Zhang et al. 2004). Arenamides are cyclohexadepsipeptides obtained from *Salinispora arenicola* found in sea sediment (Great Astrolabe Reef, Kandavu Island chain, Fiji).

Arenamides are of three main types, namely A, B, and C. A study showed that arenamides A and B exhibit anticancer activity by suppressing tumor cells. Arenamides A and B also exhibit cytotoxic activity on different human colon cancer cells as well as HCT-116 (Asolkar et al. 2009). Halolitoralins are cyclic peptides produced by bacteria called *Halobacillus litoralis* YS3106 found in the marine sediments (Huanghai Sea, China). with a molecular mass of 575 Da. These peptides exhibit in vitro anticancer activities against human gastric tumor cells BGC (Yang et al. 2002). Ieodoglucomide is found from a marine bacterium, i.e., *Bacillus licheniformis* from the marine sediment of Ieodo Reef (South Korea). Ieodoglucomides A and B showed low antimicrobial activity in vitro, but ieodoglucomide B verified cytotoxicity against stomach cancer cells (Tareq et al. 2012). Lucentamycins are isolated from the marine-derived actinomycete strain *Nocardiopsis lucentensis* CNR-712. Evidence suggests that lucentamycins A and B exhibited significant in vitro cytotoxicity against human colon carcinoma cells HCT-116 (Cho et al. 2007). Urukthapelstatin is a cyclic thiopeptide antibiotic with a molecular mass of 733 kDa obtained from the marine bacterium *Mechercharimyces asporophorigenens* YM11-542. This peptide exhibits anticancer activity against human colon carcinoma cell line HCT-116 in vitro (Matsuo et al. 2007). Fiedler et al. first reported proximicins extracted from a marine member of the rare genus *Verrucosispora*, strain MG-37 (Fiedler et al. 2008). Proximicins are of mainly three types, namely proximicins A (293 Da), B (413 kDa), and C (436 kDa). All proximicins show growth-inhibition potential against cell lines of gastric adenocarcinoma and hepatocellular carcinoma. An in vivo experiment showed that proximicin C exhibits cytotoxic activity on gastric adenocarcinoma (AGS) cells. Proximicin C can also upregulate the intracellular levels of p53 and the cyclin kinase inhibitor p21 in the different cancer cell lines such as AGS and HepG2 (Fiedler et al. 2008). Azurin is a copper-containing globular metalloprotein with redox activity with a length of 128 amino acids and a low molecular weight (14 kDa), derived from *P. aeruginosa* (Yamada et al. 2002). Azurin is one of the representative bacterial products that can show a great anticancer activity (Yamada et al. 2009). Several other studies also confirm azurin and p28 as the tumor suppressors, which prove great anticancer activity on different cancer cell lines, such as gastrointestinal cancer cell lines including liver cell line (HEPG2), colon cell line (HCT116, HT29), and pancreatic cancer (MIA-Paca2) (Mehta et al. 2011). Pep27anal2 is known as an analog of signal peptide Pep27 with a length of 27 amino acids and a molecular mass of 3.3–3.6 kDa. This peptide shows great anticancer potential, which can decrease cell proliferation in gastric cancer cell lines such as SNU-601 (Lee et al. 2005; Sung et al. 2007). Enterococcal anti-proliferative peptide (Entap) is derived from clinical strains of *Enterococcus* genus with a molecular weight of 6.2 kDa and contains 58–62 amino acids. Entap exhibits antiproliferative activity against cell lines of human gastric adenocarcinoma (AGS), colorectal adenocarcinoma (HT-29), uterine cervix adenocarcinoma (HeLa), and prostatic carcinoma (22Rv1) (Karpiński 2012; Karpiński et al. 2013).

7.3 Bacterial Toxins

Toxins expressed and secreted by bacteria play a vital role in infection and targeting the host immune systems via alteration of cellular processes, inducing apoptosis, and the inhibition of cell proliferation and differentiation. Many toxins can inhibit the function of the immune cells and disrupt the production and release of cytokines and antibodies. According to the cytotoxic action of bacterial toxins and enzymes, different bacterial toxins have anticancer potential toward different human cancer cell lines through binding to the heparin-binding EGF-like growth factor, suppressing the tumor cell growth, decreasing the angiogenesis, and inducing apoptosis (Lewis et al. 2017). *Corynebacterium diphtheria* produces an exotoxin called diphtheria toxin (DT) that exhibits anticancer activity. *Clostridium botulinum* produces botulinum neurotoxin type A (BoNT-A), which is another bacterial toxin with anticancer activity. *Pseudomonas aeruginosa* produces exotoxin A (PE) and exotoxin T (ExoT), which show great anticancer activity through inducing apoptosis in the cancer cell lines. *Clostridium perfringens* enterotoxin (CPE) is another bacterial toxin with anticancer potency. Different *E. coli* species including VT-producing *E. coli* (VTEC) and enterohemorrhagic *E. coli* (EHEC) produce a toxin, verotoxin 1, that exhibits anticancer activity through a different mechanism such as arresting the cell cycle, DNA fragmentation, and changing the cell cycle protein expression (Lutz et al. 2014; Bhattacharjee et al. 2005). Diphtheria toxin release by *Corynebacterium diphtheria* with a molecular weight of 60 kDa toxin and a length of 538 amino acids (Holmes 2000). Diphtheria toxin contains two subunits A and B and is transferred to the bacteria through bacteriophage B. The B subunit is responsible for the binding to the receptor on the cell surface, and the A subunit is responsible for inhibiting the protein synthesis via ADP-ribosylation of cytoplasmic elongation factor 2 (EF-2) (Murphy 2011). Diphtheria toxin also shows cytotoxic activity against different human carcinoma cell lines as well as gastrointestinal cancer, including adrenocortical carcinoma cell lines such as H295R, colon cancer cell lines such as SW480, SW620, HCT116, CaCo-2, and HT-29 (Vallera et al. 2002; Martarelli et al. 2009). Verotoxin 1 (VT1) is known as Shiga toxin-1 (Stx-1), secreted by different *E. coli* families such as VT-producing *E. coli* (VTEC) and enterohemorrhagic *E. coli* (EHEC) (Karmali et al. 1985). VT1 is responsible for inhibiting the synthesis of protein, cell proliferation, and apoptosis and damages endothelial cells in the hemolytic uremic syndrome (Obrig et al. 1987). VT1 has antitumor, antiangiogenic, and antineoplastic activities by blocking the protein synthesis, preventing cell growth, arresting the cell cycle at the S phase, and targeting the receptor of membrane called Gb3, which is overexpressed in several multidrug-resistant (MDR) human carcinoma cell lines such as HCT116 (Bhattacharjee et al. 2005). Exotoxin A is produced by *P. aeruginosa* with a molecular mass of 66 kDa. Exotoxin A is an ADP-ribosylation of elongation factor-2 (EF-2), which inhibits protein synthesis and leads to cell death. Exotoxin A has anticancer activity by inducing apoptosis in the tumor cells (Wood et al. 2015; Shapira et al. 2011). The *C. perfringens* enterotoxin

(CPE) is a single polypeptide with a molecular mass of 35 kDa and a length of 319 amino acids. CPE can lyse the epithelial cells via binding to the transmembrane tight junction protein claudin (Santin et al. 2007). CPE also has anticancer activity by inhibiting the proliferation of different human cancer cell lines including HCT116 and CaCo-2 cell lines of human colon cancer and SGC7901 cell lines of human gastric cancer (Black et al. 2015; Pahle et al. 2017). *C. difficile* produces two types of toxins, known as cytotoxin (TcdB) and enterotoxin (TcdA). TcdB has anticancer activity through producing the pro-inflammatory chemokines and cytokines, inhibiting cell proliferation and inducing necrosis as well as apoptosis (Eckert et al. 2014). TcdB is highly immunogenic, inducing long-term anti-tumor immunity toward different cancer cell lines as well as the colorectal cancer cell line CT26 and can be used as the effective anti-tumor vaccine or immunotherapy agent in the treatment of cancer (Huang et al. 2014).

7.4 Bacterial Enzymes

Enzymes produced by some bacteria exhibit the ability of tumor suppression. Numerous studies have shown that different enzymes are vital for cellular metabolism and normal cell growth; hence, these enzymes can be a restricting factor in rapid tumor cell growth (Ni et al. 2008). Arginine deiminase (ADI) is secreted from *Mycoplasma hominis* or *M. arginine* with a molecular weight of 46.3 kDa and has anticancer activity by inducing the caspase-independent apoptosis, decreasing the cell proliferation, and inducing autophagy followed by amino acid depletion (Fiedler et al. 2015). Lasparaginase (ASNase), another enzyme secreted by *E. coli* or *Erwinia* species, has the anti-tumor activity through decreasing the asparagine concentration of blood, resulting in the selective growth inhibition of sensitive malignant cells. The enzyme "hyaluronidase (Hyals)," obtained from *Streptococcus pyogenes*, displays anticancer activity on a variety of cancers (Lokeshwar et al. 2001).

8 Bacteria as a Target Delivery Vector for Cancer Therapeutic Agents

Over the last few decades, bacteria have been increasingly explored for the delivery of anticancer agents, being used as live, attenuated, or genetically modified vectors for specific distribution or expression of antitumor genes, anti-angiogenic genes, conventional drugs, and tumoricidal molecules. The attenuated strains of *Salmonella typhimurium* can be used as a selective target delivery vector for the therapeutic genes that encode for antitumor activity (Khan et al. 1998). The attenuated strains of *S. typhimurium* show anticancer activity against different cancer cell lines such as pancreatic cancer (ASPC-1) and human colon carcinoma cell lines (C38, WiDr, and CT26) (Pawelek et al. 1997). *Listeria monocytogenes* is an intracellular pathogen, and different recombinant forms of *L. monocytogenes* like Lm-LLO-E7 and ADXS31-142 have anticancer activity against numerous human carcinomas such as the colon cancer cell line Colo205 (Shahabi et al. 2011). Lactobacillus is largely known as a major component of human microbiota in different parts of the human's body, including the intestine and the urinary systems (O'Mahony et al. 2001). *Lactobacillus acidophilus* can suppress the tumor by enhancing the host immune response by increasing IFN-γ, IL-10, CD4+, and CD8+ T cells and reducing the serum levels of tumor markers, including CEA and CA19-9 (Agah et al. 2019). *Lactobacillus brevis* SBL8803 is a species found in fermented malt, which has shown anti-colon cancer effect through an antitumor molecule called polyphosphate (polyP) that affects the ERK pathway in the target tumor (Sakatani et al. 2016). *Lactobacillus casei* BL23 is another strain, which has anti-inflammatory and anticancer activity by stimulating the host immune response, reducing the level of IL-22, and preventing the tumor cell proliferation by upregulation of caspase-7, caspase-9, and adenoma expansion (Jacouton et al. 2017). Bifidobacteria are one of the major microbiotas in the gastrointestinal tract of humans (Choi et al. 2005). Some species (*B. longum*, *B. infantis*, *B. adolescentis*, and *B. breve*) of the genus Bifidobacterium are known as probiotics that express anticancer activity by altering the host immune response, which subsequently allows the immune components to then target the tumor cells and degrade them, alter the host intestinal microflora, and produce additional antitumor substances (Schell et al. 2002). *B. adolescentis* SPM0212 expresses butanol, which can inhibit the growth of colon cancer cell lines (Caco-2, HT-29, and SW480) through enhancing TNF-α and nitric oxide. *B. longum* SPM1205 is known as a TNF-α inhibitor and can prevent the proliferation of human colon cancer cell lines, including HT-29, SW 480, and Caco-2 (Lee et al. 2008).

9 The Gut Microbiome and Cancer Immunotherapy

Immunotherapy comprises strategies that manipulate host immune system components for targeted elimination of pathogens or diseased cells. Advances that extend the understanding of systemic immune responses to malignancies and how tumor microenvironments suppress the host's local immune responses have since led to the development of novel therapeutic interventions. In a variety of cancers, the informed use of various immune-based treatments has augmented the positive outcomes of standard therapeutic strategies, namely surgical excision, chemotherapy, and radiotherapy. Immunotherapy is thus recognized as an essential pillar for cancer treatment in current medicine. Broadly, cancer immunotherapies are classified into active or passive strategies based on their mechanism of action and the status of the patient's immune system (Papaioannou et al. 2016; Kruger et al. 2019). The major subtypes have been summarized in Fig. 4.2.

Before elaborating on the studies that describe the gut microbiome's influence on immunotherapy outcomes, we first briefly provide an outline of some major

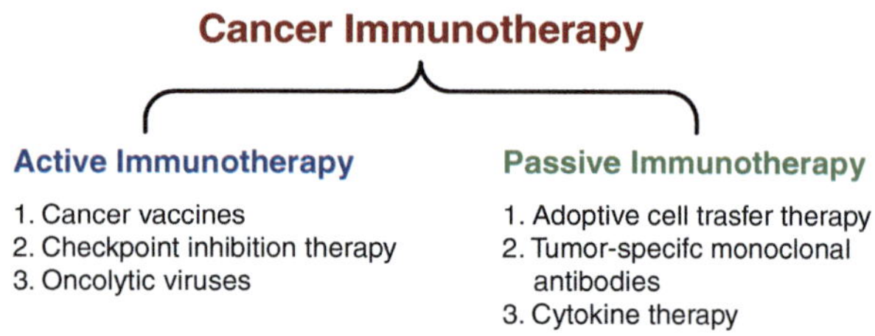

Fig. 4.2 Classification of immunotherapeutic interventions. Cancer immunotherapy and the subsets are active and passive immunotherapies

immunotherapy approaches to provide the reader with better context and to help them appreciate the nuances of patient immune modulation.

Figure 4.2 lists the major subtypes of approved immunotherapies for the treatment of various cancers. These modalities also illustrate the promising immune-based interventions that may shape the future of GI cancer therapy. Active immunotherapies serve to activate or enhance the in vivo immune responses in patients. Notable active immunotherapy strategies include therapeutic cancer vaccines, immune checkpoint inhibition (ICI) therapy, and the use of oncolytic viruses. Cancer vaccines involve the administration of multi-peptide cocktails that comprise tumor-specific antigens (TSAs) and tumor-associated antigens (TAAs). These peptides are designed to elicit stable T lymphocyte responses through the recruitment of cytotoxic T lymphocyte (CTL) or T helper (Th) cells. Use of matured dendritic cells (mDCs) is an alternative strategy for vaccine delivery, cancer antigen presentation, and activation of host immune cells (Papaioannou et al. 2016; Song et al. 2018). A number of therapeutic vaccines against GI cancers have been progressed to clinical trials, though none have received regulatory approval yet (Rahma and Khleif 2011; Fujiwara et al. 2017). ICI is currently a popular area of cancer and immunotherapy research. It involves blocking immune inhibitory receptor–ligand pair interactions through the administration of therapeutic monoclonal antibodies. Immune checkpoints are cellular mechanisms that prevent the host immunity from attacking healthy cells indiscriminately. Blocking disease-associated abnormal immune checkpoint activation restores normal immune system function by preventing T cell inhibition, promoting effector T cell activation, and antibody-mediated regulatory T cell (Treg) depletion. They subsequently permit enhanced immune responses against upregulated ligands. Widely researched checkpoint blockade targets include cytotoxic T-lymphocyte-associated protein 4 (CTLA-4), programmed cell death 1 ligand 1 (PD-L1), and programmed cell death protein 1 (PD1) (Wei et al. 2018). Several recent studies have explored the link between ICI therapy, the gut microbiome, and GI cancer. Some key findings will be described in subsequent paragraphs. Oncolytic virotherapy uses innocuous recombinant viral strains such as the Herpes Simplex Virus-1, which are engineered to specifically target tumors. Infection of tumor cells is followed by viral replication and cell lysis. The viruses used may be engineered to facilitate enhanced T effector cell recruitment to the site of infection, and the lysis of the cancerous cells may enhance the uptake and display of tumor antigens by the locally recruited antigen-presenting cells (APCs) (Papaioannou et al. 2016; Rao et al. 2019; Kruger et al. 2019).

Passive immunotherapies, on the other hand, serve to compensate for the host immune functions that are absent or inadequate. They involve the use of cells or recombinant molecules that are modified or expressed ex vivo and subsequently administered to patients. The major subtypes of passive immunotherapies are adoptive cell transfer (ACT) therapy, the use of tumor-specific monoclonal antibodies (mAbs), and cytokine therapy (Papaioannou et al. 2016).

ACT therapies involve the genetic modification of autologous lymphocytes followed by their administration to patients for tumor eradication. Modifications made to the lymphocytes aim to enhance MHC-independent recognition and killing of tumor cells aided by the tumor surface markers displayed. Among the various cell types explored, chimeric antigen receptor (CAR) T cells have garnered the most amount of interest for use in ACT-based therapies. In addition to possessing an antigen or ligand-binding domain, improvements made to CAR T cells over successive generations have resulted in CAR T cells that possess co-stimulatory domains that enhance and/or prolong T cell responses in patients. The latest generation of CAR T cells also allows inducible expression of pro-inflammatory cytokines such as interleukin (IL) 12 to improve the therapeutic benefits of treatment (Chmielewski and Abken 2015). Recombinant mAbs have the broadest range of applications in immunotherapy. Therapeutic mAbs exist in several formats based on the species of origin and the arrangement of antibody domains. Selective killing of tumors is achieved through mAbs directly causing cell death through inhibition of biological processes necessary for cancer-cell or tumor survival (e.g., inhibition of cell signaling pathways, angiogenesis, etc.). The immune-mediated killing of tumors is another mechanism through which such mAbs function (Chanier and Chames 2019; Bayer 2019). Antibody-dependent cellular cytotoxicity (ADCC), complement-dependent cytotoxicity (CDC), and/or antibody-dependent cellular phagocytosis (ADCP) are a few additional mechanisms by which therapeutic mAbs affect immune-mediated tumor cytotoxicity (Papaioannou et al. 2016). Antibody-drug conjugates (ADCs) are a subgroup of mAbs that use antibodies conjugated to chemotherapeutic agents, cytotoxins, radioactive particles, and photo-activating compounds for targeted elimination of tumors (Nejadmoghaddam et al. 2019; Bayer 2019; Kobayashi et al. 2021). Administering patients with cytokines such as ILs, interferons (IFNs), tumor necrosis factor (TNFs), colony-stimulating factors (CSFs), etc. help promote immune cell functions such as growth, differentiation, and activation, to name a few. Cytokine therapies elicit effective, durable immune responses against cancers and are thus valuable options in the armamentarium of anticancer treatments (Lee and Margolin 2011).

Historically, bacteria have been important to immunotherapy. During the early 1800s, *Clostridium perfringens* and *Streptococcus pyogens* were the first described bacterial species that caused regression of tumors in patients through activation of the immune system (McCarthy 2006; Sedighi et al. 2019). Bacillus Calmette–Guérin (BCG) is a common bladder cancer therapy approved for use since the 1970s. Intravesical administration triggers immune-mediated tumor suppression, caused by elevated cytokine levels and IFN-γ expression (Morales 2017; Rhea et al. 2021).

Among the major immunotherapy approaches described earlier, ICI therapy markedly stands out. Within a decade since the first ICI therapy was approved, they have become among the most widely prescribed anticancer agents in use as first- or second-line drugs. At present, they are used for treating more than 50 types of cancer. Active clinical trials that are evaluating ICI mAbs as mono- or combination therapies presently number over 3000 (Robert 2020). However, despite their success, ICIs suffer from certain drawbacks. The limitations include low response in patients and the development of either naturally acquired or treatment-induced resistance over time among patient subgroups. It is thus not surprising that most published studies that correlate gut microbiota and immunotherapy predominantly examine the impact of the former on ICI-based therapies. These are in part aimed at improving therapy responses since the multivariate causes that limit ICI use are yet to be completely understood. Preclinical murine model studies that were published simultaneously in 2015 examined and provided evidence supporting the ability of specific bacterial species to favorably modulate anti-CTLA-4 and anti-PD-L1 therapies. Antibiotic-treated and germ-free (GF) mice exhibited no response to CTLA-4 blockade in sarcoma, melanoma, and colon cancer (MC38) models. High-throughput pyrosequencing of 16S ribosomal RNA (rRNA) followed by principal component analysis showed how even a single dose of anti-CTLA-4 mAb could decrease the population of certain bacterial species and favor colonization by others. Inoculation of mice with *Bacteroides thetaiotaomicron*, *B. fragilis*, and *Burkholderia cepacia* improved the tumor-shrinking effects of anti-CTLA-4. The mAb treatment led to T cell-dependent apoptosis of intestinal epithelial cells (IECs). Conditions that favor the enrichment of certain bacterial taxa lead to IL12-dependent Th1 immune response against the tumor (Vétizou et al. 2015). Oral administration of *Bifidobacterium* (identified through 16 s rRNA sequencing of fecal bacteria and principal component analysis) increased antitumor immunity against melanoma in mice through the activation of dendritic cells, and the combination therapy of anti-PD-L1 mAb and *Bifidobacterium* treatment markedly abolished tumor growth. Depletion of CD8+ T cells led to the disappearance of the tumor-attenuating effect of anti-PD-L1 treatment in mice (Sivan et al. 2015).

While the above two studies predominantly drew conclusions using a melanoma model, subsequent studies described by other groups have started to expand on the role different bacterial species and taxa seem to play in the modulation of immunotherapeutic responses to GI cancers. Pushalkar and colleagues reported an increase in the efficacy of anti-PD1 immunotherapy though upregulation of PD1 in the murine model of pancreatic ductal adenocarcinoma (PDAC). Their study provided evidence that counteracting the differential increase of certain bacteria in the PDAC-associated microbiome through ablation leads to a more favorable immune signature in the tumor microenvironment. PDAC-associated dysbiosis promoted an immune-suppressive environment through activation of TLR ligands in monocytes that in turn led to T cell anergy. Bacterial ablation through antibiotic administration led to the increase in M1 macrophage differentiation, promotion of Th1 differentiation of CD4+ T cells, CD8+ T cell activation, and reduction in myeloid-derived suppressor cells (Pushalkar et al. 2018). Microbial diversity by itself however is not a predictor of poor outcomes, as demonstrated by the evidence provided by Riquelme et al., who compared the tumor microbiome signature in PDAC patients with short-term survival and long-term survival (LTS). LTS patients not only had higher tumor microbiome alpha diversity but also had a district intra-tumor microbiome signature provided by bacteria of taxa *Saccharopolyspora*, *Pseudoxanthomonas*, and *Streptomyces* in addition to enriched *Bacillus clausii*. Furthermore, the microbial signature correlated with the recruitment and activation of CD8+ T cells (Riquelme et al. 2019). Bacterial drivers of T cell recruitment to CRC were also studied in a murine NSG model that carried intraperitoneal or intracecal tumor xenografts. Tumor cells exposed to gut bacteria expressed chemokines both in vivo and in vitro. Intracecal xenografts showed higher levels of chemokines such as CCL5, CCL20, and CXCL10. The gene signatures for the described chemokines were associated with T lymphocyte subsets such as cytotoxic T cells, T helper (Th1) cells, Tregs, follicular Th cells, and IL17-producing Th cells. The abundance of *Fusobacterium nucleatum*, *E. coli*, and *B. fragilis* correlated with increased chemokine expression, T cell infiltration, and improved survival (Cremonesi et al. 2018). More recently, a defined commensal consortium composed of 11-strain mix (*Parabacteroides* spp., *Bacteroides* spp., *Alistripes senegalensis*, *Paraprevotella xylaniphila*, *Eubacterium limosum*, *Ruminococcacae bacterium* cv2, *Phascolarctobacterium faecium*, *Fusobacterium ulcerans*) was shown to induce colonic IFNγ producing CD8+ T cells. Intestinal colonization with the consortium enhanced anti-CTLA-4 ICI efficacy in a syngenic colon cancer model (MC38). The authors also noted that a subset of the IFN-γ + CD8 TIL from one of the mouse models investigated expressed TCRs specific for the MC38 TAA, p15E (Tanoue et al. 2019). The mice also did not exhibit colitis, a side effect associated with ICI therapies, thus showing promise for a safer combination therapy. The knowledge gained knowing the factors that influence T lymphocyte recruitment to tumor microenvironments opens avenues for advancing immunotherapy by exploring the possibility of using CD8+ CAR T cells to treat PDAC patients and explore ACT-based therapies for CRC treatment.

Fusobacteria have interestingly been described in a recent review to have differing and often opposing roles, which seem to be based on the molecular subtype of CRC. RNA sequencing studies of CRC tissues identified enrichment of *Fusobacteria* and *Bacteroidetes*. In microsatellite instability (MSI)-high tumors, *F. nucelatum* was negatively correlated with TILs but positively associated with TILs in non-MSI-high tumors. Fusobacteria are correlated with poor prognosis in CRC patients and bind via the Fap2 protein to the immune receptor TIGIT, which causes inhibition of NK cells and TILs. It has been suggested that *F. nucelatum* may induce expression of IL12 and transforming growth factor (TGF) β, which in turn promotes low FOX3P expressing nonsuppressive T cells that are favorable predictors of survival. Through miRNA-mediated activation of toll-like receptor (TLR) 2 and TLR4, the bacteria also increase the expression of the inflammatory mediators IL1B, IL6, and IL8 (Saito et al. 2016; Temraz et al. 2019).

10 Conclusion and Future Perspectives

By now, it might be obvious that gut bacteria are involved in several roles in relation to GI cancers, while some species are drivers of disease progression. The complexity of the gut ecosystem also means that bacterial species also have therapeutic benefits to offer either through the production of metabolites that aid in competing against conditional pathogenic species and pathobionts or through mediating mucosal immunity that keeps human hosts healthy. Genetic manipulation of bacteria allows ex vivo synthesis of microbial metabolites for therapeutic administration and the use of bacteria as delivery vehicles for therapeutics. Sequencing-based studies help lay the foundation for omics-based diagnostic methods that may help predict disease outcomes while also informing clinicians of the ideal therapies to administer, which may impact the safety and efficacy of GI cancer treatments.

The authors would like to remark that despite the vast potential the gut microbiome has to offer, there are several challenges that need to be overcome, which are not specific to GI cancer-related microbial manipulation alone. The majority of work described in the literature addressing the role that the gut microbiome plays is restricted to studying bacteria. The virome, mycobiome, and the network of interactions with the host and the gut ecology need to be uncovered to gain better perspectives of how organisms can be manipulated to better improve human health. Nucleic acid-based-omics have provided opportunities to help in the profiling of the microbiome; however, factors such as the choice of reference databases and consensus between major methods used such as metagenomic shotgun sequencing and 16 s rRNA sequencing are of concern. Whether it is for the development of monotherapies or combination therapies, establishing multiparameter models that consider patient and tumor genetics, risk factors, and microbiome composition may benefit the ability to predict response to treatments while also overcoming the inconsistencies and contrasting findings, which can be attributed to the high variability of subjects being tested. Improvements are made in the application of murine models, namely the use of GF mice or antibiotic-fed SPF mice that are then transplanted with the human gut microbiota (avatar mice), which may improve the validity of preclinical research.

A testament to the potential of uncovering the link between gut bacteria and immunotherapy is the fact that research groups studying the two have already attracted multimillion-dollar funding from biotech companies, pharmaceutical companies and investment banks that have allowed the progress of preclinical studies to move forward to clinical trials. Parallel advances made to improve immunogenicity, efficacy, and drug delivery strategies of immunotherapeutics, in addition to advancements made to technology platforms that allow patient stratification based on response outcomes, large-scale drug screening, and production of immune-based pharmaceuticals, will help usher in an era of precision medicine, which is likely to shape the future of healthcare.

References

Abdulamir AS, Hafidh RR, Abu Bakar F (2011) The association of *Streptococcus bovis/gallolyticus* with colorectal tumors: the nature and the underlying mechanisms of its etiological role. J Exp Clin Cancer Res 30(1):11. https://doi.org/10.1186/1756-9966-30-11

Abdi-Ali A, Worobec EA, Deezagi A, Malekzadeh F (2004) Cytotoxic effects of pyocin S2 produced by *Pseudomonas aeruginosa* on the growth of three human cell lines. Can J Microbiol 50(5):375–381. https://doi.org/10.1139/w04-019

Agah S, Alizadeh AM, Mosavi M, Ranji P, Khavari-Daneshvar H, Ghasemian F et al (2019) More Protection of *Lactobacillus acidophilus* than *Bifidobacterium bifidum* Probiotics on Azoxymethane-Induced Mouse Colon Cancer. Probiotics Antimicrob Proteins 11(3):857–864. https://doi.org/10.1007/s12602-018-9425-8

Ajani JA, Lee J, Sano T, Janjigian YY, Fan D, Song S (2017) Gastric adenocarcinoma. Nat Rev Dis Primers 3:17036. https://doi.org/10.1038/nrdp.2017.36

Amitay EL, Werner S, Vital M, Pieper DH, Höfler D, Gierse IJ et al (2017) Fusobacterium and colorectal cancer: causal factor or passenger? Results from a large colorectal cancer screening study. Carcinogenesis 38(8):781–788. https://doi.org/10.1093/carcin/bgx053

Arthur JC, Perez-Chanona E, Mühlbauer M, Tomkovich S, Uronis JM, Fan TJ et al (2012) Intestinal inflammation targets cancer-inducing activity of the microbiota. Science 338(6103):120–123. https://doi.org/10.1126/science.1224820

Arruebo M, Vilaboa N, Sáez-Gutierrez B, Lambea J, Tres A, Valladares M, González-Fernández A (2011) Assessment of the evolution of cancer treatment therapies. Cancers (Basel) 3(3):3279–3330. https://doi.org/10.3390/cancers3033279

Asolkar RN, Freel KC, Jensen PR, Fenical W, Kondratyuk TP, Park EJ, Pezzuto JM (2009) Arenamides A-C, cytotoxic NFkappaB inhibitors from the marine actinomycete Salinispora arenicola. J Nat Prod 72(3):396–402. https://doi.org/10.1021/np800617a

Axelrad JE, Lichtiger S, Yajnik V (2016) Inflammatory bowel disease and cancer: the role of inflammation, immunosuppression, and cancer treatment. World J Gastroenterol 22(20):4794–4801. https://doi.org/10.3748/wjg.v22.i20.4794

Baindara P, Korpole S, Grover V (2018) Bacteriocins: perspective for the development of novel anticancer drugs. Appl Microbiol Biotechnol 102(24):10393–10408. https://doi.org/10.1007/s00253-018-9420-8

Baindara P, Gautam A, Raghava GPS, Korpole S (2017) Anticancer properties of a defensin like class IId bacteriocin Laterosporulin 10. Sci Rep 7:46541. https://doi.org/10.1038/srep46541

Balamurugan R, Rajendiran E, George S, Samuel GV, Ramakrishna BS (2008) Real-time polymerase chain reaction quantification of specific butyrate-producing bacteria, Desulfovibrio and *Enterococcus faecalis* in the feces of patients with colorectal cancer. J Gastroenterol Hepatol 23:1298–1303. https://doi.org/10.1111/j.1440-1746.2008.05490.x

Balgir PP, Bhatia P, Kaur B (2010) Sequence analysis and homology-based modeling to assess structure - function relationship of pediocin CP2 of *Pediococcus acidilactici* MTCC 5101. IJBT 9:431–434

Bayer V (2019) An Overview of Monoclonal Antibodies. Semin Oncol Nurs 35(5):150927. https://doi.org/10.1016/j.soncn.2019.08.006

Bhattacharjee RN, Park KS, Uematsu S, Okada K, Hoshino K, Takeda K, Takeuchi O, Akira S, Iida T, Honda T (2005) *Escherichia coli* verotoxin 1 mediates apoptosis in human HCT116 colon cancer cells by inducing overexpression of the GADD family of genes and S phase arrest. FEBS Lett 579(29):6604–6610. https://doi.org/10.1016/j.febslet.2005.10.053

Black JD, Lopez S, Cocco E, Schwab CL, English DP, Santin AD (2015) *Clostridium perfringens* enterotoxin (CPE) and CPE-binding domain (c-CPE) for the detection and treatment of gynecologic cancers. Toxins (Basel) 7(4):1116–1125. https://doi.org/10.3390/toxins7041116

Boleij A, Muytjens CM, Bukhari SI, Cayet N, Glaser P, Hermans PW, Swinkels DW, Bolhuis A, Tjalsma H. (2011) Novel clues on the specific association of Streptococcus gallolyticus subsp gallolyticus with colorectal cancer. The Journal of infectious diseases 203(8):1101–1109. https://doi.org/10.1093/infdis/jiq169

Braun V, Pilsl H, Gross P (1994) Colicins: structures, modes of action, transfer through membranes, and evolution. Arch Microbiol 161:199–206. https://doi.org/10.1007/BF00248693

Begde D, Bundale S, Mashitha P, Rudra J, Nashikkar N, Upadhyay A (2011) Immunomodulatory efficacy of nisin–a bacterial lantibiotic peptide. J Pept Sci 17(6):438–444. https://doi.org/10.1002/psc.1341

Bray F, Ferlay J, Soerjomataram I, Siegel RL, Torre LA, Jemal A (2018) Global cancer statistics 2018: GLOBOCAN estimates of incidence and mortality worldwide for 36 cancers in 185 countries. CA Cancer J Clin 68(6):394–424. https://doi.org/10.3322/caac.21492

Buc E, Dubois D, Sauvanet P, Raisch J, Delmas J, Darfeuille-Michaud A, Pezet D, Bonnet R (2013) High prevalence of mucosa-associated *E. coli* producing cyclomodulin and genotoxin in colon cancer. PLoS One 8(2):e56964. https://doi.org/10.1371/journal.pone.0056964

Chanier T, Chames P (2019) Nanobody engineering: toward next generation immunotherapies and immunoimaging of cancer. Antibodies 8(1):13. https://doi.org/10.3390/antib8010013

Cheng WY, Wu CY, Yu J (2020) The role of gut microbiota in cancer treatment: friend or foe? Gut 69(10):1867–1876. https://doi.org/10.1136/gutjnl-2020-321153

Chmielewski M, Abken H (2015) TRUCKs: the fourth generation of CARs. Expert Opin Biol Ther 15(8):1145–1154. https://doi.org/10.1517/14712598.2015.1046430

Cho JY, Williams PG, Kwon HC, Jensen PR, Fenical W (2007) Lucentamycins A-D, cytotoxic peptides from the marine-derived actinomycete *Nocardiopsis lucentensis*. J Nat Prod 70(8):1321–1328. https://doi.org/10.1021/np070101b

Choi SS, Kang BY, Chung MJ, Kim SD, Park SH, Kim JS, Kang CY, Ha NJ (2005) Safety assessment of potential lactic acid bacteria *Bifidobacterium longum* SPM1205 isolated from healthy Koreans. J Microbiol 43(6):493–498

Chumchalová J, Smarda J (2003) Human tumor cells are selectively inhibited by colicins. Folia Microbiol (Praha) 48(1):111–115. https://doi.org/10.1007/BF02931286

Coker OO, Dai Z, Nie Y, Zhao G, Cao L, Nakatsu G, Wu WK, Wong SH, Chen Z, Sung JJY, Yu J (2018) Mucosal microbiome dysbiosis in gastric carcinogenesis. Gut 67(6):1024–1032. https://doi.org/10.1136/gutjnl-2017-314281

Cremonesi E, Governa V, Garzon JFG, Mele V, Amicarella F, Muraro MG et al (2018) Gut microbiota modulate T cell trafficking into human colorectal cancer. Gut 67(11):1984–1994. https://doi.org/10.1136/gutjnl-2016-313498

De Witte C, Schulz C, Smet A, Malfertheiner P, Haesebrouck F (2016) Other Helicobacters and gastric microbiota. Helicobacter Suppl 1:62–68. https://doi.org/10.1111/hel.12343

Dias-Jácome E, Libânio D, Borges-Canha M, Galaghar A, Pimentel-Nunes P (2016) Gastric microbiota and carcinogenesis: the role of non-*Helicobacter pylori* bacteria–A systematic review. Rev Esp Enferm Dig 108(9):530–540. https://doi.org/10.17235/reed.2016.4261/2016

D'Journo XB, Thomas PA (2014) Current management of esophageal cancer. J Thorac Dis 6:S253–S264. https://doi.org/10.3978/j.issn.2072-1439.2014.04.16

Doorakkers E, Lagergren J, Engstrand L, Brusselaers N (2016) Eradication of *Helicobacter pylori* and gastric cancer: a systematic review and meta-analysis of cohort studies. J Natl Cancer Inst 1108(9):djw132. https://doi.org/10.1093/jnci/djw132

Eckert C, Emirian A, Le Monnier A, Cathala L, De Montclos H, Goret J et al (2014) Prevalence and pathogenicity of binary toxin-positive *Clostridium difficile* strains that do not produce toxins A and B. New Microbes New Infect 3:12–17. https://doi.org/10.1016/j.nmni.2014.10.003

Elsalem L, Jum'ah AA, Alfaqih MA, Aloudat O (2020) The bacterial microbiota of gastrointestinal cancers: role in cancer pathogenesis and therapeutic perspectives. Clin Exp Gastroenterol 13:151–185. https://doi.org/10.2147/CEG.S243337

Faïs T, Delmas J, Cougnoux A, Dalmasso G, Bonnet R (2016) Targeting colorectal cancer-associated bacteria: a new area of research for personalized treatments. Gut Microbes 7(4):329–333. https://doi.org/10.1080/19490976.2016.1155020

Fan X, Alekseyenko AV, Wu J, Peters BA, Jacobs EJ, Gapstur SM et al (2018) Human oral microbiome and prospective risk for pancreatic cancer: a population-based nested case-control study. Gut 67(1):120–127. https://doi.org/10.1136/gutjnl-2016-312580

Faith JJ, Guruge JL, Charbonneau M, Subramanian S, Seedorf H, Goodman AL et al (2013) The long-term stability of the human gut microbiota. Science 341(6141):1237439. https://doi.org/10.1126/science.1237439

Farrell JJ, Zhang L, Zhou H, Chia D, Elashoff D, Akin D et al (2012) Variations of oral microbiota are associated with pancreatic diseases including pancreatic cancer. Gut 61(4):582–588. https://doi.org/10.1136/gutjnl-2011-300784

Fessler J, Matson V, Gajewski TF (2019) Exploring the emerging role of the microbiome in cancer immunotherapy. J Immunother Cancer 7(1):108. https://doi.org/10.1186/s40425-019-0574-4

Fiedler HP, Bruntner C, Riedlinger J, Bull AT, Knutsen G, Goodfellow M et al (2008) Proximicin A, B and C, novel aminofuran antibiotic and anticancer compounds isolated from marine strains of the actinomycete Verrucosispora. J Antibiot (Tokyo) 1(3):158–163. https://doi.org/10.1038/ja.2008.125

Fiedler T, Strauss M, Hering S, Redanz U, William D, Rosche Y et al (2015) Arginine deprivation by arginine deiminase of *Streptococcus pyogenes* controls primary glioblastoma growth in vitro and in vivo. Cancer Biol Ther 16(7):1047–1055. https://doi.org/10.1080/15384047.2015.1026478

Filip M, Tzaneva V, Dumitrascu DL (2018) Fecal transplantation: digestive and extradigestive clinical applications. Clujul Med 91(3):259–265. https://doi.org/10.15386/cjmed-946

Fleming M, Ravula S, Tatishchev SF, Wang HL (2012) Colorectal carcinoma: pathologic aspects. J Gastrointest Oncol 3(3):153–173. https://doi.org/10.3978/j.issn.2078-6891.2012.030

Flint HJ, Scott KP, Louis P, Duncan SH (2012) The role of the gut microbiota in nutrition and health. Nat Rev Gastroenterol Hepatol 9(10):577–589. https://doi.org/10.1038/nrgastro.2012.156

Ford AC, Yuan Y, Moayyedi P (2020) *Helicobacter pylori* eradication therapy to prevent gastric cancer: systematic review and meta-analysis. Gut 69(12):2113–2121. https://doi.org/10.1136/gutjnl-2020-320839

Francescone R, Hou V, Grivennikov SI (2014) Microbiome, inflammation, and cancer. Cancer J 20(3):181–189. https://doi.org/10.1097/PPO.0000000000000048

Fujiwara Y, Okada K, Omori T, Sugimura K, Miyata H, Ohue M et al (2017) Multiple therapeutic peptide vaccines for patients with advanced gastric cancer. Int J Oncol Res 50(5):1655–1662. https://doi.org/10.3892/ijo.2017.3955

Fulbright LE, Ellermann M, Arthur JC (2017) The microbiome and the hallmarks of cancer. PLoS Pathog 13(9):e1006480. https://doi.org/10.1371/journal.ppat.1006480

Gaida MM, Mayer C, Dapunt U, Stegmaier S, Schirmacher P, Wabnitz GH, Hänsch GM (2016) Expression of the bitter receptor T2R38 in pancreatic cancer: localization in lipid droplets and activation by a bacteria-derived quorum-sensing molecule. Oncotarget 7(11):12623–12632. https://doi.org/10.18632/oncotarget.7206

Gagnière J, Raisch J, Veziant J, Barnich N, Bonnet R, Buc E, Bringer MA, Pezet D, Bonnet M (2016) Gut microbiota imbalance and colorectal cancer. World J Gastroenterol 22(2):501–518. https://doi.org/10.3748/wjg.v22.i2.501

Garrett WS (2015) Cancer and the microbiota. Science 348:80–86. https://doi.org/10.1126/science.aaa4972

Garrett WS (2019) The gut microbiota and colon cancer. Science 21:364 (6446):1133-1135. https://doi.org/10.1126/science.aaw2367

Guevara B, Cogdill AG (2000) *Helicobacter pylori*: a review of current diagnostic and management strategies. Dig Dis Sci 65(7):1917–1931. https://doi.org/10.1007/s10620-020-06193-7

Gill SR, Pop M, Deboy RT, Eckburg PB, Turnbaugh PJ, Samuel BS et al (2006) Metagenomic analysis of the human distal gut microbiome. Science 312(5778):1355–1359. https://doi.org/10.1126/science.1124234

Holmes RK (2000) Biology and molecular epidemiology of diphtheria toxin and the tox gene. J Infect Dis 181(Suppl 1):S156–S167. https://doi.org/10.1086/315554

Huang T, Li S, Li G, Tian Y, Wang H, Shi L et al (2014) Utility of *Clostridium difficile* toxin B for inducing anti-tumor immunity. PLoS One 9(10):e110826. https://doi.org/10.1371/journal.pone.0110826

Ishaq S, Nunn L (2015) *Helicobacter pylori* and gastric cancer: a state of the art review. Gastroenterol Hepatol Bed Bench Spring 8(Suppl 1):S6–S14

Jacob F (1954) Biosynthèse induite et mode d'action d'une pyocine, antibiotique de *Pseudomonas pyocyanea* [Induced biosynthesis and mode of action of a pyocine, antibiotic produced by *Pseudomonas aeruginosa*]. Ann Inst Pasteur (Paris) 86(2):149–160

Jacouton E, Chain F, Sokol H, Langella P, Bermúdez-Humarán LG (2017) Probiotic strain *Lactobacillus casei* BL23 prevents colitis-associated colorectal cancer. Front Immunol 8:1553. https://doi.org/10.3389/fimmu.2017.01553

Jahani-Sherafat S, Alebouyeh M, Moghim S, Ahmadi Amoli H, Ghasemian-Safaei H (2018) Role of gut microbiota in the pathogenesis of colorectal cancer; a review article. Gastroenterol Hepatol Bed Bench 11(2):101–109

Jia L, Gorman GS, Coward LU, Noker PE, McCormick D, Horn TL et al (2011) Preclinical pharmacokinetics, metabolism, and toxicity of azurin-p28 (NSC745104) a peptide inhibitor of p53 ubiquitination. Cancer Chemother Pharmacol 68(2):513–524. https://doi.org/10.1007/s00280-010-1518-3

Javanmard A, Ashtari S, Sabet B, Davoodi SH, Rostami-Nejad M, Esmaeil Akbari M et al (2018) Probiotics and their role in gastrointestinal cancers prevention and treatment; an overview. Gastroenterol Hepatol Bed Bench 11(4):284–295

Joo NE, Ritchie K, Kamarajan P, Miao D, Kapila YL (2012) Nisin, an apoptogenic bacteriocin and food preservative, attenuates HNSCC tumorigenesis via CHAC1. Cancer Med 1(3):295–305. https://doi.org/10.1002/cam4.35

Kamada N, Chen GY, Inohara N, Núñez G (2013) Control of pathogens and pathobionts by the gut microbiota. Nat Immunol 14(7):685–690. https://doi.org/10.1038/ni.2608

Karmali MA, Petric M, Lim C, Fleming PC, Arbus GS, Lior H (1985) The association between idiopathic hemolytic uremic syndrome and infection by verotoxin-producing Escherichia coli. J Infect Dis 5:775–782. https://doi.org/10.1093/infdis/151.5.775

Karpiński TM (2012) New peptide (Entap) with antiproliferative activity produced by bacteria of Enterococcus genus (in Polish). Habilitation thesis. Scientific Publisher of Poznań University of Medical Sciences, 2012

Karpiński T, Szkaradkiewicz A, Gamian A (2013) New enterococcal anticancer peptide. 23rd European Congress of Clinical Microbiology and Infectious Diseases Berlin; Germany

Kaur S, Kaur S (2015) Bacteriocins as Potential Anticancer Agents. Front Pharmacol 6:272. https://doi.org/10.3389/fphar.2015.00272

Kaur B, Balgir PP, Mittu B, Kumar B, Garg N (2013) Biomedical applications of fermenticin HV6b isolated from Lactobacillus fermentum HV6b MTCC10770. Biomed Res Int 2013:168438. https://doi.org/10.1155/2013/168438

Kaur B, Balgir P, Mittu B, Chauhan A, Kumar B, Garg N (2012) Isolation and In vitro characterization of anti -Gardnerella vaginalis bacteriocin producing *Lactobacillus fermentum* HV6b isolated from human vaginal ecosystem. I 1(3):41

Khan SA, Everest P, Servos S, Foxwell N, Zähringer U, Brade H et al (1998) A lethal role for lipid A in Salmonella infections. Mol Microbiol 29(2):571–579. https://doi.org/10.1046/j.1365-2958.1998.00952.x

Kim JM, Lee JY, Yoon YM, Oh YK, Kang JS, Kim YJ, Kim KH (2006) *Bacteroides fragilis* enterotoxin induces cyclooxygenase-2 and fluid secretion in intestinal epithelial cells through NF-kappaB activation. Eur J Immunol 36(9):2446–2456. https://doi.org/10.1002/eji.200535808

Khatoon J, Rai RP, Prasad KN (2016) Role of *Helicobacter pylori* in gastric cancer: updates. World J Gastrointest Oncol 8(2):147–158. https://doi.org/10.4251/wjgo.v8.i2.147

Kobayashi H, Furusawa A, Rosenberg A, Choyke PL (2021) Near-infrared photoimmunotherapy of cancer: a new approach that kills cancer cells and enhances anti-cancer host immunity. Int Immunol 33(1):7–15. https://doi.org/10.1093/intimm/dxaa037

Kruger S, Ilmer M, Kobold S, Cadilha BL, Endres S, Ormanns S et al (2019) Advances in cancer immunotherapy 2019–latest trends. J Exp Clin Cancer Res 38(1):268. https://doi.org/10.1186/s13046-019-1266-0

Kumar B, Balgir P, Kaur B, Mittu B, Chauhan A (2012) In vitro cytotoxicity of native and rec–pediocin CP2 against cancer cell lines: a comparative study. Pharmaceutical Analytical Acta 1(6)

Kwok T, Zabler D, Urman S, Rohde M, Hartig R, Wessler S et al (2007) Helicobacter exploits integrin for type IV secretion and kinase activation. Nature 449(7164):862–866. https://doi.org/10.1038/nature06187

Laliani G, Ghasemian Sorboni S, Lari R, Yaghoubi A, Soleimanpour S, Khazaei M, Hasanian SM, Avan A (2020) Bacteria and cancer: different sides of the same coin. Life Sci 246:117398. https://doi.org/10.1016/j.lfs.2020.117398

Lakey JH, Slatin SL (2001) Pore forming colicins and their relatives. Curr Top Microbiol Immunol 257:131–161. https://doi.org/10.1007/978-3-642-56508-3_7

Lee DG, Hahm KS, Park Y, Kim HY, Lee W, Lim SC et al (2005) Functional and structural characteristics of anticancer peptide Pep27 analogues. Cancer Cell Int 5:21. https://doi.org/10.1186/1475-2867-5-21

Lee DK, Jang S, Kim MJ, Kim JH, Chung MJ, Kim KJ, Ha NJ (2008) Anti-proliferative effects of *Bifidobacterium adolescentis* SPM0212 extract on human colon cancer cell lines. BMC Cancer 8:310. https://doi.org/10.1186/1471-2407-8-310

Lee H, Kim HY (2011) Lantibiotics, class I bacteriocins from the genus Bacillus. J Microbiol Biotechnol 21(3):229–235

Lee S, Margolin K (2011) Cytokines in cancer immunotherapy. Cancers (Basel) 3(4):3856–3893. https://doi.org/10.3390/cancers3043856

Lewis DJ, Dao H Jr, Nagarajan P, Duvic M (2017) Primary cutaneous anaplastic large-cell lymphoma: complete remission for 13 years after denileukin diftitox. JAAD Case Rep 3(6):501–504. https://doi.org/10.1016/j.jdcr.2017.06.031

Li W, Deng Y, Chu Q, Zhang P (2019) Gut microbiome and cancer immunotherapy. Cancer Lett 447:41–47. https://doi.org/10.1016/j.canlet.2019.01.015

Lokeshwar VB, Rubinowicz D, Schroeder GL, Forgacs E, Minna JD, Block NL et al (2001) Stromal and epithelial expression of tumor markers hyaluronic acid and HYAL1 hyaluronidase in prostate cancer. J Biol Chem 276(15):11922–11932. https://doi.org/10.1074/jbc.M008432200

Lutz MB, Baur AS, Schuler-Thurner B, Schuler G (2014) Immunogenic and tolerogenic effects of the chimeric IL-2-diphtheria toxin cytocidal agent Ontak® on CD25+ cells. Onco Targets Ther 3:e28223. https://doi.org/10.4161/onci.28223

McCarthy EF (2006) The Toxins of William B. Coley and the treatment of bone and soft-tissue sarcomas. Iowa Orthop J 26:154–158

Maddocks OD, Short AJ, Donnenberg MS, Bader S, Harrison DJ (2009) Attaching and effacing *Escherichia coli* downregulate DNA mismatch repair protein *in vitro* and are associated with colorectal adenocarcinomas in humans. PLoS One 4(5):e5517. https://doi.org/10.1371/journal.pone.0005517

Maldonado-Contreras A, Goldfarb KC, Godoy-Vitorino F, Karaoz U, Contreras M, Blaser MJ et al (2011) Structure of the human gastric bacterial community in relation to *Helicobacter pylori* status. ISME J 5(4):574–579. https://doi.org/10.1038/ismej.2010.149

Mantovani HC, Hu H, Worobo RW, Russell JB (2002) Bovicin HC5, a bacteriocin from *Streptococcus bovis* HC5. Microbiology 148:3347–3352. https://doi.org/10.1099/00221287-148-11-3347

Manes G, Dominguez-Muñoz JE, Hackelsberger A, Leodolter A, Rössner A, Malfertheiner P (1998) Prevalence of *Helicobacter pylori* infection and gastric mucosal abnormalities in chronic pancreatitis. Am J Gastroenterol 93(7):1097–1100. https://doi.org/10.1111/j.1572-0241.1998.336_b.x

Marchesi JR, Adams DH, Fava F, Hermes GD, Hirschfield GM, Hold G et al (2016) The gut microbiota and host health: a new clinical frontier. Gut 65(2):330–339. https://doi.org/10.1136/gutjnl-2015-309990

Martin HM, Campbell BJ, Hart CA, Mpofu C, Nayar M, Singh R et al (2004) Enhanced *Escherichia coli* adherence and invasion in Crohn's disease and colon cancer. Gastroenterology 127(1):80–93. https://doi.org/10.1053/j.gastro.2004.03.054

Martarelli D, Pompei P, Mazzoni G (2009) Inhibition of adrenocortical carcinoma by diphtheria toxin mutant CRM197. Chemotherapy 55(6):425–432. https://doi.org/10.1159/000264689

Matsuo Y, Kanoh K, Yamori T, Kasai H, Katsuta A, Adachi K, Shin-Ya K, Shizuri Y (2007) Urukthapelstatin A, a novel cytotoxic substance from marine-derived *Mechercharimyces asporophorigenens* YM11-542. I. Fermentation, isolation and biological activities. J Antibiot (Tokyo) 60(4):251–255. https://doi.org/10.1038/ja.2007.30

McGuigan A, Kelly P, Turkington RC, Jones C, Coleman HG, McCain RS (2018) Pancreatic cancer: a review of clinical diagnosis, epidemiology, treatment and outcomes. World J Gastroenterol 24(43):4846–4861. https://doi.org/10.3748/wjg.v24.i43.4846

Michaud DS, Izard J, Wilhelm-Benartzi CS, You DH, Grote VA, Tjønneland A et al (2013) Plasma antibodies to oral bacteria and risk of pancreatic cancer in a large European prospective cohort study. Gut 62(12):1764–1770. https://doi.org/10.1136/gutjnl-2012-303006

Mehta RR, Yamada T, Taylor BN, Christov K, King ML, Majumdar D et al (2011) A cell penetrating peptide derived from azurin inhibits angiogenesis and tumor growth by inhibiting phosphorylation of VEGFR-2. FAK and Akt Angiogenesis 14(3):355–369. https://doi.org/10.1007/s10456-011-9220-6

Meng C, Bai C, Brown TD, Hood LE, Tian Q (2018) Human gut microbiota and gastrointestinal cancer. Genomics Proteomics Bioinformatics 16(1):33–49. https://doi.org/10.1016/j.gpb.2017.06.002

Mira-Pascual L, Cabrera-Rubio R, Ocon S, Costales P, Parra A, Suarez A, Moris F, Rodrigo L, Mira A, Collado MC (2015) Microbial mucosal colonic shifts associated with the development of colorectal cancer reveal the presence of different bacterial and archaeal biomarkers. J Gastroenterol 50(2):167–179. https://doi.org/10.1007/s00535-014-0963-x

Morales A (2017) BCG: a throwback from the stone age of vaccines opened the path for bladder cancer immunotherapy. Can J Urol 24(3):8788–8793

Murphy JR (2011) Mechanism of diphtheria toxin catalytic domain delivery to the eukaryotic cell cytosol and the cellular factors that directly participate in the process. Toxins (Basel) 3(3):294–308. https://doi.org/10.3390/toxins3030294

Narikiyo M, Tanabe C, Yamada Y, Igaki H, Tachimori Y, Kato H, Muto M, Montesano R, Sakamoto H, Nakajima Y, Sasaki H (2004) Frequent and preferential infection of *Treponema denticola*, *Streptococcus mitis*, and *Streptococcus anginosus* in esophageal cancers. Cancer Sci 95(7):569–574. https://doi.org/10.1111/j.1349-7006.2004

Nejadmoghaddam M-R, Minai-Tehrani A, Ghahremanzadeh R, Mahmoudi M, Dinarvand R, Zarnani A-H (2019) Antibody-drug conjugates: possibilities and challenges. Avicenna J Med Biotechnol 11(1):3–23

Ni Y, Schwaneberg U, Sun ZH (2008) Arginine deiminase, a potential anti-tumor drug. Cancer Lett 261(1):1–11. https://doi.org/10.1016/j.canlet.2007.11.038

Norouzi Z, Salimi A, Halabian R, Fahimi H (2018) Nisin, a potent bacteriocin and anti-bacterial peptide, attenuates expression of metastatic genes in colorectal cancer cell lines. Microb Pathog 123:183–189. https://doi.org/10.1016/j.micpath.2018.07.006

Obrig TG, Moran TP, Brown JE (1987) The mode of action of Shiga toxin on peptide elongation of eukaryotic protein synthesis. Biochem J 244(2):287–294. https://doi.org/10.1042/bj2440287

Odenbreit S, Püls J, Sedlmaier B, Gerland E, Fischer W, Haas R (2000) Translocation of *Helicobacter pylori* CagA into gastric epithelial cells by type IV secretion. Science 287(5457):1497–1500. https://doi.org/10.1126/science.287.5457.1497

Oláh A, Belágyi T, Pótó L, Romics L Jr, Bengmark S (2007) Synbiotic control of inflammation and infection in severe acute pancreatitis: a prospective, randomized, double blind study. Hepato-Gastroenterology 54(74):590–594

O'Mahony L, Feeney M, O'Halloran S, Murphy L, Kiely B, Fitzgibbon J, Lee G, O'Sullivan G, Shanahan F, Collins JK (2001) Probiotic impact on microbial flora, inflammation and tumour development in IL-10 knockout mice. Aliment Pharmacol Ther 15(8):1219–1225. https://doi.org/10.1046/j.1365-2036.2001.01027.x

Pahle J, Menzel L, Niesler N, Kobelt D, Aumann J, Rivera M, Walther W (2017) Rapid eradication of colon carcinoma by *Clostridium perfringens* Enterotoxin suicidal gene therapy. BMC Cancer 17(1):129. https://doi.org/10.1186/s12885-017-3123-x

Paiva AD, de Oliveira MD, de Paula SO, Baracat-Pereira MC, Breukink E, Mantovani HC (2012) Toxicity of bovicin HC5 against mammalian cell lines and the role of cholesterol in bacteriocin activity. Microbiology 158:2851–2858. https://doi.org/10.1099/mic.0.062190-0

Palframan SL, Kwok T, Gabriel K (2012) Vacuolating cytotoxin A (VacA), a key toxin for *Helicobacter pylori* pathogenesis. Front Cell Infect Microbiol 2:92. https://doi.org/10.3389/fcimb.2012.00092

Panebianco C, Potenza A, Andriulli A, Pazienza V (2018a) Exploring the microbiota to better understand gastrointestinal cancers physiology. Clin Chem Lab Med 56(9):1400–1412. https://doi.org/10.1515/cclm-2017-1163

Panebianco C, Andriulli A, Pazienza V (2018b) Pharmacomicrobiomics: exploiting the drug-microbiota interactions in anticancer therapies. Microbiome 6(1):92. https://doi.org/10.1186/s40168-018-0483-7

Papagianni M, Anastasiadou S (2009) Pediocins: the bacteriocins of pediococci. Sources, production, properties and applications. Microb Cell Factories 8:3. https://doi.org/10.1186/1475-2859-8-3

Papaioannou NE, Beniata OV, Vitsos P, Tsitsilonis O, Samara P (2016) Harnessing the immune system to improve cancer therapy. Ann Transl Med 4(14). https://doi.org/10.21037/atm.2016.04.01

Pawelek JM, Low KB, Bermudes D (1997) Tumor-targeted Salmonella as a novel anticancer vector. Cancer Res 57(20):4537–4544

Pillar CM, Gilmore MS (2004) Enterococcal virulence–pathogenicity island of *E. Faecalis*. Front Biosci 9:2335–2346. https://doi.org/10.2741/1400

Pormohammad A, Mohtavinejad N, Gholizadeh P, Dabiri H, Salimi Chirani A, Hashemi A, Nasiri MJ (2019) Global estimate of gastric cancer in *Helicobacter pylori*-infected population: a systematic review and meta-analysis. J Cell Physiol 234(2):1208–1218. https://doi.org/10.1002/jcp.27114

Purcell RV, Pearson J, Aitchison A, Dixon L, Frizelle FA, Keenan JI (2017) Colonization with enterotoxigenic *Bacteroides fragilis* is associated with early-stage colorectal neoplasia. PLoS One 12(2):e0171602. https://doi.org/10.1371/journal.pone.0171602

Pushalkar S, Hundeyin M, Daley D, Zambirinis CP, Kurz E, Mishra A et al (2018) The pancreatic cancer microbiome promotes oncogenesis by induction of innate and adaptive immune suppression. Cancer Discov 8(4):403–416. https://doi.org/10.1158/2159-8290.CD-17-1134

Rafter J, Bennett M, Caderni G, Clune Y, Hughes R, Karlsson PC et al (2007) Dietary synbiotics reduce cancer risk factors in polypectomized and colon cancer patients. Am J Clin Nutr 85(2):488–496. https://doi.org/10.1093/ajcn/85.2.488

Rahma OE, Khleif SN (2011) Therapeutic vaccines for gastrointestinal cancers. Gastroenterol Hepatol (NY) 7(8):517–564

Rao D, Parakrama R, Augustine T, Liu Q, Goel S, Maitra R (2019) Immunotherapeutic advances in gastrointestinal malignancies. npj Precision. Oncology 3(1):1–9. https://doi.org/10.1038/s41698-018-0076-8

Rawla P, Sunkara T, Gaduputi V (2019) Epidemiology of pancreatic cancer: global trends, etiology and risk factors. World J Oncol 10(1):10–27. https://doi.org/10.14740/wjon1166

Rezasoltani S, Asadzadeh AH, Dabiri H, Akhavan SA, Modarressi MH, Nazemalhosseini ME (2018) The association between fecal microbiota and different types of colorectal polyp as precursors of colorectal cancer. Microb Pathog 124:244–249. https://doi.org/10.1016/j.micpath.2018.08.035

Rhea LP, Mendez-Marti S, Kim D, Aragon-Ching JB (2021) Role of immunotherapy in bladder cancer. Cancer Treatment and Research Communications 26:100296. https://doi.org/10.1016/j.ctarc.2020.100296

Riquelme E, Zhang Y, Zhang L, Montiel M, Zoltan M, Dong W et al (2019) Tumor microbiome diversity and composition influence pancreatic cancer outcomes. Cell 178(4):795–806.e12. https://doi.org/10.1016/j.cell.2019.07.008

Robert C (2020) A decade of immune-checkpoint inhibitors in cancer therapy. Nat Commun 11(1):3801. https://doi.org/10.1038/s41467-020-17670-y

Rossen NG, MacDonald JK, de Vries EM, D'Haens GR, de Vos WM, Zoetendal EG, Ponsioen CY (2015) Fecal microbiota transplantation as novel therapy in gastroenterology: a systematic review. World J Gastroenterol 21(17):5359–5371. https://doi.org/10.3748/wjg.v21.i17.5359

Rubinstein MR, Wang X, Liu W, Hao Y, Cai G, Han YW (2013) *Fusobacterium nucleatum* promotes colorectal carcinogenesis by modulating E-cadherin/β-catenin signaling via its FadA adhesin. Cell Host Microbe 14(2):195–206. https://doi.org/10.1016/j.chom.2013.07.012

Rubinstein MR, Baik JE, Lagana SM, Han RP, Raab WJ, Sahoo D et al (2019) *Fusobacterium nucleatum* promotes colorectal cancer by inducing Wnt/β-catenin modulator Annexin A1. EMBO Rep 20(4):e47638. https://doi.org/10.15252/embr.201847638

Runge TM, Abrams JA, Shaheen NJ (2015) Epidemiology of barrett's esophagus and esophageal adenocarcinoma. Gastroenterol Clin N Am 44(2):203–231. https://doi.org/10.1016/j.gtc.2015.02.001

Sahl HG, Bierbaum G (1998) Lantibiotics: biosynthesis and biological activities of uniquely modified peptides from gram-positive bacteria. Annu Rev Microbiol 52:41–79. https://doi.org/10.1146/annurev.micro.52.1.41

Saito T, Nishikawa H, Wada H, Nagano Y, Sugiyama D, Atarashi K et al (2016) Two FOXP3 + CD4 + T cell subpopulations distinctly control the prognosis of colorectal cancers. Nat Med 22(6):679–684. https://doi.org/10.1038/nm.4086

Sakatani A, Fujiya M, Ueno N, Kashima S, Sasajima J, Moriichi K et al (2016) Polyphosphate derived from *Lactobacillus brevis* inhibits colon cancer progression through induction of cell apoptosis. Anticancer Res 36(2):591–598

Santin AD, Bellone S, Marizzoni M, Palmieri M, Siegel ER, McKenney JK et al (2007) Overexpression of claudin-3 and claudin-4 receptors in uterine serous papillary carcinoma: novel targets for a type-specific therapy using *Clostridium perfringens* enterotoxin (CPE). Cancer 109(7):1312–1322. https://doi.org/10.1002/cncr.22536

Sawada A, Fujiwara Y, Nagami Y, Tanaka F, Yamagami H, Tanigawa T et al (2016) Alteration of esophageal microbiome by antibiotic treatment does not affect incidence of rat esophageal adenocarcinoma. Dig Dis Sci 61(11):3161–3168. https://doi.org/10.1007/s10620-016-4263-6

Sedighi M, Zahedi BA, Hamblin MR, Ohadi E, Asadi A, Halajzadeh M et al (2019) Therapeutic bacteria to combat cancer; current advances, challenges, and opportunities. Cancer Med 8(6):3167–3181. https://doi.org/10.1002/cam4.2148

Shapira S, Shapira A, Starr A, Kazanov D, Kraus S, Benhar I, Arber N (2011) An immunoconjugate of anti-CD24 and Pseudomonas exotoxin selectively kills human colorectal tumors in mice. Gastroenterology 3:935–946. https://doi.org/10.1053/j.gastro.2010.12.004

Sivan A, Corrales L, Hubert N, Williams JB, Aquino-Michaels K, Earley ZM et al (2015) Commensal Bifidobacterium promotes antitumor immunity and facilitates anti-PD-L1 efficacy. Science 350(6264):1084–1089. https://doi.org/10.1126/science.aac4255

Smarda J, Smajs D (1998) Colicins-exocellular lethal proteins of *Escherichia coli*. Folia Microbiol (Praha) 43:563–582. https://doi.org/10.1007/BF02816372

Schell MA, Karmirantzou M, Snel B, Vilanova D, Berger B, Pessi G et al (2002) The genome sequence of *Bifidobacterium longum* reflects its adaptation to the human gastrointestinal tract. Proc Natl Acad Sci U S A 99(22):14422–14427. https://doi.org/10.1073/pnas.212527599

Schwabe RF, Jobin C (2013) The microbiome and cancer. Nat Rev Cancer 13(11):800–812. https://doi.org/10.1038/nrc3610

Sethi V, Kurtom S, Tarique M, Lavania S, Malchiodi Z et al (2018) Gut microbiota promotes tumor growth in mice by modulating immune response. Gastroenterology 155(1):33–37.e6

Siegel RL, Miller KD, Goding Sauer A, Fedewa SA, Butterly LF, Anderson JC et al (2020) Colorectal cancer statistics. CA Cancer J Clin 70(3):145–164. https://doi.org/10.3322/caac.21601

Siegel RL, Miller KD, Jemal A (2017) Cancer statistics. CA Cancer J Clin 67(1):7–30. https://doi.org/10.3322/caac.21387

Simons A, Alhanout K, Duval R (2020) Bacteriocins, antimicrobial peptides from bacterial origin: overview of their biology and their impact against multidrug-resistant bacteria. Microorganisms 8(5):639. https://doi.org/10.3390/microorganisms8050639

Shahabi V, Seavey MM, Maciag PC, Rivera S, Wallecha A (2011) Development of a live and highly attenuated Listeria monocytogenes-based vaccine for the treatment of Her2/neu-overexpressing cancers in human. Cancer Gene Ther 18(1):53–62. https://doi.org/10.1038/cgt.2010.48

Snider EJ, Freedberg DE, Abrams JA (2016) Potential role of the microbiome in Barrett's esophagus and esophageal adenocarcinoma. Dig Dis Sci 61(8):2217–2225. https://doi.org/10.1007/s10620-016-4155-9

Sohda M, Kuwano H (2017) Current status and future prospects for esophageal cancer treatment. Ann Thorac Cardiovasc Surg 23(1):1–11. https://doi.org/10.5761/atcs.ra.16-00162

Soleimanpour S, Hasanian SM, Avan A, Yaghoubi A, Khazaei M (2020) Bacteriotherapy in gastrointestinal cancer. Life Sci 254:117754. https://doi.org/10.1016/j.lfs.2020.117754

Song Q, Zhang C, Wu X (2018) Therapeutic cancer vaccines: from initial findings to prospects. Immunol Lett 196:11–21. https://doi.org/10.1016/j.imlet.2018.01.011

Sukri A, Hanafiah A, Mohamad Zin N, Kosai NR (2020) Epidemiology and role of Helicobacter pylori virulence factors in gastric cancer carcinogenesis. APMIS 128(2):150–161. https://doi.org/10.1111/apm.13034

Sung WS, Park Y, Choi CH, Hahm KS, Lee DG (2007) Mode of antibacterial action of a signal peptide, Pep27 from *Streptococcus pneumoniae*. Biochem Biophys Res Commun 363(3):806–810. https://doi.org/10.1016/j.bbrc.2007.09.041

Swidsinski A, Khilkin M, Kerjaschki D, Schreiber S, Ortner M, Weber J, Lochs H (1998) Association between intraepithelial *Escherichia coli* and colorectal cancer. Gastroenterology 115(2):281–286. https://doi.org/10.1016/s0016-5085(98)70194-5

Tang B, Tang L, Huang C, Tian C, Chen L, He Z et al (2020) The effect of probiotics supplementation on gut microbiota after *Helicobacter pylori* eradication: a multicenter randomized controlled trial. Infect Dis Ther 10(1):317–333. https://doi.org/10.1007/s40121-020-00372-9

Tan J, Chen YX (2016) Dietary and lifestyle factors associated with colorectal cancer risk and interactions with microbiota: fiber, red or processed meat and alcoholic drinks. Gastrointest Tumors 3(1):17–24. https://doi.org/10.1159/000442831

Tanoue T, Morita S, Plichta DR, Skelly AN, Suda W, Sugiura Y et al (2019) A defined commensal consortium elicits CD8 T cells and anti-cancer immunity. Nature 565(7741):600–605. https://doi.org/10.1038/s41586-019-0878-z

Tareq FS, Kim JH, Lee MA, Lee HS, Lee YJ, Lee JS, Shin HJ (2012) Ieodoglucomides A and B from a marine-derived bacterium *Bacillus licheniformis*. Org Lett 14(6):1464–1467. https://doi.org/10.1021/ol300202z

Temraz S, Nassar F, Nasr R, Charafeddine M, Mukherji D, Shamseddine A (2019) Gut microbiome: a promising biomarker for immunotherapy in colorectal cancer. Int J Mol Sci 20(17). https://doi.org/10.3390/ijms20174155

Tomita K, Ogawa T, Uozumi T, Watanabe K, Masaki H (2000) A cytotoxic ribonuclease which specifically cleaves four isoaccepting arginine tRNAs at their anti-codon loops. Proc Natl Acad Sci U S A 97:8278–8283. https://doi.org/10.1073/pnas.140213797

Toprak NU, Yagci A, Gulluoglu BM, Akin ML, Demirkalem P, Celenk T, Soyletir G. (2006) A possible role of Bacteroides fragilis enterotoxin in the aetiology of colorectal cancer. Clinical microbiology and infection : the official publication of the European Society of Clinical Microbiology and Infectious Diseases 12(8):782–786. https://doi.org/10.1111/j.1469-0691.2006.01494.x.

Vallera DA, Li C, Jin N, Panoskaltsis-Mortari A, Hall WA (2002) Targeting urokinase-type plasminogen activator receptor on human glioblastoma tumors with diphtheria toxin fusion protein DTAT. J Natl Cancer Inst 94(8):597–606. https://doi.org/10.1093/jnci/94.8.597

Vétizou M, Pitt JM, Daillère R, Lepage P, Waldschmitt N, Flament C et al (2015) Anticancer immunotherapy by CTLA-4 blockade relies on the gut microbiota. Science 350(6264):1079–1084. https://doi.org/10.1126/science.aad1329

Villarante KI, Elegado FB, Iwatani S, Zendo T, Sonomoto KE, de Guzman E (2011) Purification, characterization and in vitro cytotoxicity of the bacteriocin from *Pediococcus acidilactici* K2a2-3 against human colon adenocarcinoma (HT29) and human cervical carcinoma (HeLa) cells. World J Microbiol Biotechnol 27:975–980. https://doi.org/10.1007/s11274-010-0541-1

Virtue AT, McCright SJ, Wright JM, Jimenez MT, Mowel WK, Kotzin JJ et al (2019) The gut microbiota regulates white adipose tissue inflammation and obesity via a family of microRNAs. Sci Transl Med 11(496):eaav1892. https://doi.org/10.1126/scitranslmed.aav1892

Wang J, Zhao L, Yan H, Che J, Huihui L, Jun W, Liu B, Cao B (2016) A meta-analysis and systematic review on the association between human papillomavirus (Types 16 and 18) infection and esophageal cancer worldwide. PLoS One 11(7):e0159140. https://doi.org/10.1371/journal.pone.0159140

Wei MQ, Ellem KA, Dunn P, West MJ, Bai CX, Vogelstein B (2007) Facultative or obligate anaerobic bacteria have the potential for multimodality therapy of solid tumours. Eur J Cancer 43(3):490–496. https://doi.org/10.1016/j.ejca.2006.10.005

Wei SC, Duffy CR, Allison JP (2018) Fundamental mechanisms of immune checkpoint blockade therapy. Cancer Discov 8(9):1069–1086. https://doi.org/10.1158/2159-8290.CD-18-0367

Wood SJ, Goldufsky JW, Bello D, Masood S, Shafikhani SH (2015) *Pseudomonas aeruginosa* ExoT induces mitochondrial apoptosis in target host cells in a manner that depends on its GTPase-activating protein (GAP) domain activity. J Biol Chem 290(48):29063–29073. https://doi.org/10.1074/jbc.M115.689950

Wong SH, Zhao L, Zhang X, Nakatsu G, Han J, Xu W et al (2017) Gavage of fecal samples from patients with colorectal cancer promotes intestinal carcinogenesis in germ-free and conventional mice. Gastroenterology 153(6):1621–1633.e6. https://doi.org/10.1053/j.gastro.2017.08.022

World Health Organization. Global health observatory. Geneva: world Health Organization; 2018. who.int/gho/database/en/. Accessed June 21, 2018

Wroblewski LE, Peek RM Jr, Wilson KT (2010) *Helicobacter pylori* and gastric cancer: factors that modulate disease risk. Clin Microbiol Rev 23(4):713–739. https://doi.org/10.1128/CMR.00011-10

Yahiro K, Akazawa Y, Nakano M, Suzuki H, Hisatune J, Isomoto H, et al (2015) *Helicobacter pylori* VacA induces apoptosis by accumulation of connexin 43 in autophagic vesicles via a Rac1/ERK-dependent pathway. Cell Death Discov 1:15035. doi:https://doi.org/10.1038/cddiscovery.2015.35

Yaghoubi A, Khazaei M, Jalili S, Hasanian SM, Avan A, Soleimanpour S, Cho WC (2020) Bacteria as a double-action sword in cancer. Biochim Biophys Acta Rev Cancer 1874(1):188388. https://doi.org/10.1016/j.bbcan.2020.188388

Yamada T, Goto M, Punj V, Zaborina O, Chen ML, Kimbara K et al (2002) Bacterial redox protein azurin, tumor suppressor protein p53, and regression of cancer. Proc Natl Acad Sci U SA 99(22):14098–14103. https://doi.org/10.1073/pnas.222539699

Yamada T, Mehta RR, Lekmine F, Christov K, King ML, Majumdar D et al (2009) A peptide fragment of azurin induces a p53-mediated cell cycle arrest in human breast cancer cells. Mol Cancer Ther 8(10):2947–2958. https://doi.org/10.1158/1535-7163.MCT-09-0444

Yang Y, Weng W, Peng J, Hong L, Yang L, Toiyama Y et al (2017) Fusobacterium nucleatum increases proliferation of colorectal cancer cells and tumor development in mice by activating toll-like receptor 4 signaling to nuclear factor-κB, and up-regulating expression of MicroRNA-21. Gastroenterology 152(4):851–866.e24. https://doi.org/10.1053/j.gastro.2016.11.018

Yang L, Tan R-x, Wang Q, Huang W-y, Yin Y-x (2002) Antifungal cyclopeptides from *Halobacillus litoralis* YS3106 of marine origin. Tetrahedron Lett 43(37):6545–6548. https://doi.org/10.1016/S0040-4039(02)01458-2

Yang L, Lu X, Nossa CW, Francois F, Peek RM, Pei Z (2009) Inflammation and intestinal metaplasia of the distal esophagus are associated with alterations in the microbiome. Gastroenterology 137(2):588–597. https://doi.org/10.1053/j.gastro.2009.04.046

Yang X, Tang B, Li BS, Xie R, Hu CJ, Luo G, Qin Y, Dong H, Yang SM (2015) *Helicobacter pylori* virulence factor CagA promotes tumorigenesis of gastric cancer via multiple signaling pathways. Cell Commun Signal 13:30. https://doi.org/10.1186/s12964-015-0111-0

Zabaleta J (2012) Multifactorial etiology of gastric cancer. Methods Mol Biol 863:411–435. https://doi.org/10.1007/978-1-61779-612-8_26

Zaidi AH, Kelly LA, Kreft RE, Barlek M, Omstead AN, Matsui D et al (2016) Associations of microbiota and toll-like receptor signaling pathway in esophageal adenocarcinoma. BMC Cancer 16:52. https://doi.org/10.1186/s12885-016-2093-8

Zhang HL, Hua HM, Pei YH, Yao XS (2004) Three new cytotoxic cyclic acylpeptides from marine *Bacillus* sp. Chem Pharm Bull (Tokyo) 52(8):1029–1030. https://doi.org/10.1248/cpb.52.1029

Zhang JJ, Wu HS, Wang L, Tian Y, Zhang JH, Wu HL (2010) Expression and significance of TLR4 and HIF-1alpha in pancreatic ductal adenocarcinoma. World J Gastroenterol 16(23):2881–2888. https://doi.org/10.3748/wjg.v16.i23.2881

Chapter 5
Genetic and Epigenetic Regulation by Gut Microbe-Modulated Metabolites in Chronic Metabolic Diseases

S. Sumi and Chandrasekharan C. Kartha

1 Introduction

The gastrointestinal tract is one of the most densely populated anatomical sites with trillions of bacteria, archaea, viruses, and eukaryotes residing in the intestinal mucosa (Savage 1977). These bacteria, collectively known as microbiota, have a significant role in the physiological processes in the host. Symbiosis and commensalism reign in this host–microbe-based human holobiont. Even minute imbalance in this equation promotes various disease conditions. Research focus on host–microbiota interactions and their health implications have increased during the last two decades.

Several metagenomic studies reveal that a healthy human gut is populated predominantly by members of Firmicutes, Bacteroidetes, Actinobacteria, and Proteobacteria phyla, with relatively few other phyla such as Fusobacteria and Verrucomicrobia (Kaur et al. 2020; De et al. 2020). Around 3.8×10^{13} bacteria are estimated to be present in an adult human body, which is approximately equal to the number of total human cells (3×10^{13}) (Sender et al. 2016). Their combined genomes, generally termed as "gut microbiome," are presumed to contain more than three million genes (Human Microbiome Project Consortium 2012). The microbiome expression products complement host physiological and metabolic mechanisms. It is now established that microbial dysbiosis causes altered gut metagenome and their microbial functions, which, when present with classic genetic and lifestyle

S. Sumi (✉)
Cardiovascular Diseases and Diabetes Biology, Rajiv Gandhi Centre for Biotechnology, Thiruvananthapuram, Kerala, India
e-mail: sumis@rgcb.res.in

C. C. Kartha
Society for Continuing Medical Education & Research (SOCOMER), Kerala Institute of Medical Sciences, Thiruvananthapuram, Kerala, India

S. Thomas (ed.), *Human Microbiome*,
https://doi.org/10.1007/978-981-16-7672-7_5

factors, promote the pathogenesis of metabolic disorders (Vijay-Kumar et al. 2010) and cancers (Hope et al. 2005).

The majority of studies until now have been focusing on the interaction of gut microbiota with gut homeostasis (Kawamoto et al. 2012), intestinal mucosa, and development of colon cancer (Hope et al. 2005). A significantly higher bacterial load was demonstrated in the colorectal cancer biopsies (Zhou et al. 2016). As the bacterial metabolites enter the host's circulation via intestinal epithelial mucosa, recent focus is changing toward the role of gut microbiota and its myriad of metabolites in systemic diseases such as diabetes, atherosclerosis, obesity, and cancers (Vinjé et al. 2014). Besides their role in host metabolism, microbiota-secreted multiple low-molecular-weight (LMW) substances closely interact with various cellular targets. Many of these microbial metabolites can interfere in the genomic, epigenomic, and host metabolic processes (Bhat and Kapila 2017).

Recent research indicates an altered microbial LMW pattern in patients with obesity and metabolic diseases (Arora and Bäckhed 2016). Exactly how these molecules affect genetic and epigenetic alterations in biological signaling pathways is unexplained. The capability of the intestinal microbiota to produce folic acid, Vitamin B6, and S-adenosyl methionine (SAM) (Jacob 2000; LeBlanc et al. 2013) could affect host DNA methylation patterns, while short-chain fatty acids (SCFAs) produced from bacterial fermentation (Louis and Flint 2009) may alter chromatin organization and genetic transcription through histone de/acetylation. An insight into the mode of interaction between microbiota-derived small molecules and host epigenome will allow the design of novel epigenetics-based interventions in complex diseases such as type 2 diabetes mellitus (T2DM) and obesity. In this context, we discuss the role of gut microbiota-derived chemical moieties in the genetic and epigenetic regulation of biological signaling in the pathophysiology of lifestyle diseases such as obesity and diabetes.

2 Gut Microbial Dysbiosis in Metabolic Diseases

Recent reports suggest that gut microbiota pattern between normal individuals and patients with obesity or T2DM differs significantly. Qin et al. (2012) performed a two-step metagenome-wide association analysis using deep shotgun sequencing of the gut microbiome from 345 Chinese individuals. They reported a very significant role of the intestinal microbial community in the pathogenesis of T2DM. The gut microbiome of patients with diabetes had very less butyrate-producing bacteria, including *Faecalibacterium prausnitzii*, *Eubacterium rectale*, etc. Besides serving as an energy resource for colonocytes, butyrate effectively reduces inflammation, carcinogenesis, and oxidative stress and improves gut barrier integrity. The investigators also indicated that gut dysbiosis, characterized by a higher number of opportunistic microbes, increased ability for sulfate reduction, and low numbers of SCFA producing bacteria, is a significant characteristic of metabolic diseases such as T2DM.

How the gut microbiome regulates glucose homeostasis and insulin sensitivity is a significant question. Several research groups have found a positive correlation between host hyperglycemia and intestinal microbial dysbiosis (Gérard and Vidal 2019; Lim et al. 2016). Lim et al. (2016) observed that *Lactobacillus sakei* OK67 reduces inflammation and induces tight junction protein expression in high-fat-diet (HFD)-induced hyperglycemia and obesity in mice models. Gut microbiota also converts tyrosine to 4-cresol, which reduces hyperglycemia and fatty liver in animal disease models (Brial et al. 2018).

Insulin resistance is another major pathogenetic factor in metabolic diseases. The outcomes of SCFA administration in mice were studied in detail, and butyrate was demonstrated to have a protective effect against diet-dependent obesity and insulin resistance (Gao et al. 2009). Some groups have explored the role of diet-supplemented butyrate in pancreatic β cell mass, function, and insulin sensitivity in target tissues of animal models for obesity (Li et al. 2013). Gut microbe-derived SCFAs have a major role in maintaining host physiological homeostasis in the pancreas, skeletal muscles, liver, and adipose tissues (Table 5.1).

In our studies on pancreatic beta cells, we observed the effect of butyrate in increasing cell proliferation and differentiation (unpublished data). Treatment with butyrate significantly reduced the oxidative stress in pancreatic beta cells exposed to hyperglycemic conditions. Microbial dysbiosis causes higher gut permeability that results in chronic low-grade inflammation that reduces insulin sensitivity (Fig. 5.1). When there is an imbalance of gut microbiota species and gut barrier disruption, bacteria and their metabolites, lipopolysaccharide (LPS), phenylacetic acid,

Table 5.1 Major effects of SCFAs in the pancreas, skeletal muscles, liver, and adipose tissues

Organ/tissue	Effects
Pancreas	Higher beta cell mass Increased insulin secretion Less glucagon secretion
Adipose tissue	Reduced adipose tissue mass Higher leptin production Decreased inflammation Enhanced differentiation Browning of white adipose tissue Increased beta-oxidation Increased lipolysis
Liver	Less inflammation Increased beta-oxidation Less lipogenesis Less glucose production Lower bile acid synthesis
Skeletal muscle	High fatty acid oxidation Less lipid accumulation Increased beta-oxidation Increased lean body mass Higher insulin response

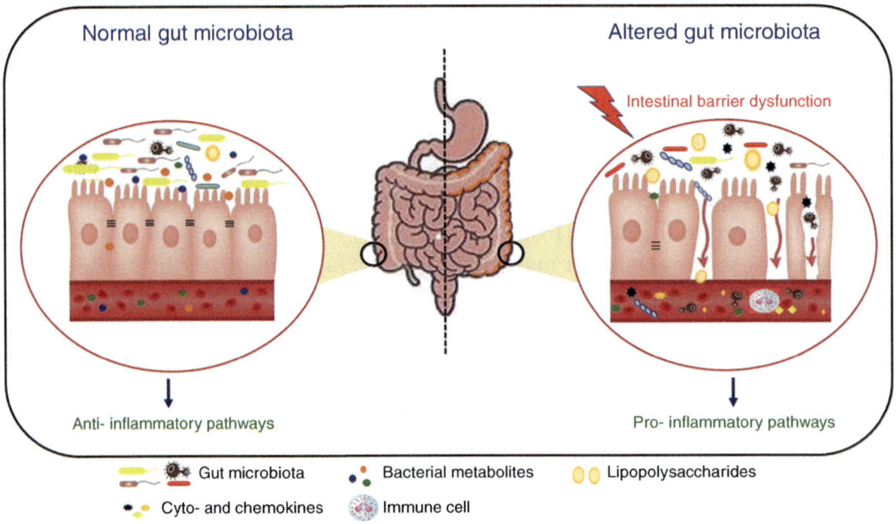

Fig. 5.1 Schematic representation of altered gut microbiota-induced intestinal barrier dysfunction. *Left panel.* In healthy individuals, there is a predominance of symbiotic bacteria, their metabolites, and an intact intestinal barrier. *Right panel.* Microbial dysbiosis induces inflammation and loss of barrier function that in turn allows increased translocation of bacterial components and bacteria into systemic circulation

Table 5.2 Primary producers of various SCFAs

Acetate (C3)	Propionate (C3)	Butyrate (C4)
Bifidobacterium longum	*Pelotomaculum schinkii*	*Clostridium leptum*
Clostridium ljungdahlii	*Syntrophobacter*	*Eubacterium rectale*
Bacteroides thetaiotaomicron	*Coprococcus catus*	*Faecalibacterium prausnitzii*
Prevotella spp	*Veillonella gazogenes*	*Eubacterium hallii*
Streptococcus spp	*Megasphaera elsdenii*	*Faecalibacterium prausnitzii*
Lactobacillus spp	*Roseburia inulinivorans*	*Anaerostipes caccae*
Bifidobacterium spp	*Ruminococcus obeum*	*Coprococcus eutactus*
Blautia hydrogenotrophica	*Akkermansia muciniphila*	*Roseburia inulinivorans*
	Phascolarctobacterium succinatutens	*Ruminococcus bromii*

imidazole propionate, and endotoxins, enter the systemic circulation and cause low-grade inflammation through pro-inflammatory cytokine expression.

SCFAs such as acetate (C2) and propionate (C3) are mainly produced by Bacteroidetes, while butyrate (C4) is secreted by Firmicutes (Ríos-Covián et al. 2016; den Besten et al. 2013). Gut microbial species produce all SCFAs, but some species contribute more of one SCFA type than others. Some such species predominantly producing either acetate or propionate or butyrate are described in Table 5.2.

Four-carbon SCFA butyrate acts as the main energy source of epithelial cells lining the gut and maintains the integrity of the gut barrier. SCFAs such as butyrate induce gluconeogenesis and lipogenesis in the liver (Ji et al. 2019). Gluconeogenesis activates gut–brain neural connections and maintains glucose homeostasis. SCFAs such as butyrate elevate the expression of genes associated with gluconeogenesis via a cAMP-dependent mechanism. It was also reported that gut microbiota metabolizes histidine to imidazole propionate that prevents insulin receptor substrate signaling (Koh et al. 2018).

Another class of gut microbiota-derived peptides is incretin hormones. Incretins such as glucagon-like peptide-1 (GLP-1) and gastric inhibitory polypeptide (GIP) are significant gut peptides that stimulate insulin secretion from beta cells (Pais et al. 2016). Microbial metabolites directly regulate incretin secretion from enteroendocrine cells. Sulfate-reducing bacteria in the colon produce hydrogen sulfide that induces intestinal GLP-1 secretion. Diet supplementation with chondroitin sulfate in mice models has shown higher levels of sulfate-reducing bacteria in the gut (Pichette et al. 2017). These mice had high GLP-1, and good insulin response improved blood glucose control after 4 weeks of diet supplementation. Studies in colonic enteroendocrine L cells have shown that gut microbiota-induced indole regulates GLP-1 secretion. This observation indicates the role of gut microbiota in host glycemic control (Chimerel et al. 2014). Further comprehensive studies elucidating the mechanistic details of these metabolites in response to specific phyla of microbiota are essential to translate these findings into nutrition management strategies. Conventionalized mice models will be of great benefit in such studies (Nicaise et al. 1993; Druart et al. 2015). Germ-free mice in which microbiota was re-colonized are also called conventionalized mice models. Conventionalized mice-based studies help to precisely study the effect of specific bacterial species on host metabolism and genetic and epigenetic regulation.

3 Gut Microbiota and Epigenetic Regulation of Metabolic Diseases

Epigenetic regulation refers to hereditable changes that alter inherent gene expression without changing the DNA sequence. The most commonly described mechanisms of epigenetics are DNA methylation and histone alterations and noncoding RNAs (ncRNA). The earlier research on the gut microbiota–host axis focused more on the microbiota-induced biochemical signaling and differential expression of genes involved in glycemic homeostasis. Very recently, a novel concept of "microbiota-nutrient metabolism-host epigenetics" in physiological homeostasis is evolving (Miro-Blanch and Yanes 2019; Lee 2019). Gut microbiome-induced epigenetic alterations in the host can plausibly occur due to microbial metabolites,

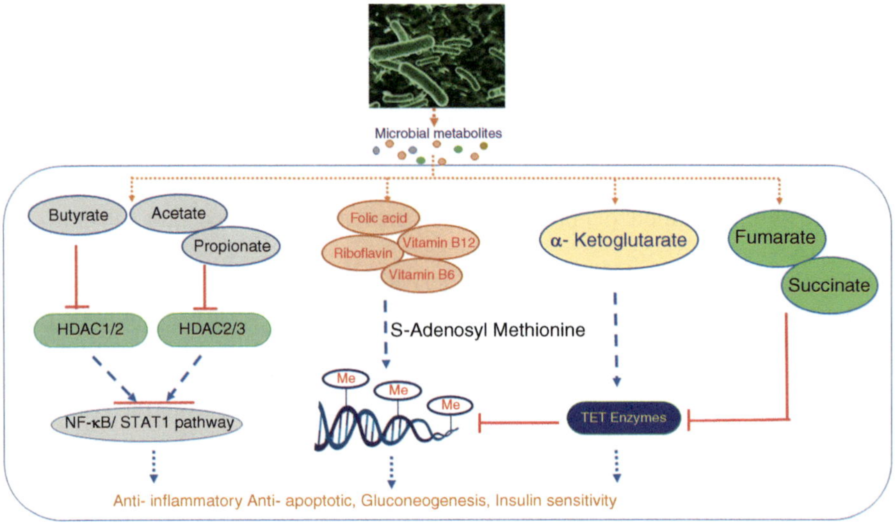

Fig. 5.2 Epigenetic modifications regulated by gut microbiota and their metabolites. Short-chain fatty acids produced by the gut microbiota could regulate histone modification by inhibiting various HDACs. Similarly, microbial metabolites such as folate, riboflavin, and vitamin co-factors induce the formation of S-adenosylmethionine, which is a methyl donor in methylation reactions. α-ketoglutarate enhances TET enzymes, which cause demethylation of gene promoters, maintaining a molecular homeostasis in healthy individuals. Fumarate and succinate, also produced from microbiota, have the potential to inhibit TET enzymes

which serve as chemical donors for promoter methylation or histone modifications or due to its interaction with enzymes responsible for epigenetic modifications (Fig. 5.2).

3.1 DNA Methylation Programming by Gut Microbiota

Methylation refers to an epigenomic process where a methyl group from SAM is added to the fifth position of cytosine to generate 5-methylcytosine in CpG islands in genomic DNA. Methylation has major regulatory effects on gene transcription, mainly by affecting the DNA binding transcription factors. The microbial metabolites-induced altered methylome hypothesis was experimentally substantiated by studies that observed a significant association of differential methylation with specific bacterial predominance in the host gut (Al Akeel 2013). In a study reported from Finland, the methylome pattern was found to be highly altered in pregnant women in whom Firmicute species were predominant (Kumar et al. 2014). Methylation analysis revealed 568 hypermethylated genes and 254 hypomethylated genes in these women, and most of these altered genes were associated with cardiovascular disease lipid metabolism, inflammation, and obesity. Recently, studies were also conducted by classifying individuals based on low and high Bacteroidetes to Firmicutes ratio, where it was found that 258 genes were differentially methylated, of which genes such as *HDAC7* and insulin-like growth factor 2 mRNA binding protein 2 (*IGF2BP2)* are significant in glucose homeostasis (Ramos-Molina et al. 2019). Most importantly, studies done in obese and T2DM patients have reported low numbers of butyrate-producers as well as lower promoter methylation of the SCFA receptor *(FFAR3)* gene compared to lean individuals (Remely et al. 2014).

Precisely how the gut microbiota induces firm control over host gene methylation is still not described. Gut microbiota-induced folic acids and vitamins generate SAM (AdoMet), a key methyl-donating substrate for the action of methyltransferase enzymes such as DNA methyltransferases (DNMTs) and histone methyltransferases (HMTs) (Abbasi et al. 2018). Bacterial species, especially *Lactobacillus* and *Bifidobacterium*, secrete significant levels of folic acid from p-aminobenzoic acid and dihydropterin pyrophosphate (Rossi et al. 2011). Vitamin B6, B12, riboflavin, and choline act along with folic acid to ensure homocysteine balance, SAM generation, and DNA methylation (Abbasi et al. 2018). The reduced numbers of folate and vitamin co-factors producing microbiota results in lower SAM and global DNA hypomethylation in hosts resulting in overexpression of associated genes. However, folate deficiency is implicated both in global DNA hypomethylation and hypermethylation too (Crider et al. 2012). This process may happen due to the low SAM-induced hypomethylation of CpGs within the promoters of *DNMT1* and *DNMT3A* enzyme-coding genes.

Microbial LMW also contains several other components that have specific roles in maintaining homeostasis in methylation in the host. Microbial alpha-ketoglutarate is a co-substrate of TET dioxygenase enzyme, which results in DNA demethylation (Zdzisińska et al. 2017) as well as other metabolites such as fumarate and succinate promote methylation by inhibiting TET enzymes (Rowland et al. 2018; Xiao et al. 2012).

Methionine is the substrate for SAM. Contrary to the general belief that microbiota decides the methylation status, studies have also shown that dietary methionine can affect the microbiota communities (Schaible et al. 2011). Interestingly, studies have delineated the role of bacterial infection in host DNA methylation too. Individuals with pathogenic *Helicobacter pylori* infection were demonstrated to have higher methylation levels in eight regions of CpG islands, including p16 core and p16 noncore regions as well as *LOX* (Maekita et al. 2006).

3.2 Histone Modifications

Histone modification is a type of epigenetic change that affects chromatin macrostructure, organization, and even the binding of effector molecules during transcription. Most of the histone modifications are reversible and respond to environmental changes. They can induce or prevent transcription. Histone acetyltransferase (HAT) enzymes induce transcriptional activation whereby they catalyze acetylation of amino-terminal lysine residues on histone proteins (Roth et al. 2001). Histone deacetylases (HDACs) prevent transcriptional activation by deacetylating the amino-terminal lysine residues of histones (Yuille et al. 2018). The association between microbial LMW SCFA and histone acetylation is already known (Natarajan and Pluznick 2014). Studies have demonstrated that microbial metabolites affect host histone modifications in tissues apart from the intestine too (Krautkramer et al. 2016). This study substantiated the role of microbial SCFAs in histone modifications such as acetylation and methylation in the colon, liver, and white adipose tissues. Microbial SCFA also reduces glucose-mediated histone modifications in the host. Röth et al. (2019) reported that probiotic *Lactobacillus reuteri* strain 6475 produces 2-carbon-transporting folate called 5,10-ethenyltetrahydrofolyl polyglutamate. EtTHF transfers two carbons to homocysteine, producing immunomodulatory amino acid ethionine that was found to have inhibitory roles on methylation and ethylation of histone lysine residues.

Microbial LMWs, including SCFAs, SAM, acetyl-CoA, etc., have the potential to regulate histone modifications. The most prominent SCFA secreted by microbiota, butyrate, is a well-known histone deacetylase inhibitor (HDACi) (Steliou et al. 2012). Propionate and acetate also increase histone acetylation and chromatin relaxation (Licciardi et al. 2011). All SCFAs, especially butyrate and propionate, hinder the functioning of class I, IIa HDACs (Schilderink et al. 2013; Davie 2003). There are also reports that butyrate inhibits the activity of HDAC1 and 2 while propionate and acetate negatively regulate HDAC2 and 3 (Davie 2003). These metabolites have a wide range of anti-inflammatory roles in the host. In hematopoietic cells, it has been shown that SCFAs promote histone acetylation, thereby inducing *FOXP3* gene expression in CD4 + T cells (Smith et al. 2013). This results in the differentiation of Treg cells that prevent the immune response and maintain homeostasis. SCFAs also repress the transcription of *NF-kB* and *STAT1* genes that induce pro-inflammatory cytokine production (Liu et al. 2012; Vinolo et al. 2011; Martin-Gallausiaux et al. 2018). Treatment of macrophages with butyrate has been demonstrated to cause histone acetylation and reduced expression of LPS-induced pro-inflammatory cytokines too (Jiang et al. 2020). Similarly, specific microbiota species in conventionalized mice were shown to upregulate genes coding for major histocompatibility complex class II, plausibly via epigenetic modifications (Kubinak et al. 2015; D'Aquila et al. 2020).

Studies on conventionalized mice suggest that the bacterial composition in microbiota specifically modulates the histone codes. Studies have shown that higher levels of H3K27me3 and H3K36 and low H3K18me1, H3K23, K27me2, and K36me1 are seen in the colon, liver, and adipose tissues of such mice (D'Aquila et al. 2020). Interestingly, histone codes such as H3K27me1 and H3K36me2 were found to be abundant in adipose tissue and less elsewhere.

Pathogenic bacteria such as *Listeria monocytogenes* and *Helicobacter pylori* were also shown to have the potential to regulate global histone acetylation in the host and regulate downstream signaling pathways. *L. monocytogenes* promote acetylation of histone H4 and phosphorylation/acetylation of histone H3 in infected human endothelial cells and very higher levels of cytokine expression (Schmeck et al. 2005). Patients with metabolic diseases such as obesity are highly susceptible to infections. In normal individuals with proper gut microbiome and immune system, the gut *microbiota* prevents *L. monocytogenes* colonization in the *gut* lumen and thereby averts systemic dissemination. However, patients with an altered gut microbial profile may fail in this purpose and result in altered epigenetic signaling and aberrant gene expression profiles.

3.3 Noncoding RNAs

MicroRNAs (miRNA) are ncRNAs, which are approximately 21–24 nucleotides in length with no protein-coding function. They are mostly engaged in post-transcriptional regulation. Recent studies indicate that host–microbiota interactions are also regulated by miRNAs-based epigenetic regulation (Yuan et al. 2019). There was a distinct difference in miRNA profiles of germ-free mice colonized with specific gut microbiota. Host miR-141-3p was found associated with the abundance of Bacteroidetes and Firmicutes and the miR-200a-3p level with an abundance of *Actinobacteria*, *Bacteroidetes*, *Firmicutes*, and *Proteobacteria* (Moloney et al. 2018). The mode of miRNA–microbiome interaction is not properly described, yet it can be assumed that miRNAs expressed by the host may transcriptionally regulate microbiome mRNA.

Microbial dysbiosis in the gut results in intestinal epithelial barrier dysfunction and ensuing inflammation. The probiotic bacterium *Escherichia coli* Nissle 1917 (EcN) possesses an intestinal barrier enhancing effect and induced a differential expression of miR-203, miR-483-3p, and miR-595. Further analysis of these miRNAs indicates their role in tight junction protein function (Guo et al. 2019). Pathogenic *E. coli* (EPEC) caused tight junction disruption by inhibiting these miRNAs. Treatment with a specific inhibitor to these miRNAs reduced the disturbances in tight junctions, which were previously caused by EPEC strain.

Gut microbiota-derived metabolites were also found to have a role in the regulation of miR-181. Studies in germ-free mice colonized with conventional mice microbiota demonstrated higher levels of miR-181a/b in epididymal white adipocytes (Virtue et al. 2019). Their further studies also indicate that tryptophan-derived metabolites reduce miR-181 levels in these adipocytes. The role of miR-181 in adiposity and insulin sensitivity needs to be further studied.

Host intestinal epithelial cell-derived miRNAs may target microbial mRNA and thereby control microbiota via transcriptional regulation or deregulation of gene expression. Human miRNAs such as miR-515-5p and miR-1226-5p were shown to regulate 16S/23S rRNA in *Fusobacterium nucleatum* and yegH in *Escherichia coli*, respectively (Liu et al. 2016).

It is known that host miRNA can regulate microbiota. Interestingly, gut microbiota can also affect host miRNA expression. Xue et al. (2011) observed that microbiota downregulate the miR-10a expression in intestinal epithelial and dendritic cells and TLR–ligand interactions via MyD88-dependent pathway.

Microbiota-based downregulation of miR-10a increases the expression of IL-12 and IL-23, which promotes intestinal immune homeostasis. Nakata et al. (2017) have demonstrated in intestinal epithelial cell (IEC) culture models that miR-21-5p levels in intestinal endothelial cells are induced by commensal bacteria. Elevated miR-21-5p was shown to augment intestinal epithelial permeability, indicating a therapeutic target for preventing intestinal epithelial barrier dysfunction. Moreover, altered intestinal microbiota can influence the host gene expression by miRNAs modulation in various diseases. As miRNA modulation via antagomirs and agomirs are mechanistically possible, gut microbiota–miRNA–host gene expression axis can be a potential therapeutic target in metabolic diseases.

Other types of ncRNAs such as long ncRNAs (lncRNAs) were also found to be involved in gut microbiome–host interaction. LncRNAs are ncRNAs with more than 200 nucleotides length and which do not encode for any proteins. Very few studies have been done to understand the precise role of gut microbiota in the expression of lncRNA in hosts. Researchers from the University of Washington performed RNA sequencing-based screening of tissue-specific lncRNA expression in germ-free and conventionalized mice (Dempsey et al. 2018). They found that majority of lncRNAs were co-regulated with adjacent protein-coding genes. They observed a ubiquitous expression of lncRNAs in the enterohepatic and the peripheral metabolic tissues of conventionalized mice. While in germ-free mice, the lncRNAs were differentially expressed in various tissues with predominant expression in the jejunum. This study reported the first evidence of the concept of gut microbiome regulating host lncRNA expression.

Much earlier to this study, Liang et al. (2015) characterized intestinal microbiome-regulated lncRNAs in the gut epithelium of germ-free mice with the conventional and *E. coli*-colonized gnotobiotic mice. By bioinformatic analysis, they found the overexpression of six lncRNAs involved in immune processes. The lncRNAs were different in mice models based on their gut microbiota. Further studies are warranted in this context to understand microbiota-induced host lncRNAs in homeostasis and disease.

4 Probiotics and Host Epigenetic Regulation

Gut microbial dysbiosis is associated with various facets of metabolic diseases, including diabetes, obesity, and insulin resistance. Modulation of microbiota composition, either directly (diet, bacterial metabolites, prebiotics, and probiotics) or indirectly (e.g., immunotherapeutics) may help in the management of patients with lifestyle metabolic diseases. The last two decades have seen several research groups focusing on the antidiabetic effects of probiotics that increase LMW production, as well as bacterial colonization and prebiotics. Various randomized and placebo-controlled clinical trials studied the outcome of probiotic administration on glucose and lipid parameters in T2DM. In one study, the participants in the "probiotic intervention group" had 300 g/day of probiotic yogurt with 10^6 CFU/ml *Lactobacillus acidophilus La5* and 10^6 CFU/ml *Bifidobacterium lactis Bb12* strains, whereas the control group participants consumed 300 g/day of conventional yogurt (Ejtahed et al. 2011). After 6 weeks of daily consumption, it was shown that the probiotic treatment with *L. acidophilus La5* and *B. lactis Bb12* yielded an increase in the antioxidant potential and low fasting blood glucose in the study subjects. Similarly, Tong et al. (2018) in a study of 450 patients demonstrated that T2DM with high lipid profile was controlled when enriched with beneficial *Blautia* and *Faecalibacterium.*

Studies have also tried to look at the association of diet and probiotics with epigenetic regulation in study subjects. It was found that black raspberries boost *Anaerostipes* which generates butyrate. Black raspberries also enrich anti-inflammatory bacteria, such as *Akkermansia* and *Desulfovibrio*. These microbes reduced DNMT1 levels and promoter methylation of genes present in the WNT-signaling pathway in tumors (Wang et al. 2011a, b). In HFD-fed mice, sodium butyrate supplementation was found to reduce fasting glucose and induce greater insulin sensitivity (Gao et al. 2009).

The role of prebiotics in the gut microbial composition is also being studied. Prebiotics, being a source of microbial SCFAs, may improve glycemic control and induce the growth or functioning of beneficial microorganisms. In HFD-fed mice, supplementation of oligofructose prebiotics resulted in larger numbers of *Bifidobacterium* species with improved glucose tolerance and glucose-induced insulin secretion (Cani et al. 2007). Prebiotics including inulins, oligodextrans, lactose, etc. were reported to have hypocholesterolemic effects in T2DM patients (Yoo and Kim 2016). Hald et al. (2016) focused on the role of two dietary fibers such as arabinoxylan and resistant starch type 2, on the gut microbiome and SCFAs in 19 patients with metabolic diseases. They observed that a diet rich in these dietary fibers increased *Bifidobacterium* numbers and butyrate levels in patients with metabolic syndrome.

Gut microbiota metabolizes several dietary bioactive compounds that modulate epigenetic enzymatic activity. Microbial degradation of epigallocatechin-3-gallate (EGCG) generates phenolic acids that reduce 80% of DNMTs' activity in the host (Remely et al. 2017). Supplementation of EGCG increases *DNMT1* and reduces inflammatory *IL-6* in the colon. EGCG supplementation also results in hypermethylation of *MLH1* and *DNMT1* gene promoters and thereby reduces their expression in the liver.

Resveratrol is a beneficial polyphenol that helps in glycemic and lipid control. It also reduces fat mass, blood pressure, chronic inflammation, and oxidative stress in several investigations. *Bifidobacteria infantis* and *Lactobacillus acidophilus* produce resveratrol from stilbenoid glucoside, piceid, which is a major resveratrol derivative (Basholli-Salihu et al. 2016). The current consensus is that resveratrol may prevent intestinal inflammation via gut microbiota modulation (Chaplin et al. 2018). More studies are required to understand the significant role of gut microbiota in nutritional interventions such as resveratrol supplementation.

5 Gut Microbiota Markers for Metabolic Diseases

Studies are at present focusing on obtaining a T2DM-associated gut metagenome profile in patients. There is a possibility of using gut microbiota-derived metabolites as novel markers for early prediction of T2DM, which will help in the nutritional modulation. For instance, researchers are exploring the possibility of using trimethylamine-N-oxide (TMAO), which is generated in the liver in response to the gut microbiota-derived metabolite, trimethylamine (TMA), as a biomarker for cardiovascular diseases (Yang et al. 2019). Plasma levels of TMAO were found to be positively correlated with adverse cardiovascular events (Dong et al. 2018).

Another alternative is LPS that is produced by Gram-negative gut bacteria and may act as a good marker for inflammation, insulin resistance, and increased fat mass (Cani et al. 2007). But, the use of LPS is limited as a marker due to its less half-life and increased susceptibility to interfering substances. LPS-binding protein (LBP), on the other hand, is a reliable circulatory that can be detected by ELISA-based kits. LBP helps in the recognition and host immune response to LPS and amplifies host immune responses to LPS and hence may reflect the LPS level in the host. Zhang et al. (2013) have reported reduced serum LBP in mice that had a higher abundance of beneficial bacteria and lower opportunistic pathogens. Circulating LBP was found to be higher in patients with coronary artery disease (Lepper et al. 2007) and obesity-related insulin resistance (Moreno-Navarrete et al. 2012). These studies indicate the potential of using LBP as an accurate biomarker for the early detection of metabolic syndrome in patients.

Gut microbiota produces essential amino acids such as leucine, isoleucine, and valine, which are classified as branched-chain amino acids (BCAAs). Plasma BCAA levels positively correlate with obesity and serum insulin (Newgard et al. 2009). Quite a few studies have looked at the association of plasma BCAAs with insulin resistance in patients with metabolic diseases. Wang et al. (2011a, b) conducted a study in 2422 normoglycemic individuals who were further followed up for 12 years to understand if metabolite profiles can be used as predictive markers for the development of T2DM. They observed that BCAA levels individually or in combination have the potential to predict the development of T2DM in patients.

Glutamate, a secondary product of the catabolism of BCAAs, was reported to be significantly higher in obese individuals than in non-obese lean individuals (Liu et al. 2017; Ejtahed et al. 2020). Glutamate expression was also found correlated to the abundance of *Ruminococcus*, *Coprococcus*, and *Dorea* species. Liu et al. (2017) have found a notable reduction in glutamate-fermenting *Bacteroides thetaiotaomicron* in obese study subjects. Other metabolite-based marker candidates are *Bifidobacterium* spp.-derived conjugated linoleic acid and *Eubacterium ventriosum*- and *Lactobacillus*-secreted conjugated linolenic acid (Park et al. 2012; Devillard et al. 2007). These polyunsaturated fatty acids-derived metabolites demonstrate beneficial effects in preventing obesity (Druart et al. 2014; Ejtahed et al. 2020).

Even though several studies have focused on microbial markers for metabolic diseases, it is still not known if these alterations in gut microbiota are just a bystander effect in patients with diseases. Yassour et al. (2016), in their study of 36 healthy Korean monozygotic twins with subclinical values of higher body mass index (BMI), observed that BMI values were inversely proportional to the number of *Akkermansia muciniphila* and directly correlated with riboflavin and NAD biosynthesis. This observation indicates the role of altered microbiota in the onset of disease.

6 Microbiological Memory in Epigenetic Regulation

Diet, nutrition, and microbiota are all demonstrated to be associated with the specific methylome and histone modifications in an individual. Prenatal and postnatal nutrition and probiotic supplementation induce stable, heritable modifications in offspring and are termed "fetal programming." Recently, another term, "microbiological memory," is coined to understand the inherited epigenetic programming in an offspring subjected to the cell microenvironment (microbiome, microbial LMWs supplemented through the diet) (Devaux and Raoult 2018). It is hypothesized that the microbiological memory of an individual remains stable when both diet and microbiota are constant. As per this model, the association of microbiome and host cells results in a continuous modification of host genes resulting in a specific epigenetic signature. Studies are essential to delineate the inherent epigenetic programming of an individual to attain a comprehensive understanding of the gut microbe–host interaction in both homeostasis and diseases.

7 Conclusions

Genetic susceptibility to metabolic diseases such as obesity and T2DM is generally studied based on gene variations. Epigenomics is the interface between the genetic predisposition and the influence of environmental factors in metabolic diseases. A gut-centric theory of metabolic syndrome started to evolve in 2000–2010, when studies in animal models and humans demonstrated the HFD-induced intestinal barrier and entry of contents in intestinal luminal into the systemic circulation. Gut microbiota is an epigenetic modulator in the host and contributes significantly to homeostasis. Altered gut microbiota influences the pathogenesis of chronic diseases such as obesity and insulin resistance. This indigenous microbiota produces multiple low-molecular-weight metabolites that can interact with different targets in the cells and tissues. Studies in human and animal models indicate the protective role of the gut microbiota that produces metabolites such as SCFAs and SAM in maintaining the integrity of the gut barrier. SCFAs are prominent in regulating the host epigenetic processes via DNA methylation and histone modifications. Even though research groups focused on specific microbiota-induced epigenetic changes, their works faced many challenges as the results obtained were masked by the effect of microbiota already present in animals. The advent of conventionalized animals greatly solves this conundrum. Further in-depth studies focusing on specific gut microbial species and their metabolites in shaping the host epigenome are essential for the early nutritional management of metabolic diseases.

References

Abbasi IHR, Abbasi F, Wang L et al (2018) Folate promotes S-adenosyl methionine reactions and the microbial methylation cycle and boosts ruminants production and reproduction. AMB Express 8(1):65. https://doi.org/10.1186/s13568-018-0592-5

Al Akeel R (2013) Role of epigenetic reprogramming of host genes in bacterial pathogenesis. Saudi J Biol Sci 20(4):305–309. https://doi.org/10.1016/j.sjbs.2013.05.003

Arora T, Bäckhed F (2016) The gut microbiota and metabolic disease: current understanding and future perspectives. J Intern Med 280(4):339–349. https://doi.org/10.1111/joim.12508

Basholli-Salihu M, Schuster R, Mulla D et al (2016) Bioconversion of piceid to resveratrol by selected probiotic cell extracts. Bioprocess Biosyst Eng 39(12):1879–1885. https://doi.org/10.1007/s00449-016-1662-1

Bhat MI, Kapila R (2017) Dietary metabolites derived from gut microbiota: critical modulators of epigenetic changes in mammals. Nutr Rev 75(5):374–389. https://doi.org/10.1093/nutrit/nux001

Brial F, Alzaid F, Sonomura K et al (2018) The microbial metabolite 4-cresol improves glucose homeostasis and enhances β-cell function. bioRxiv 1:444893. https://doi.org/10.1101/444893

Cani PD, Neyrinck AM, Fava F et al (2007) Selective increases of bifidobacteria in gut microflora improve high-fat-diet-induced diabetes in mice through a mechanism associated with endotoxaemia. Diabetologia 50(11):2374–2383. https://doi.org/10.1007/s00125-007-0791-0

Chaplin A, Carpéné C, Mercader J (2018) Resveratrol, metabolic syndrome, and gut microbiota. Nutrients 10(11):1651. https://doi.org/10.3390/nu10111651

Chimerel C, Emery E, Summers DK et al (2014) Bacterial metabolite indole modulates incretin secretion from intestinal enteroendocrine L cells. Cell Rep 9(4):1202–1208. https://doi.org/10.1016/j.celrep.2014.10.032

Crider KS, Yang TP, Berry RJ et al (2012) Folate and DNA methylation: a review of molecular mechanisms and the evidence for folate's role. Adv Nutr 3(1):21–38. https://doi.org/10.3945/an.111.000992

D'Aquila P, Carelli LL, De Rango F et al (2020) Gut microbiota as important mediator between diet and DNA methylation and histone modifications in the host. Nutrients 12(3):597. https://doi.org/10.3390/nu12030597

Davie JR (2003) Inhibition of histone deacetylase activity by butyrate. J Nutr 133(7 Suppl):2485S–2493S. https://doi.org/10.1093/jn/133.7.2485S

De R, Mukhopadhyay AK, Dutta S (2020) Metagenomic analysis of gut microbiome and resistome of diarrheal fecal samples from Kolkata, India, reveals the core and variable microbiota including signatures of microbial dark matter. Gut Pathog 12:32. https://doi.org/10.1186/s13099-020-00371-8

Dempsey J, Zhang A, Cui JY (2018) Coordinate regulation of long non-coding RNAs and protein-coding genes in germ-free mice. BMC Genomics 19(1):834. https://doi.org/10.1186/s12864-018-5235-3

den Besten G, van Eunen K, Groen AK et al (2013) The role of short-chain fatty acids in the interplay between diet, gut microbiota, and host energy metabolism. J Lipid Res 54(9):2325–2340. https://doi.org/10.1194/jlr.R036012

Devaux CA, Raoult D (2018) The microbiological memory, an epigenetic regulator governing the balance between good health and metabolic disorders. Front Microbiol 9:1379. https://doi.org/10.3389/fmicb.2018.01379

Devillard E, McIntosh FM, Duncan SH et al (2007) Metabolism of linoleic acid by human gut bacteria: different routes for biosynthesis of conjugated linoleic acid. J Bacteriol 189(6):2566–2570. https://doi.org/10.1128/JB.01359-06

Dong Z, Liang Z, Guo M et al (2018) The association between plasma levels of trimethylamine N-oxide and the risk of coronary heart disease in chinese patients with or without type 2 diabetes mellitus. Dis Markers 2018:1578320. https://doi.org/10.1155/2018/1578320

Druart C, Dewulf EM, Cani PD et al (2014) Gut microbial metabolites of polyunsaturated fatty acids correlate with specific fecal bacteria and serum markers of metabolic syndrome in obese women. Lipids 49(4):397–402. https://doi.org/10.1007/s11745-014-3881-z

Druart C, Bindels LB, Schmaltz R et al (2015) Ability of the gut microbiota to produce PUFA-derived bacterial metabolites: proof of concept in germ-free versus conventionalized mice. Mol Nutr Food Res 59(8):1603–1613. https://doi.org/10.1002/mnfr.201500014

Ejtahed HS, Mohtadi-Nia J, Homayouni-Rad A et al (2011) Effect of probiotic yogurt containing *lactobacillus acidophilus* and *Bifidobacterium lactis* on lipid profile in individuals with type 2 diabetes mellitus. J Dairy Sci 94(7):3288–3294. https://doi.org/10.3168/jds.2010-4128

Ejtahed HS, Angoorani P, Soroush AR et al (2020) Gut microbiota-derived metabolites in obesity: a systematic review. Biosci Microbiota Food Health 39(3):65–76. https://doi.org/10.12938/bmfh.2019-026

Gao Z, Yin J, Zhang J et al (2009) Butyrate improves insulin sensitivity and increases energy expenditure in mice. Diabetes 58(7):1509–1517. https://doi.org/10.2337/db08-1637

Gérard C, Vidal H (2019) Impact of gut microbiota on host glycemic control. Front Endocrinol (Lausanne) 10:29. https://doi.org/10.3389/fendo.2019.00029

Guo S, Chen S, Ma J et al (2019) *Escherichia coli* Nissle 1917 protects intestinal barrier function by inhibiting NF-κB-mediated activation of the MLCK-P-MLC signaling pathway. Mediat Inflamm 2019:5796491. https://doi.org/10.1155/2019/5796491

Hald S, Schioldan AG, Moore ME et al (2016) Effects of arabinoxylan and resistant starch on intestinal microbiota and short-chain fatty acids in subjects with metabolic syndrome: a randomised crossover study. PLoS One 11(7):e0159223. https://doi.org/10.1371/journal.pone.0159223

Hope ME, Hold GL, Kain R et al (2005) Sporadic colorectal cancer–ole of the commensal microbiota. FEMS Microbiol Lett 244(1):1–7. https://doi.org/10.1016/j.femsle.2005.01.029

Human Microbiome Project Consortium (2012) Structure, function and diversity of the healthy human microbiome. Nature 486(7402):207–214. https://doi.org/10.1038/nature11234

Jacob RA (2000) Folate, DNA methylation, and gene expression: factors of nature and nurture. Am J Clin Nutr 72(4):903–904. https://doi.org/10.1093/ajcn/72.4.903

Ji X, Zhou F, Zhang Y et al (2019) Butyrate stimulates hepatic gluconeogenesis in mouse primary hepatocytes. Exp Ther Med 17(3):1677–1687. https://doi.org/10.3892/etm.2018.7136

Jiang L, Wang J, Liu Z et al (2020) Sodium butyrate alleviates lipopolysaccharide-induced inflammatory responses by down-regulation of NF-κB, NLRP3 signaling pathway, and activating histone acetylation in bovine macrophages. Front Vet Sci 7:579674. https://doi.org/10.3389/fvets.2020.579674

Kaur K, Khatri I, Akhtar A et al (2020) Metagenomics analysis reveals features unique to Indian distal gut microbiota. PLoS One 15(4):e0231197. https://doi.org/10.1371/journal.pone.0231197

Kawamoto S, Tran TH, Maruya M et al (2012) The inhibitory receptor PD-1 regulates IgA selection and bacterial composition in the gut. Science 336(6080):485–489. https://doi.org/10.1126/science.1217718

Koh A, Molinaro A, Ståhlman M et al (2018) Microbially produced imidazole propionate impairs insulin signaling through mTORC1. Cell 175(4):947–961.e17. https://doi.org/10.1016/j.cell.2018.09.055

Krautkramer KA, Kreznar JH, Romano KA et al (2016) Diet-microbiota interactions mediate global epigenetic programming in multiple host tissues. Mol Cell 64(5):982–992. https://doi.org/10.1016/j.molcel.2016.10.025

Kubinak JL, Stephens WZ, Soto R et al (2015) MHC variation sculpts individualized microbial communities that control susceptibility to enteric infection. Nat Commun 6:8642. https://doi.org/10.1038/ncomms9642

Kumar H, Lund R, Laiho A et al (2014) Gut microbiota as an epigenetic regulator: pilot study based on whole-genome methylation analysis. mBio 5(6):e02113–e02114. https://doi.org/10.1128/mBio.02113-14

LeBlanc JG, Milani C, de Giori GS et al (2013) Bacteria as vitamin suppliers to their host: a gut microbiota perspective. Curr Opin Biotechnol 24(2):160–168. https://doi.org/10.1016/j.copbio.2012.08.005

Lee HS (2019) The interaction between gut microbiome and nutrients on development of human disease through epigenetic mechanisms. Genomics Inform 17(3):e24. https://doi.org/10.5808/GI.2019.17.3.e24

Lepper PM, Schumann C, Triantafilou K et al (2007) Association of lipopolysaccharide-binding protein and coronary artery disease in men. J Am Coll Cardiol 50(1):25–31. https://doi.org/10.1016/j.jacc.2007.02.070

Li HP, Chen X, Li MQ (2013) Butyrate alleviates metabolic impairments and protects pancreatic β cell function in pregnant mice with obesity. Int J Clin Exp Pathol 6(8):1574–1584

Liang L, Ai L, Qian J et al (2015) Long noncoding RNA expression profiles in gut tissues constitute molecular signatures that reflect the types of microbes. Sci Rep 5:11763. https://doi.org/10.1038/srep11763

Licciardi PV, Ververis K, Karagiannis TC (2011) Histone deacetylase inhibition and dietary short-chain fatty acids. ISRN Allergy 2011:869647. https://doi.org/10.5402/2011/869647

Lim SM, Jeong JJ, Woo KH et al (2016) Lactobacillus sakei OK67 ameliorates high-fat diet-induced blood glucose intolerance and obesity in mice by inhibiting gut microbiota lipopolysaccharide production and inducing colon tight junction protein expression. Nutr Res 36(4):337–348. https://doi.org/10.1016/j.nutres.2015.12.001

Liu R, Hong J, Xu X et al (2017) Gut microbiome and serum metabolome alterations in obesity and after weight-loss intervention. Nat Med 23(7):859–868. https://doi.org/10.1038/nm.4358

Liu S, da Cunha AP, Rezende RM et al (2016) The host shapes the gut microbiota via fecal microrna. Cell Host Microbe 19(1):32–43. https://doi.org/10.1016/j.chom.2015.12.005

Liu T, Li J, Liu Y et al (2012) Short-chain fatty acids suppress lipopolysaccharide-induced production of nitric oxide and proinflammatory cytokines through inhibition of NF-κB pathway in RAW264.7 cells. Inflammation 35(5):1676–1684. https://doi.org/10.1007/s10753-012-9484-z

Louis P, Flint HJ (2009) Diversity, metabolism and microbial ecology of butyrate-producing bacteria from the human large intestine. FEMS Microbiol Lett 294(1):1–8. https://doi.org/10.1111/j.1574-6968.2009.01514.x

Maekita T, Nakazawa K, Mihara M et al (2006) High levels of aberrant DNA methylation in helicobacter pylori-infected gastric mucosae and its possible association with gastric cancer risk. Clin Cancer Res 12(3 Pt 1):989–995. https://doi.org/10.1158/1078-0432.CCR-05-2096

Martin-Gallausiaux C, Larraufie P, Jarry A et al (2018) Butyrate produced by commensal bacteria down-regulates indolamine 2,3-dioxygenase 1 (ido-1) expression via a dual mechanism in human intestinal epithelial cells. Front Immunol 9:2838. https://doi.org/10.3389/fimmu.2018.02838

Miro-Blanch J, Yanes O (2019) Epigenetic regulation at the interplay between gut microbiota and host metabolism. Front Genet 10:638. https://doi.org/10.3389/fgene.2019.00638

Moloney GM, Viola MF, Hoban AE et al (2018) Faecal microRNAs: indicators of imbalance at the host-microbe interface? Benef Microbes 9(2):175–183. https://doi.org/10.3920/BM2017.0013

Moreno-Navarrete JM, Ortega F, Serino M et al (2012) Circulating lipopolysaccharide-binding protein (LBP) as a marker of obesity-related insulin resistance. Int J Obes 36(11):1442–1449. https://doi.org/10.1038/ijo.2011.256

Nakata K, Sugi Y, Narabayashi H et al (2017) Commensal microbiota-induced microRNA modulates intestinal epithelial permeability through the small GTPase ARF4. J Biol Chem 292(37):15426–15433. https://doi.org/10.1074/jbc.M117.788596

Natarajan N, Pluznick JL (2014) From microbe to man: the role of microbial short chain fatty acid metabolites in host cell biology. Am J Physiol Cell Physiol 307(11):C979–C985. https://doi.org/10.1152/ajpcell.00228.2014

Newgard CB, An J, Bain JR et al (2009) A branched-chain amino acid-related metabolic signature that differentiates obese and lean humans and contributes to insulin resistance [published correction appears in Cell Metab 9(6):565–6]. Cell Metab 9(4):311–326. https://doi.org/10.1016/j.cmet.2009.02.002

Nicaise P, Gleizes A, Forestier F et al (1993) Influence of intestinal bacterial flora on cytokine (IL-1, IL-6 and TNF-alpha) production by mouse peritoneal macrophages. Eur Cytokine Netw 4(2):133–138

Pais R, Gribble FM, Reimann F (2016) Stimulation of incretin secreting cells. Ther Adv Endocrinol Metab 7(1):24–42. https://doi.org/10.1177/2042018815618177

Park HG, Cho HT, Song MC et al (2012) Production of a conjugated fatty acid by Bifidobacterium breve LMC520 from α-linolenic acid: conjugated linolenic acid (CLnA). J Agric Food Chem 60(12):3204–3210. https://doi.org/10.1021/jf2041559

Pichette J, Fynn-Sackey N, Gagnon J (2017) Hydrogen sulfide and sulfate prebiotic stimulates the secretion of GLP-1 and improves glycemia in male mice. Endocrinology 158(10):3416–3425. https://doi.org/10.1210/en.2017-00391

Qin J, Li Y, Cai Z et al (2012) A metagenome-wide association study of gut microbiota in type 2 diabetes. Nature 490(7418):55–60. https://doi.org/10.1038/nature11450

Ramos-Molina B, Sánchez-Alcoholado L, Cabrera-Mulero A et al (2019) Gut microbiota composition is associated with the global dna methylation pattern in obesity. Front Genet 10:613. https://doi.org/10.3389/fgene.2019.00613

Remely M, Aumueller E, Merold C et al (2014) Effects of short chain fatty acid producing bacteria on epigenetic regulation of FFAR3 in type 2 diabetes and obesity. Gene 537(1):85–92. https://doi.org/10.1016/j.gene.2013.11.081

Remely M, Ferk F, Sterneder S et al (2017) EGCG prevents high fat diet-induced changes in gut microbiota, decreases of DNA strand breaks, and changes in expression and DNA methylation of Dnmt1 and MLH1 in C57BL/6J male mice. Oxidative Med Cell Longev 2017:3079148. https://doi.org/10.1155/2017/3079148

Ríos-Covián D, Ruas-Madiedo P, Margolles A et al (2016) Intestinal short chain fatty acids and their link with diet and human health. Front Microbiol 7:185. https://doi.org/10.3389/fmicb.2016.00185

Rossi M, Amaretti A, Raimondi S (2011) Folate production by probiotic bacteria. Nutrients 3(1):118–134. https://doi.org/10.3390/nu3010118

Röth D, Chiang AJ, Hu W et al (2019) Two-carbon folate cycle of commensal lactobacillus reuteri 6475 gives rise to immunomodulatory ethionine, a source for histone ethylation. FASEB J 33(3):3536–3548. https://doi.org/10.1096/fj.201801848R

Roth SY, Denu JM, Allis CD (2001) Histone acetyltransferases. Annu Rev Biochem 70:81–120. https://doi.org/10.1146/annurev.biochem.70.1.81

Rowland I, Gibson G, Heinken A et al (2018) Gut microbiota functions: metabolism of nutrients and other food components. Eur J Nutr 57(1):1–24. https://doi.org/10.1007/s00394-017-1445-8

Savage DC (1977) Microbial ecology of the gastrointestinal tract. Annu Rev Microbiol 31:107–133. https://doi.org/10.1146/annurev.mi.31.100177.000543

Schaible TD, Harris RA, Dowd SE et al (2011) Maternal methyl-donor supplementation induces prolonged murine offspring colitis susceptibility in association with mucosal epigenetic and microbiomic changes. Hum Mol Genet 20(9):1687–1696. https://doi.org/10.1093/hmg/ddr044

Schilderink R, Verseijden C, de Jonge WJ (2013) Dietary inhibitors of histone deacetylases in intestinal immunity and homeostasis. Front Immunol 1(4):226. https://doi.org/10.3389/fimmu.2013.00226

Schmeck B, Beermann W, van Laak V et al (2005) Intracellular bacteria differentially regulated endothelial cytokine release by MAPK-dependent histone modification. J Immunol 175(5):2843–2850. https://doi.org/10.4049/jimmunol.175.5.2843

Sender R, Fuchs S, Milo R (2016) Revised estimates for the number of human and bacteria cells in the body. PLoS Biol 19(14(8)):e1002533. https://doi.org/10.1371/journal.pbio.1002533

Smith PM, Howitt MR, Panikov N et al (2013) The microbial metabolites, short-chain fatty acids, regulate colonic Treg cell homeostasis. Science 341(6145):569–573. https://doi.org/10.1126/science.1241165

Steliou K, Boosalis MS, Perrine SP et al (2012) Butyrate histone deacetylase inhibitors. Biores Open Access 1(4):192–198. https://doi.org/10.1089/biores.2012.0223

Tong X, Xu J, Lian F et al (2018) Structural alteration of gut microbiota during the amelioration of human type 2 diabetes with hyperlipidemia by metformin and a traditional chinese herbal formula: a multicenter, randomized, open label clinical trial. mBio 9(3):e02392–e02317. https://doi.org/10.1128/mBio.02392-17

Vijay-Kumar M, Aitken JD, Carvalho FA et al (2010) Metabolic syndrome and altered gut microbiota in mice lacking toll-like receptor 5. Science 328(5975):228–231. https://doi.org/10.1126/science.1179721

Vinjé S, Stroes E, Nieuwdorp M et al (2014) The gut microbiome as novel cardio-metabolic target: the time has come! Eur Heart J 35(14):883–887. https://doi.org/10.1093/eurheartj/eht467

Vinolo MA, Rodrigues HG, Hatanaka E et al (2011) Suppressive effect of short-chain fatty acids on production of proinflammatory mediators by neutrophils. J Nutr Biochem 22(9):849–855. https://doi.org/10.1016/j.jnutbio.2010.07.009

Virtue AT, McCright SJ, Wright JM et al (2019) The gut microbiota regulates white adipose tissue inflammation and obesity via a family of microRNAs. Sci Transl Med 11(496):eaav1892. https://doi.org/10.1126/scitranslmed.aav1892

Wang LS, Arnold M, Huang YW et al (2011a) Modulation of genetic and epigenetic biomarkers of colorectal cancer in humans by black raspberries: a phase I pilot study. Clin Cancer Res 17(3):598–610. https://doi.org/10.1158/1078-0432.CCR-10-1260

Wang TJ, Larson MG, Vasan RS et al (2011b) Metabolite profiles and the risk of developing diabetes. Nat Med 17(4):448–453. https://doi.org/10.1038/nm.2307

Xiao M, Yang H, Xu W et al (2012) Inhibition of α-KG-dependent histone and DNA demethylases by fumarate and succinate that are accumulated in mutations of FH and SDH tumor suppressors. Genes Dev 26(12):1326–1338. https://doi.org/10.1101/gad.191056.112

Xue X, Feng T, Yao S et al (2011) Microbiota downregulates dendritic cell expression of miR-10a, which targets IL-12/IL-23p40. J Immunol 187(11):5879–5886. https://doi.org/10.4049/jimmunol.1100535

Yang S, Li X, Yang F et al (2019) Gut microbiota-dependent marker TMAO in promoting cardiovascular disease: inflammation mechanism, clinical prognostic, and potential as a therapeutic target. Front Pharmacol 10:1360. https://doi.org/10.3389/fphar.2019.01360

Yassour M, Lim MY, Yun HS et al (2016) Sub-clinical detection of gut microbial biomarkers of obesity and type 2 diabetes. Genome Med 8(1):17. https://doi.org/10.1186/s13073-016-0271-6

Yoo JY, Kim SS (2016) Probiotics and prebiotics: present status and future perspectives on metabolic disorders. Nutrients 8(3):173. https://doi.org/10.3390/nu8030173

Yuan C, Steer CJ, Subramanian S (2019) Host-microrna-microbiota interactions in colorectal cancer. Genes (Basel) 10(4):270. https://doi.org/10.3390/genes10040270

Yuille S, Reichardt N, Panda S et al (2018) Human gut bacteria as potent class I histone deacetylase inhibitors *in vitro* through production of butyric acid and valeric acid. PLoS One 13(7):e0201073. https://doi.org/10.1371/journal.pone.0201073

Zdzisińska B, Żurek A, Kandefer-Szerszeń M (2017) Alpha-ketoglutarate as a molecule with pleiotropic activity: well-known and novel possibilities of therapeutic use. Arch Immunol Ther Exp 65(1):21–36. https://doi.org/10.1007/s00005-016-0406-x
Zhang C, Li S, Yang L et al (2013) Structural modulation of gut microbiota in life-long calorie-restricted mice. Nat Commun 4:2163. https://doi.org/10.1038/ncomms3163
Zhou Y, He H, Xu H et al (2016) Association of oncogenic bacteria with colorectal cancer in South China. Oncotarget 7(49):80794–80802. https://doi.org/10.18632/oncotarget.13094

Chapter 6
Gut Microbiota-Related Clinical Events and Therapeutic Interventions in Alcohol-Associated Liver Disease

Cyriac Abby Philips, Juan Pablo Arab, Xingshun Qi, Norihiro Imai, and Philip Augustine

1 Introduction

Alcohol use remains a major cause of socioeconomic burden, morbidity, and mortality globally and is linked to more than 60 acute and chronic diseases. Alcohol use is also a leading risk factor for premature death and disability among people in their most productive years. Over the last decade, the prevalence of high-risk drinking, defined by the National Institute on Alcohol Abuse and Alcoholism (NIAAA) as more than four drinks per day or 14 in a week for men, and more than three drinks a day or seven per week for women as well as AUD, has increased globally. Even though global trends showed an overall increase, it was also observed that alcohol

C. A. Philips (✉)
Clinical and Translational Hepatology, The Liver Institute, Center of Excellence in Gastrointestinal Sciences, Rajagiri Hospital, Aluva, Kerala, India
e-mail: abbyphilips@theliverinst.in

J. P. Arab
Medicine Departmento Gastroenterología y Hepatología, Escuela de Medicina, Pontificia Universidad Católica de Chile, Santiago, Chile
e-mail: jparab@uc.cl

X. Qi
Department of Gastroenterology, General Hospital of Northern Theater Command, Shenyang, China

N. Imai
Gastroenterology and Hepatology, Nagoya University Graduate School of Medicine, Nagoya, Aichi, Japan
e-mail: norihiro@med.nagoya-u.ac.jp

P. Augustine
Gastroenterology and Advanced GI Endoscopy, Center of Excellence in Gastrointestinal Sciences, Rajagiri Hospital, Aluva, Kerala, India

S. Thomas (ed.), *Human Microbiome*,
https://doi.org/10.1007/978-981-16-7672-7_6

use declined in western and, more recently, Eastern European countries. Nonetheless, alcohol consumption increased substantially in several Asian countries such as India, Vietnam, China, and sub-Saharan Africa, where consumption is nearly equal to the global average. In this regard, the leading cause of alcohol-attributable deaths was cirrhosis in countries with a low human development index. Alcohol-associated cirrhosis was also the commonest cause of death in people aged 40–59 years (GBD 2016 Alcohol Collaborators 2018; Shield et al. 2020).

Alcohol-associated liver disease (AALD) ranges from isolated alcohol-associated steatosis to cirrhosis with episodes of alcohol-associated hepatitis (AH) that may present with acute decompensation of cirrhosis or acute on chronic liver failure (ACLF). Patients with AH have a 40% mortality at 6 months and a nine-times risk of cirrhosis progression than those with only alcohol-associated steatosis. Two percent of patients with alcohol-associated cirrhosis develop hepatocellular carcinoma annually (Avila et al. 2020; Crabb et al. 2020).

In patients with AALD, multiple mechanisms act at the host level that defines clinical events. Gender, genetic predisposition (probably polygenic), associated metabolic syndrome, concomitant liver diseases such as chronic hepatitis B or C virus infections, the pattern of drinking, other drug use, and nutritional status define the risk of and time to progression of AALD. Among the genetic factors associated with causation and progression of AALD, it was shown that single nucleotide polymorphisms of alcohol and acetaldehyde dehydrogenase protected against alcohol use disorders and subsequent disease development; polymorphism of patatin-like phospholipase domain-containing 3 (PNPLA3) rs738409 polymorphism was significantly associated with cirrhosis; and hepatocellular carcinoma in patients with AALD and several other genetic polymorphisms in HSD17B13, TM6SF2, MBOAT7, and SERPINA1 genes and heterogenous carriage of the alpha-1-antitrypsin Pi*Z variant were found to be associated with the development of cirrhosis in AALD (Stickel and Hampe 2012; Atkinson et al. 2017; Kolla et al. 2018; Abul-Husn et al. 2018). A recent multinational genome-wide association study identified a new locus at *Fas-Associated Factor family member 2* (FAF2) implicating lipid droplets, associated with reduced risk for developing cirrhosis in heavy drinkers (Schwantes-An et al. 2020).

Current evidence sheds light on the central role of intestinal microbiota in plausible causation and progression of AALD and associated clinical events (Bajaj 2019; Arab et al. 2020). In this regard, dysbiosis, or the disruption of key intestinal microbial communities and their function leading to loss of host-microbe symbiosis, results in disturbed crosstalk at the molecular level that ultimately leads to the causation of specific clinical events associated with AALD or progression of cirrhosis and its complications. In a dysbiotic environment, there occurs a reduction in diversity and richness of microbial communities; loss of beneficial microbes that produce advantageous metabolites that maintain or promote homeostatic effects at the local, organ, or systems level; and the emergence of pathobiont expansion. Acute and chronic alcohol consumption triggers pro-inflammatory cascades within the gut lumen and at the mucosal level through the generation of alcohol metabolism generated byproducts that further incite intestinal bacterial dysbiosis with the diversification of pathogenic species such as Gram-negative bacteria. The direct effects of alcohol and its metabolism within the intestines and at the hepatocyte level and changes in gut bacteria and its functional metabolism promote liver injury that promotes disease development or culminate into severe AH events (Adolph et al. 2018; Philips et al. 2019a, b; Bajaj 2019; Bajaj et al. 2020).

2 Gut Microbiota in Alcohol-Associated Liver Disease

2.1 Lessons from the Animal Models

Three weeks of intragastric alcohol feeding in mice increased *Bacteroidetes* and *Verrucomicrobia* with a reduction in *Firmicutes* compared to pair-feeding (feeding two groups of experimental animals with a diet identical for the item (alcohol) whose effects are being tested on one group). An overgrowth of *Akkermansia muciniphila*, a bacterium that degrades mucin, was observed in the alcohol-fed group along with depletion of *Lactobacilli* (Yan et al. 2011).

In a Lieber-DiCarli alcohol feeding mouse model (exclusive ethanol-containing liquid diet formula to simulate early AALD; no fibrosis or inflammation), expansion in Gram-negative *Proteobacteria* and Gram-positive *Actinobacteria* phyla was noted, including a higher proportion of alkaline tolerant *Alcaligenes* and *Corynebacterium*, associated with increased colonic pH and liver steatosis (Bull-Otterson et al. 2013). In a chronic ethanol feeding (for 8 weeks) mouse model, authors noted changes in bacterial alpha-diversity in the ileum and liver that led to compositional changes especially in the ileum due to an increase in endotoxin-producing Gram-negative phyla—*Prevotella* was specifically increased in the mucus layer of ileum as well as in liver tissue, demonstrating the loss of gut barrier integrity in the presence of alcohol leading to bacterial translocation (Bluemel et al. 2019). When rhesus monkeys were fed alcohol, fecal composition analysis revealed an increase in *Firmicutes*, *Proteobacteria,* and *Verrucomicrobia*, associated with a reduction in *Bacteroidetes* and *Actinobacteria*. Lactobacillus and Streptococcus were reduced in the alcohol group at the genus level compared to controls fed water (Wang et al. 2019).

In medaka (Japanese rice fish), exposure to alcohol for 2 months lead to liver steatosis and inflammation associated with a reduction in *Fusobacterium*, *Tenericutes,* and *Firmicutes* and an increase in *Bacteroides* and *Proteobacteria*. At the genera level, *Alcaligenes incertae sedis* and *Cloacibacterium incertae sedis* decreased while *Cetobacterium incertae sedis* and *Erysipelotrichaceae incertae sedis* increased with ethanol administration, similar to observations in human AALD (Fujisawa et al. 2019). Alcohol feeding for 7 days in germ-free Swiss mice led to reduced liver injury compared to conventional mice due to reduced intestinal permeability and lower neutrophil accumulation (Canesso et al. 2014). In contrast, the C57BL/6 germ-free mice developed a more pronounced liver injury, inflammation, and steatosis than conventional mice when fed alcohol. This differential response to alcohol feeding was probably due to differences in rodent strains (Chen et al. 2015a, b).

2.2 Lessons from AALD in Humans

Patients with AALD without cirrhosis develop dysbiosis demonstrating lower *Bacteroidetes* and higher *Proteobacteria*. Similarly, in another study, dysbiosis associated with a lower abundance of taxa *Ruminococcaceae* was notable in patients with AALD without chronic liver disease and active drinking, which reversed after abstinence. AALD patients with high intestinal permeability had lower levels of *Faecalibacterium prausnitzii*, a bacterial species known for its anti-inflammatory properties (Mutlu et al. 2012; Leclercq et al. 2014).

In patients with alcohol-associated cirrhosis who were abstinent, the quality and quantity of dysbiosis were similar to other etiologies of cirrhosis. However, hepatic encephalopathy, rather than cirrhosis's etiology, was found to be associated with specific dysbiosis patterns, probably as a marker of disease severity. *Alcaligenaceae*, *Porphyromonadaceae*, *Enterobacteriaceae* were positively and *Ruminococcaceae* negatively associated with cognition and inflammation in alcohol-associated cirrhosis and hepatic encephalopathy (Chen et al. 2011; Bajaj et al. 2012; Kakiyama et al. 2013).

Fecal sample next-generation sequencing among patients with alcohol-associated cirrhosis with active drinking showed that cirrhosis patients had a median of 27 times more bacterial DNA of *Enterobactericaea* in feces as translocated to the liver, compared to the healthy volunteers. When ascites samples from the volunteers with cirrhosis were analyzed, around half of them contained bacterial DNA from *Enterobactericaea, Clostridium leptum,* or *Lactobacillus* species (Tuomisto et al. 2014). In another study, depletion of commensal bacteria associated with increases in *Lactobacillus* and *Bifidobacterium* species were seen in patients with alcohol-associated cirrhosis and active drinking (Dubinkina et al. 2017). A study on the stool, duodenal, ileal, and colonic microbiota analysis between healthy control individuals, actively drinking patients with cirrhosis, and abstinent patients with cirrhosis demonstrated that the proportion of autochthonous taxa (*Lachnospiraceae, Ruminococcaeae*, and *Clostridiales* Cluster XIV) were significantly lower in all alcohol-associated cirrhosis tissues compared to non-alcoholic tissues (Bajaj et al. 2017). In a study on patients with AH, it was shown that the circulating bacterial DNA was found to be higher among those with heavy alcohol drinking, which was also associated with significantly decreased *Bacteroidetes* and enrichment of *Fusobacteria* (Puri et al. 2018). A study on gut microbiota analysis in patients with or without cirrhosis and the presence or absence of AH showed that cirrhosis patients with AH had a different microbiota than cirrhotics without AH. The abundance of *Actinobacteria* was higher, and that of *Bacteroidetes* was lower in the former. Cirrhotic patients with AH had a higher abundance of *Actinomyces*, *Rothia*, and *Bifidobacterium* and a lower abundance of *Bilophila* and *Parabacteroides* than those without AH. In a more recent study from India in patients with severe AH, the authors analyzed gut microbiota association with clinical events. At admission, 27%, 42%, and 58% had acute kidney injury, hepatic encephalopathy, and infections, respectively; and 38.5% died at 90 days follow-up: *Veillonellaceae*, *Prevotellaceae* with sepsis; *Dehalobacteriaceae*, *Turicibacteraceae* with the model for end-stage liver disease >25 and *Enterobacteriaceae*, *Peptococcaceae* with death on follow-up. *Lachnobacterium*, *Catenibacterium* was notably associated with hepatic encephalopathy at admission, while *Eubacterium*, *Capnocytophaga* were associated with Child-Pugh score > 10. Prevotella was associated with survival post steroid treatment. Co-occurrence between *Christensenella*, *Prevotella*, and mutual exclusion between *Megamonas*, *Citrobacter* was associated with hepatic encephalopathy at admission, and *Enterococcus cecorum*, *Acinetobacter schindleri*, and *Mitsuokella* were associated with acute kidney injury at admission (Philips et al. 2020a, b). Deviating from bacterial community profiling, a collaborative multicenter study on fungal microbiota (mycobiome) in patients with AALD demonstrated that fungal diversity was lower in subjects with AUD than healthy controls with overgrowth of *Candida* and concomitant reduction in *Epicoccum*, *Galactomyces*, and *Debaryomyces* in the former (Ho et al. 2017; Yang et al. 2017) (Fig. 6.1).

3 Gut Microbial Metabolites in Alcohol-Associated Liver Disease

In small animal studies, it was elegantly shown that alcohol feeding led to increased intestinal permeability, dependent on gut microbiota-modulated reduction in intestinal levels of hypoxia-induced factor 1-alpha activity expression, which also led to hepatic steatosis. The endotoxin producing *Enterobacteriaceae* was preferentially upregulated in alcohol consumption, resulting in reduced beneficial gut-derived short-chain fatty acids (SCFAs), the latter due to reduction in *Lachnospiraceae* and *Ruminococcaceae* (Bull-Otterson et al. 2013; Wang et al. 2011; Chen et al. 2015a, b). Levels of secondary bile acids were higher in patients with AALD and active drinking, which was also associated with increased intestinal permeability changes. Furthermore, in persons with AUD, fecal bacterial metabolome demonstrated higher SCFAs and sulfides, reduced antioxidant fatty acids, and inverse association of alcohol dependence with butyrate-producing bacterial communities. In patients with AH, reduction in *Akkermansia muciniphila* was associated with loss of gut barrier function via mucosal disruption that resulted in liver injury. In mouse models of AH, administration of *A. muciniphila* led to an improvement in alcohol-associated

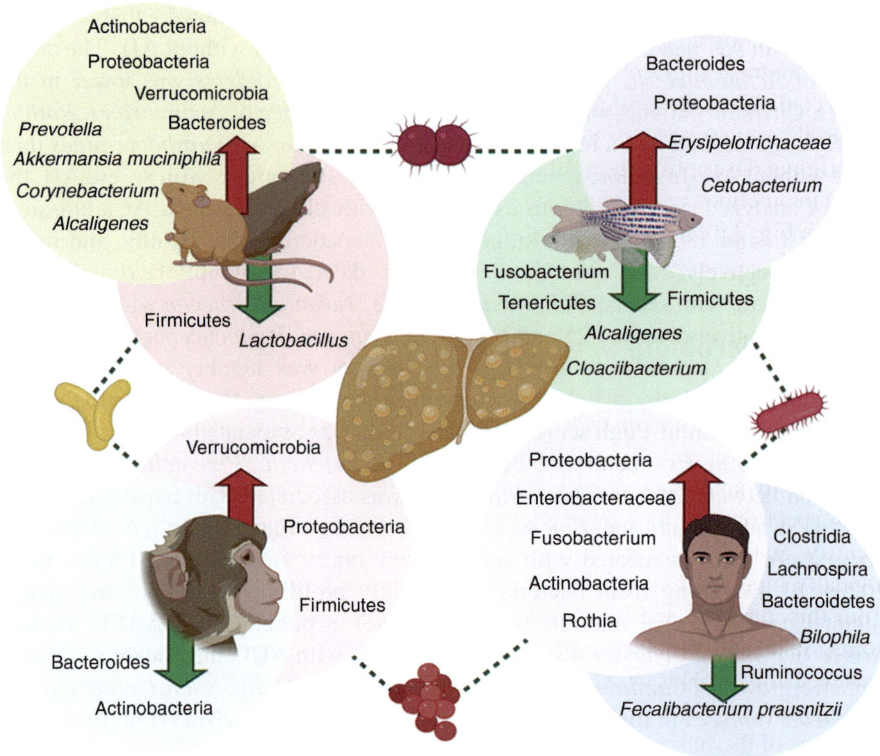

Fig. 6.1 Schematic representation of gut microbiota changes and pertinent similarities at the family, class, and genus levels in rodent, fish, mammal, and human models of alcohol-associated liver disease

liver injury and gut mucosal integrity. Intestinal dysbiosis is also associated with a reduction in de-novo production of bile acids with high serum levels of secondary bile acids (Ciocan et al. 2018). Reduction in *Bacteroidetes* and increase in *Actinobacteria* were associated with upregulation of glutathione metabolism, reduction in biotin metabolism in patients with cirrhosis and AH (Bajaj et al. 2012; Bajaj et al. 2017; Hartmann et al. 2019; Dubinkina et al. 2017). Heavy alcohol drinking was associated with higher plasma levels of threonine, glutamine, guanidinosuccinate, higher urinary 3-hydroxytetradecanedioic acid and isocitric acid levels, and a lower concentration of urinary sebacic acid. Propionate and isobutyrate and microbial decompensation products such as dimethyl di- and trisulfides were found altered in the fecal metabolome of alcohol users with AALD. Apart from the beneficial SCFAs, intestinal metabolites such as caryophyllene, a natural alcohol use suppressant, and camphene, the hepatic steatosis attenuator, were reduced, and tetradecane the oxidative stress biomarker was elevated in patients with AALD (Couch et al. 2015).

Actively drinking cirrhotics had lower stool concentrations of citrate, malate, and phosphate (bioenergetics pathways); threonine, ornithine, and serine (amino acid metabolism); and ribosine, orotic acid, and hexanoate (pyrimidine intermediate metabolism pathway), which were significantly related to increasing total serum bile acids as well as secondary bile acids, the former highest in patients with severe AH. Further, patients with AH also show major changes in metabolites such as higher eicosapentaenoic and docosapentaenoic acids and lower mono glycerol associated with lipolysis oxidative stress (Rachakonda et al. 2014; Liang et al. 2015; Bajaj et al. 2020).

Endogenous tryptophan metabolites in humans (kynurenines, serotonin, and melatonin) and bacterial metabolites (indole, indole derivatives, skatole, and tryptamine) play an important role in the regulation of local (intestinal) and systemic immune homeostasis. Tryptophan is converted into indole and indole derivatives by tryptophanase expressed in many bacterial species, including *Escherichia coli*, *Clostridium* spp., and *Bacteroides* spp. The indole derivatives, such as indole-3-aldehyde, indole-3-acetic acid, indole-3-propionic acid, indole-3-acetaldehyde, and indole acrylic acid through the aryl hydrocarbon receptor (AhR, cytosolic ligand-activated transcription factor), promote xenobiotic metabolism to regulate adaptive immunity and gut barrier function. In a mouse model of alcohol-induced liver disease, dysbiosis reduced the intestinal levels of indole-3-acetic acid that decreased expression of interleukin-22 and Reg3γ in the intestine. Indole-3-acetic acid oral supplementation protected mice from ethanol-induced steatohepatitis via induction of intestinal expression of IL-22 and Reg3γ, which prevented bacterial translocation to the liver.

Furthermore, *Bifidobacterium* and *Lactobacillus* produce indole-3-lactic acid, which induces immunoregulatory T cells, suppresses inflammatory T cells, and prevents LPS-mediated detrimental effects on the liver. Fecal and serum levels of tryptophan and tryptophan-derived metabolites, indole-3-acetic acid, indole-3-propionic acid, and indole-3-lactic acid decreased in AH patients (Mendes and Schnabl 2020). In patients with severe AH, pathways associated with lipopolysaccharide protein biosynthesis, glycosyltransferase activity, branch-chain amino acid degradation, and riboflavin metabolism were significantly upregulated compared to healthy controls in whom downregulation of alanine, aspartate, and glutamate metabolism was more significant (Philips et al. 2018). Stool samples from patients with AH had about 2700-fold more *Enterococcus faecalis* compared to non-alcoholic controls. The exotoxin cytolysin, secreted by *E. faecalis*, was demonstrated to cause hepatocyte injury in mice, and cytolysin-positive/cytolytic *E. faecalis* correlated with liver disease severity and mortality in patients with AH, but not in those with non-alcoholic fatty liver disease (Duan et al. 2019). Therefore, gut microbiota-associated metabolites, in addition to dysbiosis, have significant relevance to progression and clinical events in patients with AALD (Philips et al. 2019a, b). Pertinent bacterial interactions, dysbiotic events, and functional metabolomic changes in patients with AALD are shown in Fig. 6.2.

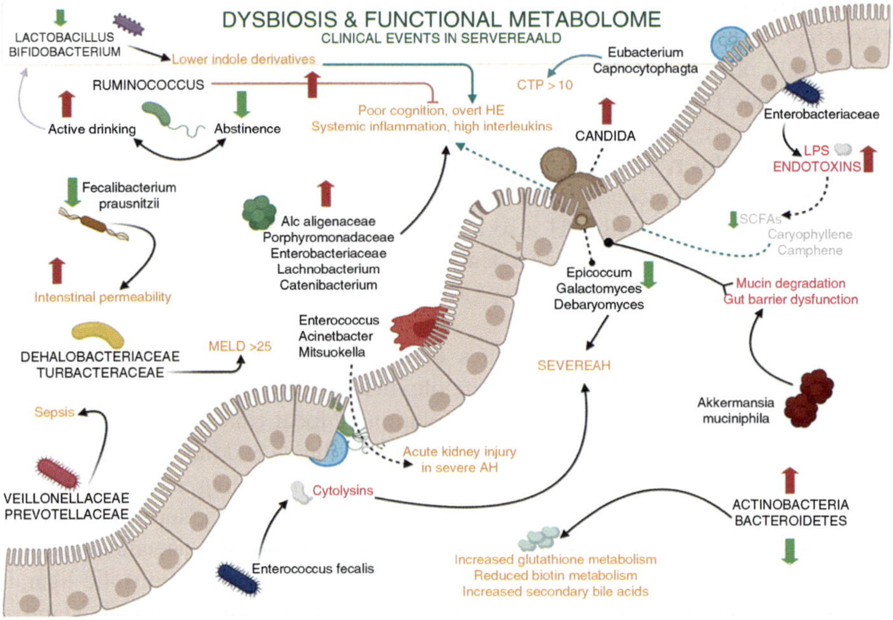

Fig. 6.2 Schematic representation of dysbiotic changes and functional metabolomic differences affecting the gut microbiota in patients with severe alcohol-associated liver disease including severe alcohol-associated hepatitis. Changes in microbiota also affect the functional metabolism within the intestine that results in the disruption of balance between beneficial and damaging metabolites. *AH* alcohol-associated hepatitis, *SCFA* short-chain fatty acid, *LPS* lipopolysaccharide, *CTP* Child-Pugh score, *MELD* model for end-stage liver disease

4 Fecal Microbiota Transplantation for Alcohol-Associated Liver Disease

The term fecal microbiota transplantation (FMT) has seen major changes in terminology over the last year. To improve patient acceptance, industry attentiveness, and grant/funding options, various terminologies have been proposed, including intestinal microbiota transplantation, microbiome restoration therapy, and, recently, the more reasonable, intestinal microbiota reinstitution therapy or IMRT (Philips et al. 2020a, b). Nonetheless, a globally accepted terminology in this regard is currently lacking, and hence the term FMT has stood the test of time.

The role of modulating the microbiota through FMT has been demonstrated initially in rodent models followed by human pilot trials. Germ-free mice transplanted with intestinal microbiota from patients with severe AH developed severe hepatocyte inflammation and necrosis, greater intestinal permeability, and more bacterial translocation to the liver when compared to mice transplanted with intestinal microbiota from an AALD patient without AH. In conventional mice humanized with the intestinal microbiota from a severe AH patient, a second subsequent transfer of intestinal microbiota from patients without AH improved alcohol-induced liver injury (Llopis et al. 2016).

The further significance of gut microbiota in AALD was further demonstrated by FMT between mice housed in different facilities. Two groups of mice were housed in two nearby facilities. The mice in one facility developed AALD (alcohol-sensitive), and the mice in other facilities did not (alcohol-resistant), despite a similar alcohol intake. Alcohol-induced hepatic steatosis and liver inflammation were associated with gut dysbiosis in the alcohol-sensitive mice. Importantly, transplantation of gut microbiota from the resistant mice to the sensitive mice restored gut microbiota 'eubiosis' and prevented the development of AALD. The authors concluded that gut microbiota modulation could prevent alcohol-induced liver injury and has therapeutic potential (Ferrere et al. 2017). Cytolytic *E. faecalis* in fecal samples was linked to more severe clinical outcomes and increased mortality in patients with AH. Bacteriophages of the virulent *Picovirinae* group were demonstrated to specifically target cytolytic *E. faecalis*, resulting in reduced liver injury, a method of precision-editing of intestinal microbiota to improve host-disease outcomes (Duan et al. 2019).

In the first human trial on FMT (Fig. 6.3) in AALD—AH, an open-label study with 1 year follow-up in male patients with steroid ineligible severe alcohol-associated injury, daily FMT from several healthy donors through a nasojejunal tube for 7 days improved survival when compared with historical controls associated with a reduction in potentially pathogenic species and improved liver functions. Phyla *Firmicutes* dominated in donors and recipients at 1 year after FMT. Pathogenic *Proteobacteria* were reduced while *Actinobacteria* increased post-FMT in recipients. The relative abundance of pathogenic species, such as *Klebsiella pneumonia,* was reduced, and nonpathogenic species (*Enterococcus villorum* and *Bifidobacterium longum*—6 months and *Megasphaera elsdenii*—at 12 months) were remarkable after FMT. *Pseudomonas* and *Escherichia coli* mediated methane metabolism and fluorobenzoate acid degradation pathways and

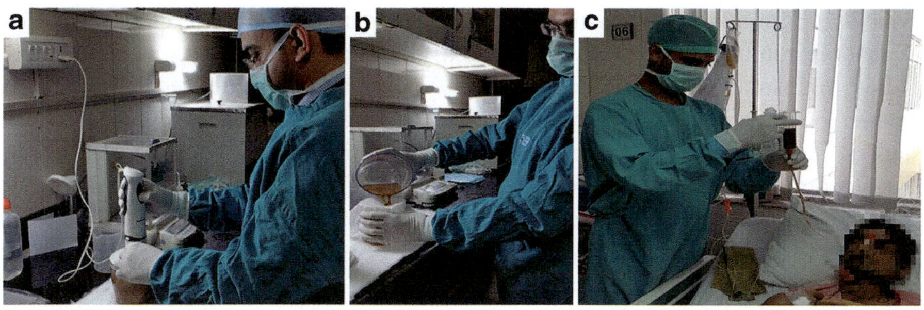

Fig. 6.3 Procedure of fecal microbiota transplantation in a patient with severe alcohol-associated hepatitis. The freshly collected (minimum 30 g) stool samples are blended lightly for 2 minutes (**a**) admixed with 100 ml normal saline and then filtered (**b**) thrice through sterile gauze material to remove solid and vegetable matter. It is then packed into sterile bottles and then infused at bedside through a nasoduodenal tube (**c**) which is placed under fluoroscopic guidance

bacterial invasion of the epithelial cells that were upregulated at baseline ameliorated after FMT at 1 year. The bile secretion, carotenoid biosynthesis, and pantothenate biosynthesis pathways, which were low at the baseline in severe AH patients, improved to near normal levels following FMT (Philips et al. 2017).

The same group conducted an open-label study with 3 months of follow-up in male patients treated with FMT (100 ml daily through the nasojejunal tube for 7 days) compared to pentoxifylline, corticosteroids, and nutritional therapy. They found that 90 days of survival was highest in the FMT group, which was also associated with bacterial communities and functional composition changes. Pathogenic bacteria such as *Bilophila, Citrobacter, Enterobacter*, and *Klebsiella*, which were significant at baseline, were preferentially modulated to beneficial taxa such as *Bacteroides, Parabacteroides*, and *Porphyromonas* at the end of 1 week and *Roseburia* and *Micrococcus* beyond 30 days after FMT. Fecal predictive functional metabolome analysis showed that, at the end of 1 week, glycine, serine, and threonine metabolism and tropane, piperidine, and pyridine alkaloid biosynthesis were significantly upregulated. Beyond 30 days after FMT, the amino acid biosynthesis, nitrogen and thiamine metabolism, and peroxisome and peroxisome proliferator-activated receptor signaling pathways were significantly active while lipopolysaccharide signaling pathways remained dormant (Philips et al. 2018).

In patients with severe AH who develop ACLF, in the presence of <4 organ failures and the Chronic Liver Failure Consortium (CLIF-C) ACLF score < 64, standard medical care was associated with survival in only 39% with 100% mortality in those with >4 OFs or CLIF-C ACLF score > 64 (Gustot et al. 2015). Severe AH-related ACLF is also associated with a poor outcome in the presence of poor response to corticosteroids (based on the Lille model). In Lille-non-responders with ACLF grades 0, 1, and (2 + 3) the 90-day survival rates are 68.1%, 45.8%, and 36.7%. In this regard, a retrospective study on FMT in patients with AH-related ACLF ineligible for steroid therapy demonstrated that at 548 days follow-up, the overall survival rate was 66% with overall mean survival of 389 days. In the lower and higher grades of ACLF, the proportion of patients surviving at the end of 548 days follow-up was 72.7% and 58.3%, respectively, which was higher than what was currently known with the medical standard care (Philips et al. 2019a, b). A randomized trial of FMT versus pentoxifylline in patients with severe AH, published in abstract form, demonstrated 83% survival in patients receiving healthy donor stool transplants compared to 40% in the latter. Clinically evident ascites and hepatic encephalopathy were more among patients receiving pentoxifylline. *Firmicutes* and *Bacteroides* increased post-FMT with a gradual reduction in Proteobacteria. The genus *Bifidobacterium* was most abundant after FMT between 90 and 180 days (Philips et al. 2017). Two recently concluded studies, published in abstract form, from two different centers on FMT's role in severe AALD demonstrated short and intermediate-term benefits with FMT. In the first study from Chandigarh, it was demonstrated that FMT, performed in a single setting with freshly prepared (within 6 hours of collection) stool suspension constituted from 30 g of donor stool homogenized with 100 mL of normal saline delivered through a nasojejunal tube, was feasible, tolerable, and safe in patients with severe AH. FMT was associated with a lower one-month mortality rate and a higher rate of ascites resolution. There was a trend toward improved three-month survival and resolution of hepatic encephalopathy among patients receiving FMT (Dhiman et al. 2020). A randomized controlled trial from New Delhi on severe AH compared patients on FMT with corticosteroid therapy and found that post-FMT, survival was better in those receiving healthy donor stool, and bacterial taxa correlated with three-month survival. *Lactobacillus ruminis*, *Collinsella aerofaciens*, *Butyricicoccus pullicaecorum*, *Bifidobacterium adolescentis*, *Bifidobacterium bifidum*, *Roseburia faecis*, and *Dorea formicigenerans,* which were absent at baseline in patients with severe AH, were introduced post-FMT, taking 14–28 days to establish. In the FMT arm, *Tenericutes* was significantly associated with mortality. Pathogenic genera such as *Klebsiella*, *Salmonella*, and *Mycoplasma* showed a significant increase among those who died. The phylum *Firmicutes*, family *Lachnospiraceae*, and genera *Veillonella* and *Prevotella* were associated with good outcomes. A more recent randomized clinical trial of FMT for AUD showed that the microbial diversity increased with higher *Ruminococcaceae* and other SCFA producing taxa post-FMT but not placebo and that FMT was safe and associated with short-term reduction in alcohol craving and consumption, reduction in AUD-related events over 6 months, in patients with alcohol-associated cirrhosis and alcohol misuse (Bajaj et al. 2020). A summary of human trials on FMT in patients with alcohol-associated liver disease is shown in Table 6.1.

5 Other Therapies of Microbiota Modulation in Alcohol-Associated Liver Disease

Prior to direct gut microbiota modulation using FMT, a large body of evidence hinted on the utility of antibiotics, dietary and prebiotic interventions (mostly in preclinical studies) and probiotics for improving clinical outcomes in alcohol and non-alcohol associated fatty liver disease, chronic liver disease, and cirrhosis. Short courses of antibiotics in patients with AALD have demonstrated improvement in liver disease severity scores and hepatic encephalopathy, which was also proven in small animal models. Nonetheless, the effect of such interventions specific to clinical endpoints in the natural history of AALD including reduction in mortality is not yet demonstrated (Sung et al. 2016; Zhou and Zhong 2017; Gu et al. 2019). The use of non-absorbable antibiotics such as rifaximin was shown to reduce endotoxemia in pre-clinical and clinical studies and improve minimal and overt hepatic encephalopathy. However, strong evidence for its benefit on mortality in AALD is not yet proven (Sarin et al. 2019). Similarly, a variety of probiotic strains such as *Lactobacillus rhamnosus* GG, *Lactobacillus acidophilus*, *Lactobacillus helveticus*, *Bifidobacterium*, VSL#3 (proprietary generic multi-strain product), heat-killed *Lactobacillus brevis* SBC8803, and *Lactobacillus rhamnosus* GG supernatant were demonstrated to show beneficial effects on systemic inflammation, leaky gut,

Table 6.1 Studies (fully published and abstract forms) on healthy donor fecal microbiota transplantation in humans with alcohol-associated liver disease

Study	Methods	Patients	Summary
Philips et al. (2016)	Randomized controlled trial, 100 ml saline blended fresh donor feces via nasoduodenal tube for 7 days ($N = 15$) compared to pentoxifylline 400 mg thrice daily for 28 days ($N = 30$)	Severe alcoholic hepatitis patients not eligible for corticosteroid treatment, survival at 90 days post-treatment	Patients in FMT arm had better 90-day survival. *Bacteroides* increased and *Proteobacteria* reduced postFMT. Ascites and hepatic encephalopathy higher in pentoxifylline arm
Philips et al. (2017)	Open-label pilot study, 100 ml freshly prepared, saline blended stool sample via nasoduodenal tube, once daily for 7 days ($N = 8$)	Severe AH, steroid ineligible compared with matched historical controls, followed up to 1 year	FMT improved patient survival compared to controls on standard care; beneficial bacterial changes and predictive functional metabolism showed deleterious pathways downregulated from baseline. Non-pathogenic *Enterococcus villorum, Bifidobacterium longum, Megasphaera elsdenii* increased post FMT at 6–12 months after FMT
Philips et al. (2018)	Open-label trial, FMT daily for 7 days compared with steroids, pentoxifylline, and nutritional therapy	FMT, $N = 16$; pentoxifylline, $N = 10$; steroids, $N = 8$; nutritional therapy, $N = 17$; end point was survival at 90-day post treatment	Survival highest in FMT group, specific beneficial bacterial taxa such as *Roseburia* and *Micrococcus* increased at 1 month's post FMT; lipopolysaccharide synthesis pathways ameliorated beyond 1 month after FMT
Philips et al. (2019a, b)	Open-label, retrospective analysis of patients with severe AH and ACLF; FMT 100 ml daily once for 7 days through nasoduodenal tube	$N = 23$, 548 days follow-up. The first study to look at the role of FMT in AH-related ACLF	Overall survival rate was 66% with overall mean survival of 389.3 days. The commonest cause of death on follow-up was sepsis in 62.5%. The proportion of patients surviving the end of 548 days follow-up in lower grades of ACLF was 72.7% and in higher grades, 58.3%

Table 6.1 (continued)

Study	Methods	Patients	Summary
Bajaj et al. (2020)	Randomized, double-blind trial, 1:1 allocation, single time FMT enema from a donor enriched in *Lachnospiraceae* and *Ruminococcaceae*	$N = 10$; craving reduced significantly in 90% of FMT versus 30% in placebo at day15, improved cognition and psychosocial quality of life; serum interleukin-6 and lipopolysaccharide binding protein reduced, butyrate/isobutyrate increased. Microbial diversity increased	Alcohol misuse and associated poor quality of life indices reduced in patients with cirrhosis active alcohol consumption after FMT compared to placebo as early as 15 days to 6 months
Dhiman et al. (2020)	Patients with severe AH, underwent pre-FMT 5 days treatment with broad-spectrum antibiotics, 100 ml of freshly prepared stool (in saline) delivered via nasojejunal tube once	$N = 13$ in FMT arm compared with matched controls, $N = 20$ who underwent standard care	1- and 3-month survival rates were higher in patients in FMT arm. Resolution of hepatic encephalopathy and ascites higher in patients receiving FMT. Rates of spontaneous bacterial peritonitis and upper gastrointestinal bleed similar in both groups. Excessive flatulence, gastroesophageal reflux, nausea commonest side effects in FMT arm
Sharma et al. (2020)	Randomized control trial of FMT versus corticosteroids in patients with severe AH	Severe AH patients, DF score > 32 and alcohol intake within last 30 days randomized to receive steroids ($n = 57$) or FMT ($n = 55$)	Healthy donor FMT safe in severe AH patients, establishes beneficial flora within 4 weeks and improves 90-day survival. Among non-survivors pathogenic Tenericutes phyla predominated while phylum Firmicutes, family Lachnospiraceae, and genera Veillonella and Prevotella, favored survival at end of follow-up (90 days)

FMT fecal microbiota transplantation, *AH* alcohol-related hepatitis, *ACLF* acute-on-chronic liver failure, *DF* discriminant function.

endotoxemia, and liver functions in pre-clinical models and patients with alcohol-associated cirrhosis (Li et al. 2016; Hong et al. 2019). However, even though the prevention and treatment of AALD using probiotics is an attractive area of research, further high-quality studies are warranted to identify beneficial strains, combinations, and timing of treatment for reduction in specific adverse clinical events. Studies on specific dietary interventions and prebiotics in the context of human AALD are limited and require further large trials.

6 Conclusions and Future Directions

Intestinal microbiota contributes toward the development and progression of AALD-related clinical events. The role of microbiota and their functional metabolome in promoting severe AH has been well demonstrated in rodent and human studies. Manipulation of a dysbiotic microbiota in healthy donor FMT, toward a beneficial one, to improve clinical outcomes as a proof of concept has been made clear through elegant animal studies and pilot trials in human AH. Nonetheless, the timing of such therapies, the ideal route, the efficacious dose and duration, and the role of repeatability currently remain unknown. Future studies on the human gut microbiome should focus on specific groups of microbial communities and their functional metabolome that drive defined clinical events in patients with AALD, such as steatosis, development, and severe AH progression promote sepsis or extrahepatic organ failures. Our understanding of how microbial communities interact with each other and interactions between different microbial communities such as bacterial and fungi or viruses and phages' role in such interactions remain fetal. The microbial composition and their overall functionality and the beneficial interaction that can be simulated or modulated remain an unmet need in current metagenomic studies. Precision medicine to target the microbiota or modulating the microbiota-derived metabolites toward host benefit and disease amelioration would truly improve on clinical outcomes in this difficult to treat patient group in whom no recommended treatments exist.

References

Abul-Husn NS, Cheng X, Li AH, Xin Y, Schurmann C, Stevis P et al (2018) A protein-truncating HSD17B13 variant and protection from chronic liver disease. N Engl J Med 378(12):1096–1106. https://doi.org/10.1056/NEJMoa1712191

Adolph TE, Grander C, Moschen AR, Tilg H (2018) Liver-microbiome Axis in health and disease. Trends Immunol 39(9):712–723. https://doi.org/10.1016/j.it.2018.05.002

Arab JP, Arrese M, Shah VH (2020) Gut microbiota in non-alcoholic fatty liver disease and alcohol-related liver disease: current concepts and perspectives. Hepatol Res 50(4):407–418. https://doi.org/10.1111/hepr.13473

Atkinson SR, Way MJ, McQuillin A, Morgan MY, Thursz MR (2017) Homozygosity for rs738409: G in PNPLA3 is associated with increased mortality following an episode of severe alcoholic hepatitis. J Hepatol 67(1):120–127. https://doi.org/10.1016/j.jhep.2017.01.018

Avila MA, Dufour JF, Gerbes AL, Zoulim F, Bataller R, Burra P et al (2020) Recent advances in alcohol-related liver disease (ALD): summary of a gut round table meeting. Gut 69(4):764–780. https://doi.org/10.1136/gutjnl-2019-319720

Bajaj JS, Ridlon JM, Hylemon PB, Thacker LR, Heuman DM, Smith S et al (2012) Linkage of gut microbiome with cognition in hepatic encephalopathy. Am J Physiol Gastrointest Liver Physiol 302(1):G168–G175. https://doi.org/10.1152/ajpgi.00190.2011

Bajaj JS, Kakiyama G, Zhao D, Takei H, Fagan A, Hylemon P et al (2017) Continued alcohol misuse in human cirrhosis is associated with an impaired gut-liver Axis. Alcohol Clin Exp Res 41(11):1857–1865. https://doi.org/10.1111/acer.13498

Bajaj JS (2019) Alcohol, liver disease and the gut microbiota. Nat Rev Gastroenterol Hepatol 16(4):235–246. https://doi.org/10.1038/s41575-018-0099-1

Bajaj JS, Gavis EA, Fagan A, Wade JB, Thacker LR, Fuchs M et al (2020) A randomized clinical trial of fecal microbiota transplant for alcohol use disorder. Hepatology Epub ahead of print. https://doi.org/10.1002/hep.31496

Bull-Otterson L, Feng W, Kirpich I, Wang Y, Qin X, Liu Y et al (2013) Metagenomic analyses of alcohol induced pathogenic alterations in the intestinal microbiome and the effect of *lactobacillus rhamnosus* GG treatment. PLoS One 8(1):e53028. https://doi.org/10.1371/journal.pone.0053028

Bluemel S, Wang L, Kuelbs C, Moncera K, Torralba M, Singh H et al (2019) Intestinal and hepatic microbiota changes associated with chronic ethanol administration in mice. Gut Microbes 11(3):265–275. https://doi.org/10.1080/19490976.2019.1595300

Canesso MCC, Lacerda NL, Ferreira CM, Gonçalves JL, Almeida D, Gamba C et al (2014) Comparing the effects of acute alcohol consumption in germ-free and conventional mice: the role of the gut microbiota. BMC Microbiol 14:240. https://doi.org/10.1186/s12866-014-0240-4

Chen P, Torralba M, Tan J, Embree M, Zengler K, Stärkel P et al (2015b) Supplementation of saturated long-chain fatty acids maintains intestinal eubiosis and reduces ethanol-induced liver injury in mice. Gastroenterology 148(1):203–214.e16. https://doi.org/10.1053/j.gastro.2014.09.014

Chen P, Miyamoto Y, Mazagova M, Lee KC, Eckmann L, Schnabl B et al (2015a) Microbiota protects mice against acute alcohol-induced liver injury. Alcohol Clin Exp Res 39(12):2313–2323. https://doi.org/10.1111/acer.12900

Chen Y, Yang F, Lu H, Wang B, Chen Y, Lei D et al (2011) Characterization of fecal microbial communities in patients with liver cirrhosis. Hepatology 54(2):562–572. https://doi.org/10.1002/hep.24423

Ciocan D, Voican CS, Wrzosek L, Hugot C, Rainteau D, Humbert L et al (2018) Bile acid homeostasis and intestinal dysbiosis in alcoholic hepatitis. Aliment Pharmacol Ther 48(9):961–974. https://doi.org/10.1111/apt.14949

Couch RD, Dailey A, Zaidi F, Navarro K, Forsyth CB, Mutlu E et al (2015) Alcohol induced alterations to the human fecal VOC metabolome. PLoS One 10(3):e0119362. https://doi.org/10.1371/journal.pone.0119362

Crabb DW, Im GY, Szabo G, Mellinger JL, Lucey MR (2020) Diagnosis and treatment of alcohol-associated liver diseases: 2019 practice guidance from the American Association for the Study of Liver Diseases. Hepatology 71(1):306–333. https://doi.org/10.1002/hep.30866

Dhiman RK, Sharma A, Roy A, Premkumar M, Taneja S, Kumar A et al (2020) Role of fecal microbiota transplantation in severe alcoholic hepatitis: assessment of impact on prognosis and short-term outcomes. J Hepatol 73:S179

Duan Y, Llorente C, Lang S, Brandl K, Chu H, Jiang L et al (2019) Bacteriophage targeting of gut bacterium attenuates alcoholic liver disease. Nature 575(7783):505–511. https://doi.org/10.1038/s41586-019-1742-x

Dubinkina VB, Tyakht AV, Odintsova VY, Yarygin KS, Kovarsky BA, Pavlenko AV et al (2017) Links of gut microbiota composition with alcohol dependence syndrome and alcoholic liver disease. Microbiome 5(1):141. https://doi.org/10.1186/s40168-017-0359-2

Ferrere G, Wrzosek L, Cailleux F, Turpin W, Puchois V, Spatz M et al (2017) Fecal microbiota manipulation prevents dysbiosis and alcohol-induced liver injury in mice. J Hepatol 66(4):806–815. https://doi.org/10.1016/j.jhep.2016.11.008

Fujisawa K, Takami T, Nagatomo T, Fukui Y, Hoshida H, Saeki I et al (2019) Usefulness of adult medaka fish as a model for the evaluation of alcoholic fatty liver. Alcohol 77:147–154. https://doi.org/10.1016/j.alcohol.2019.01.005

GBD (2016) Alcohol collaborators (2018) Alcohol use and burden for 195 countries and territories, 1990–2016: a systematic analysis for the global burden of disease study 2016. Lancet 392(10152):1015–1035. https://doi.org/10.1016/S0140-6736(18)31310-2

Gu Z, Liu Y, Hu S, You Y, Wen J, Li W, Wang Y (2019) Probiotics for alleviating alcoholic liver injury. Gastroenterol Res Pract 27(2019):9097276. https://doi.org/10.1155/2019/9097276
Gustot T, Fernandez J, Garcia E, Morando F, Caraceni P, Alessandria C et al (2015) CANONIC study investigators of the EASL-CLIF consortium. Clinical course of acute-on-chronic liver failure syndrome and effects on prognosis. Hepatology 62(1):243–252. https://doi.org/10.1002/hep.27849
Hartmann P, Chu H, Duan Y, Schnabl B. (2019) Gut microbiota in liver disease: too much is harmful, nothing at all is not helpful either. Am J Physiol Gastrointest Liver Physiol. 316(5):563–573. https://doi.org/10.1152/ajpgi.00370.2018
Ho SB, Bataller R, Stärkel P, Fouts DE, Schnabl B (2017) Intestinal fungi contribute to development of alcoholic liver disease. J Clin Invest 127(7):2829–2841. https://doi.org/10.1172/JCI90562
Hong M, Han DH, Hong J, Kim DJ, Suk KT (2019) Are probiotics effective in targeting alcoholic liver diseases? Probiotics Antimicrob Proteins 11(2):335–347. https://doi.org/10.1007/s12602-018-9419-6
Kakiyama G, Pandak WM, Gillevet PM, Hylemon PB, Heuman DM, Daita K et al (2013) Modulation of the fecal bile acid profile by gut microbiota in cirrhosis. J Hepatol 58(5):949–955. https://doi.org/10.1016/j.jhep.2013.01.003
Kolla BP, Schneekloth TD, Biernacka J, Shah V, Lazaridis KN, Geske J et al (2018) PNPLA3 association with alcoholic liver disease in a cohort of heavy drinkers. Alcohol Alcohol 53(4):357–360. https://doi.org/10.1093/alcalc/agy007
Leclercq S, Matamoros S, Cani PD, Neyrinck AM, Jamar F, Stärkel P et al (2014) Proc Natl Acad Sci U S A 111(42):E4485–E4493. https://doi.org/10.1073/pnas.1415174111
Li F, Duan K, Wang C, McClain C, Feng W (2016) Probiotics and alcoholic liver disease: treatment and potential mechanisms. Gastroenterol Res Pract 2016:5491465. https://doi.org/10.1155/2016/5491465
Liang Q, Wang C, Li B, Zhang A (2015) Metabolomics of alcoholic liver disease: a clinical discovery study. RSC Adv 5:80381–80387
Llopis M, Cassard AM, Wrzosek L, Boschat L, Bruneau A, Ferrere G et al (2016) Intestinal microbiota contributes to individual susceptibility to alcoholic liver disease. Gut 65(5):830–839. https://doi.org/10.1136/gutjnl-2015-310585
Mendes BG, Schnabl B (2020) From intestinal dysbiosis to alcohol-associated liver disease. Clin Mol Hepatol 26(4):595–605. https://doi.org/10.3350/cmh.2020.0086
Mutlu EA, Gillevet PM, Rangwala H, Sikaroodi M, Naqvi A, Engen PA et al (2012) Colonic microbiome is altered in alcoholism. Am J Physiol Gastrointest Liver Physiol 302(9):G966–G978. https://doi.org/10.1152/ajpgi.00380.2011
Philips CA, Shasthry SM, Pande A, Jamwal KD, Khillan V, Hussain MS et al (2016) Outcomes of fecal microbiota transplantation in steroid ineligible severe alcoholic hepatitis-a randomized control trial (NCT02458079). Indian J Gastroenterol 35:1–111. https://doi.org/10.1007/s12664-016-0715-3
Philips CA, Pande A, Shasthry SM, Jamwal KD, Khillan V, Chandel SS et al (2017) Healthy donor fecal microbiota transplantation in steroid-ineligible severe alcoholic hepatitis: a pilot study. Clin Gastroenterol Hepatol 15(4):600–602. https://doi.org/10.1016/j.cgh.2016.10.029
Philips CA, Augustine P, Padsalgi G, Ahamed R, Jose A, Rajesh S (2019a) Only in the darkness can you see the stars: severe alcoholic hepatitis and higher grades of acute-on-chronic liver failure. J Hepatol 70(3):550–551. https://doi.org/10.1016/j.jhep.2018.10.004
Philips CA, Phadke N, Ganesan K, Ranade S, Augustine P (2018) Corticosteroids, nutrition, pentoxifylline, or fecal microbiota transplantation for severe alcoholic hepatitis. Indian J Gastroenterol 37(3):215–225. https://doi.org/10.1007/s12664-018-0859-4
Philips CA, Augustine P, Yerol PK, Rajesh S, Mahadevan P (2019b) Severe alcoholic hepatitis: current perspectives. Hepat Med 11:97–108. https://doi.org/10.2147/HMER.S197933

Philips CA, Ahamed R, Rajesh S, Augustine P (2020a) 'You know my name, but not my story'–deciding on an accurate nomenclature for fecal microbiota transplantation. J Hepatol 72(6):1212–1213. https://doi.org/10.1016/j.jhep.2020.02.004

Philips C, Ganesan K, Ranade S, Chopra V, Patil K, Shende S et al (2020b) The role of gut microbiota in clinical complications and treatment response in alcoholic hepatitis–a CIRCOS®, linear discriminant analysis effect size biomarker and CONET® co-occurrence network analysis (abstract). Gut 69:A23–A24

Puri P, Liangpunsakul S, Christensen JE, Shah VH, Kamath PS, Gores GJ et al (2018) TREAT consortium. The circulating microbiome signature and inferred functional metagenomics in alcoholic hepatitis. Hepatology 67(4):1284–1302. https://doi.org/10.1002/hep.29623

Rachakonda V, Gabbert C, Raina A, Bell LN, Cooper S, Malik S et al (2014) Serum metabolomic profiling in acute alcoholic hepatitis identifies multiple dysregulated pathways. PLoS One 9(12):e113860. https://doi.org/10.1371/journal.pone.0113860

Sarin SK, Pande A, Schnabl B (2019) Microbiome as a therapeutic target in alcohol-related liver disease. J Hepatol 70(2):260–272. https://doi.org/10.1016/j.jhep.2018.10.019

Schwantes-An TH, Darlay R, Mathurin P, Masson S, Liangpunsakul S, Mueller S et al (2020) Genome-wide association study and meta-analysis on alcohol-related liver cirrhosis identifies novel genetic risk factors. Hepatology. https://doi.org/10.1002/hep.31535

Sharma S, Pande A, Khillan V, Rastogi A, Arora V, Shasthry SM et al (2020) Post-fecal microbiota transplant taxa correlate with 3-month survival in severe alcoholic hepatitis patients. J Hepatol 73:S137

Shield K, Manthey J, Rylett M, Probst C, Wettlaufer A, Parry CDH et al (2020) National, regional, and global burdens of disease from 2000 to 2016 attributable to alcohol use: a comparative risk assessment study. Lancet Public Health 5(1):e51–e61. https://doi.org/10.1016/S2468-2667(19)30231-2

Stickel F, Hampe J (2012) Genetic determinants of alcoholic liver disease. Gut 61(1):150–159. https://doi.org/10.1136/gutjnl-2011-301239

Sung H, Kim SW, Hong M, Suk KT (2016) Microbiota-based treatments in alcoholic liver disease. World J Gastroenterol 22(29):6673–6682. https://doi.org/10.3748/wjg.v22.i29.6673

Tuomisto S, Pessi T, Collin P, Vuento R, Aittoniemi J, Karhunen PJ (2014) Changes in gut bacterial populations and their translocation into liver and ascites in alcoholic liver cirrhotics. BMC Gastroenterol 14:40. https://doi.org/10.1186/1471-230X-14-40

Wang H, Yan Y, Yi X, Duan Y, Wang J, Li S et al (2019) Histopathological features and composition of gut microbiota in rhesus monkey of alcoholic liver disease. Front Microbiol 10:165. https://doi.org/10.3389/fmicb.2019.00165

Wang Y, Kirpich I, Liu Y, Ma Z, Barve S, McClain CJ et al (2011) *Lactobacillus rhamnosus* GG treatment potentiates intestinal hypoxia-inducible factor, promotes intestinal integrity and ameliorates alcohol-induced liver injury. Am J Pathol 179(6):2866–2875. https://doi.org/10.1016/j.ajpath.2011.08.039

Yan AW, Fouts DE, Brandl J, Stärkel P, Torralba M, Schott E et al (2011) Enteric dysbiosis associated with a mouse model of alcoholic liver disease. Hepatology 53(1):96–105. https://doi.org/10.1002/hep.24018

Yang AM, Inamine T, Hochrath K, Chen P, Wang L, Llorente C et al (2017) Intestinal fungi contribute to development of alcoholic liver disease. J Clin Invest 127(7):2829–2841. https://doi.org/10.1172/JCI90562

Zhou Z, Zhong W (2017) Targeting the gut barrier for the treatment of alcoholic liver disease. Liver Res 1(4):197–207. https://doi.org/10.1016/j.livres.2017.12.004

Chapter 7
Microbiota-Gut-Brain Axis in Neurological Disorders

Soumya Sundaram, Dinoop Korol Ponnambath, and Sruthi S. Nair

1 Introduction

The human body accommodates a multitude of microorganisms that inhabit various anatomical sites, such as skin, mucosa, gastrointestinal tract, respiratory tract, urogenital tract and mammary glands, and are collectively defined as the microbiome, the composition and functions of which are crucial to health and survival (Moos et al. 2016). Among the different organ systems that coordinate the functions of the human body, central nervous system comprising the brain and spinal cord, plays a primary role in controlling awareness, movements, sensations, thoughts, speech and memory by integrating sensory information and responding accordingly. Furthermore, the microbiome engages in complex interactions with the organ systems of the human body thereby regulating the functions of both the entities. Such relationships between the central nervous system (CNS) and the gut microbiome have been explored in detail and is termed as microbiome-gut-brain (MGB) axis, a complex bidirectional inter-communication that exists between the gut microbiome and the crucial areas of the CNS (Malan Muller et al. 2018; Cryan 2019). Various neuroactive compounds such as neurotransmitters, metabolites, cytokines, and hormones are synthesised by the gut microbiota and the host as a result of this interaction. These neuromodulatory substances gain access to the brain by different pathways, thus affecting the local homeostasis. Dysregulation of the MGB axis has been implicated in the pathogenesis of various

S. Sundaram (✉) · S. S. Nair
Department of Neurology, Sree Chitra Tirunal Institute for Medical Sciences and Technology, Trivandrum, Kerala, India
e-mail: drsoumya@sctimst.ac.in

D. K. Ponnambath
Department of Microbiology, Sree Chitra Tirunal Institute for Medical Sciences and Technology, Trivandrum, Kerala, India

S. Thomas (ed.), *Human Microbiome*,
https://doi.org/10.1007/978-981-16-7672-7_7

neuroinflammatory, neurodevelopmental and neurodegenerative diseases which is mediated by either a direct line of communication through vagus nerve or immune system activation or both (Cryan 2019). In turn, gut microbiota composition is influenced by the brain in response to stress and endocrine factors (Tremlett et al. 2017).

The specific aetiology and trigger for many of the neurodevelopmental and neurodegenerative diseases remain elusive to the scientific community. An intricate interplay of genetic and environmental factors is attributed to their pathogenesis (Tremlett et al. 2017). There is accumulating evidence suggesting a definitive role of human gut microbiome in the pathogenesis of Parkinson's disease (PD), multiple sclerosis (MS) and autism spectrum disorders (ASD) whereas animal and human studies are ongoing to elucidate the role of microbiome in Alzheimer disease (AD), stroke, traumatic brain injury, amyotrophic lateral sclerosis and glioma (Cryan 2019). Prebiotics, probiotics, diet modifications, aerobic exercises and microbial transplantation are the few interventions employed for targeting gut dysbiosis in neurological disorders, subject to critical screening and evaluation (Tremlett et al. 2017; Kang et al. 2017; Gubert et al. 2018; Yang et al. 2019).

2 Microbiome-Gut-Brain Axis

The bidirectional interactions in the MGB axis are perpetuated through direct and indirect pathways. These pathways are mediated by the following components, (i) endocrine (hypothalamic-pituitary-adrenal [HPA] axis); (ii) immune-reactive molecules (cytokines and chemokines); (iii) metabolic pathways; (iv) limbic system; and (v) neural system (afferent and efferent pathways) (Malan-Muller et al. 2018). The communication is also dependent on the permeability of the blood-brain barrier (BBB) and the intestinal epithelial barrier (IEB) (Lyte and Cryan 2014).

2.1 Gut Microbiota

The microbiota colonising the human gastrointestinal tract is a collection of bacteria, archaea and eukaryotes which have co-evolved in a mutual relationship over the years. An estimated 10^{14} organisms inhabit the gastrointestinal tract which supersedes the number of human cells in the body (Cani 2018). Recent genomic studies suggest that the ratio of bacterial to human cells is 1.3:1 (Gill et al. 2006; Sender et al. 2016). The inherited human genome is non-modifiable, while the microbiome composition is highly dynamic and diverse at various stages of life, with the changes being driven by extraneous factors like diet, physical activity and stress. Earlier studies had classified the gut microbiota into 3 enterotypes—*Bacteriodes* (Enterotype 1), *Prevotella* (Enterotype 2) and *Ruminococcus* (Enterotype 3) which is currently considered obsolete (MetaHIT Consortium et al. 2011). The gut microbiota is primarily composed of 2 major phyla, Bacteroidetes (largely composed of

Bacteroides and *Prevotella* species) and Firmicutes (comprised of *Clostridium*, *Lactobacillus* and *Ruminococcus* species). Bacteria in the human gut belong to either of these phyla, although a few gut pathogens from phylum Proteobacteria are also present (Gill et al. 2006).

2.2 Bi-directional Pathways between Gut and Brain

The bidirectional interaction pathways existing between the CNS and the gut microbiome have also been termed and referred to as microbial endocrinology (Lyte and Cryan 2014). Many neuroactive compounds (gamma-aminobutyric acid [GABA], acetylcholine, serotonin, dopamine and norepinephrine) are produced by both the host and the microbiota. Food consumed by humans contains pre-existing functional neuroactive components as well as substrates required to produce these compounds. Gut microbiota can directly produce neuroactive compounds from these substrates (Wall et al. 2014; Strandwitz 2018).

Neuroactive compounds produced in the gut are known to influence the host behavioural functions via two pathways. These compounds are either taken up from the gut into the portal circulation and then to systemic circulation, ultimately reaching the brain or they directly interact with the G protein-coupled receptors (GPCR-43)/neurokinin (NK)1R or NK2R receptors expressed on the enteric nervous system (ENS) which innervates the gastrointestinal tract (Aresti Sanz and El Aidy 2019). The brain functions influenced by these pathways can result in alteration of behaviour or cognition and also determine the food preferences and appetite. The brain also regulates the composition of the microbiota through the specific release of certain neurochemicals like substance P, calcitonin gene-related peptide, neuropeptide Y, somatostatin and vasoactive intestinal polypeptide into the gut lumen. This leads to the bidirectional vicious axis between the CNS and the gut microbiome (Fig. 7.1) (Holzer and Farzi 2014; Lyte and Cryan 2014).

2.3 Role of Neurotransmitters in MGB Axis Homeostasis

Numerous neurotransmitters secreted by a wide spectrum of microbial species have been identified in the human gut which can impact the CNS. These neurotransmitters can potentially enter the brain from gut through portal circulation, but is limited by the intact BBB. Neurotransmitters like serotonin, dopamine, GABA and norepinephrine cannot typically breach the BBB. Yet, their precursors have the capability to cross the BBB and subsequently get converted to active neurotransmitters. For example, tryptophan is the serotonin precursor and tryptophan concentration in the blood plasma has been shown to correlate with brain serotonin levels (Strandwitz 2018). The alternate hypothesis is that the microorganism-derived neurotransmitters affect the brain through the vagus nerve and its afferent neurons. Vagus nerve

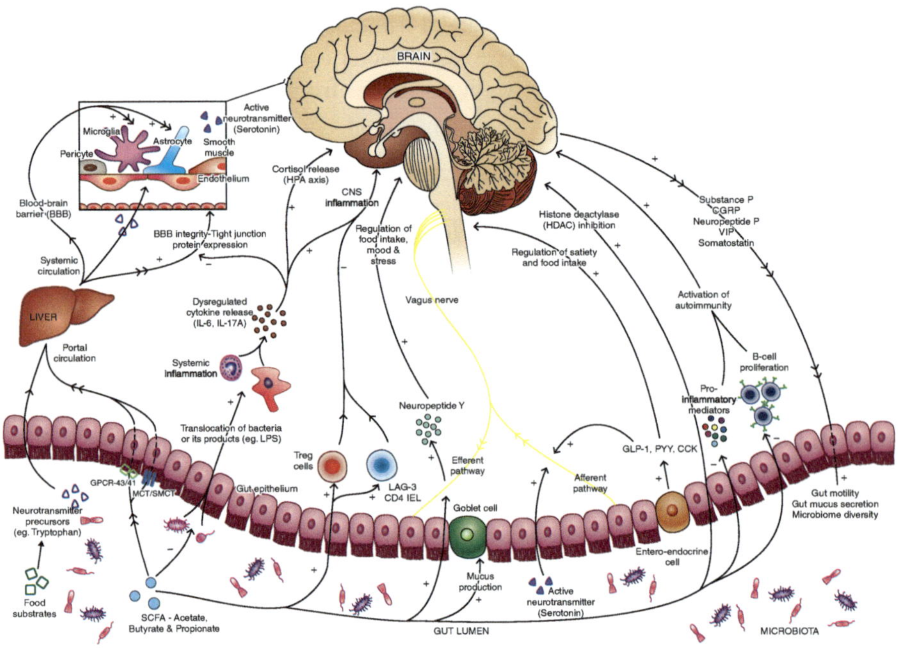

Fig. 7.1 Microbiome-gut-brain (MGB) axis. *BBB* blood-brain barrier, *CCK* cholecystokinin, *CD4* cluster differentiation-4, *CGRP* calcitonin gene-related peptide, *CNS* central nervous system, *GLP-1* glucagon-like peptide-1, *GPCR43/41* G protein-coupled receptor 43/41, *HDAC* histone deacetylase, *HPA* hypothalamic-pituitary-adrenal axis, *IEL* intra-epithelial lymphocyte, *IL* interleukin, *LAG-3* lymphocyte activation gene-3, *LPS* lipopolysaccharide, *MCT* monocarboxylate transporter, *PYY* peptide YY, *SCFA* short-chain fatty acid, *SMCT* sodium-coupled monocarboxylate transporter, *Treg* regulatory T cell, *VIP* vasoactive intestinal peptide

signalling plays a crucial role in mediating stress, mood and satiety (Browning et al. 2017).

2.4 Role of Gut Neuroactive Metabolites in MGB Axis

Among the various bacterial metabolic by-products, short-chain fatty acids (SCFAs) have been identified as the key mediators of the brain-gut interaction. SCFAs are composed of up to 6 carbon atoms and are produced by the anaerobic fermentation of dietary fibres and resistant starches by the large intestine microbiota (Pascale et al. 2018). The predominant SCFAs produced are acetate (60%), propionate (20%) and butyrate (20%). They are absorbed by colonocytes via H^+ or sodium-dependent monocarboxylate transporters into the portal circulation and are metabolised in the liver (Vijay and Morris 2014). Butyrate is primarily consumed as a preferred fuel source by colonocytes. A minor fraction reaches the systemic circulation and crosses the BBB thereby impacting CNS functions (Silva et al. 2020).

SCFAs play a key role in maintaining the gut-brain axis which includes preserving the integrity of colonocytes and IEB, preventing the translocation of bacteria or their toxic products, maintaining colonic mucus production, regulation of gut mucosal immunity and systemic inflammation via inhibition of histone deacetylation and T regulatory cells (Tregs) differentiation and regulation of appetite, sleep and whole-body homeostasis (Arpaia et al. 2013; Smith et al. 2013). Histone deacetylase (HDAC) inhibitors like butyrate and other SCFAs are emerging as promising therapeutic options due to their effects on transcription activation and gene regulation in the CNS (Dietz and Casaccia 2010).

2.5 *Role of Gut Microbiota in Modulation of Microglia and Astrocyte Biology*

Early microbiome studies to determine the influence of gut microbiota on CNS modulation was performed on germ-free (GF) or gnotobiotic animals. These gnotobiotic animals have been a powerful tool to determine causation and to study the effect of a particular bacteria or a dietary intervention on the MGB axis. A few animal studies have shown the role of gut microbiota on biological changes in microglia and astrocyte (Erny et al. 2015; Fung et al. 2017). Microglia constitute 10–15% of all cells in the CNS and as scavenger cells of brain, remove apoptotic nerve cells, kill invasive pathogens and trim redundant synaptic connections (Nayak et al. 2014). Microglial maturation and function are affected by the gut microbiota. Compared to conventionally colonised (specific pathogen-free, SPF) controls, GF mice have microglia with abnormal morphology, altered gene expression and impaired functional response to stimulation. The microglial changes were observed to revert to normal morphology once the diet for GF mice was supplemented with SCFAs (Erny et al. 2015). Similar changes in astrocyte morphology and functions due to the gut microbiota have been reported. Astrocytes are the major glial cells in the CNS that provide physical and metabolic support to the neurons. They take part in synaptogenesis and maintain the integrity of BBB by supporting the lining endothelial cells. Animal studies have shown that microbial metabolites bind to the aryl hydrocarbon receptor (AHR) in astrocytes, reducing symptoms of experimental autoimmune encephalitis (EAE) by inhibiting type I interferon signalling. This effect was reversed when mice were treated with antibiotics. Microbiota-dependent metabolism of tryptophan into AHR ligands engages AHR on astrocytes, thus leading to an increase in astrocyte expression of the inhibitor protein SOCS2 and consequently inhibiting activation of the transcription factor NF-κB and thereby limiting CNS inflammation (Rothhammer et al. 2016).

2.6 Role of Immune System in MGB Axis

Inflammation in the gut has been shown to mirror systemic and CNS inflammation. Peripheral gut insults like antibiotic use and unbalanced diet disrupt the IEB leading to translocation of bacteria and bacterial products like lipopolysaccharide (LPS) into the bloodstream creating a systemic inflammatory response by cytokine release. Dysregulated cytokine release can impact the permeability of BBB, thereby allowing entry of preformed neurotransmitters and pro-inflammatory cells and molecules to initiate a CNS inflammatory milieu (Pan et al. 2011).

Several studies suggest that local and systemic immune activation has critical effects on neural plasticity and behavioural effects, thereby confirming the contribution of immune system in the MGB axis homeostasis (Blander et al. 2017). Recognition of microbe-associated molecular patterns (MAMPs) in microbes and their metabolic by-products (SCFAs) can activate distinct immunological pathways for maintaining the CNS homeostasis (Thursby and Juge 2017). The mode of delivery (caesarean section) and antibiotic use have been linked with immune activation phenomenon in the newborn due to gut dysbiosis (Moos et al. 2016). These early gut dysbiotic changes have been positively correlated with defective neuronal plasticity and behavioural development, classically observed in ASD (Dzidic et al. 2018). The gut colonisers, Lactobacillus and Bifidobacterium, produce the neurotransmitter, acetylcholine, modulating the vagal nerve signalling, thereby attenuating the release of harmful cytokines (IL-1β, IL-6 and TNF-α) (El Aidy et al. 2014). A potential mechanism by which immune activation affects physiology and behaviour is via actions on the serotonergic system. As observed in mice studies by Lowry et al., serotonergic neurons in the dorsal raphe regulate mood and behaviour in response to the immune stimuli. Vagal afferent nerve fibres are also involved in transferring signals from peripheral immune activation to the CNS (Lowry et al. 2007). In addition, microbial infections can trigger antibody-mediated CNS autoimmunity and consequent neurodegeneration (Wu and Wu 2012). Altogether, peripheral immune response regulation by the gut microbiota and exogenous microbial challenges are critical in moulding CNS function and behaviour.

2.7 Composition and Correlation of Dysbiosis with Brain Behaviour

Increase in the members of the phylum Firmicutes or decrease in the members of phylum Bacteroidetes or their subclasses (altered Bacteroidetes/Firmicutes ratio) has been consistently observed in individuals with CNS disorders. In addition to the ratio, relative abundance or deficiency of certain genera in gut microbiota have been strongly linked with CNS disorders (Gill et al. 2006). Colonic bacterial spp. *Faecalibacterium prausnitzii* (Clostridial cluster IV), *Coprococcus*, *Roseburia* (Clostridial cluster XIV), *Eubacterium* and *Anaerostipes* have been identified to produce high concentrations of butyrate, an HDAC inhibitor. Butyrate-producing

Faecalibacterium and *Coprococcus* were consistently associated with high mental quality-of-life (QOL) indicators (Cheung et al. 2019). The genera *Akkermansia*, *Oscillospira* and *Lactococcus* were found to be relatively abundant in individuals with higher sociability, less anxiety or depression (Johnson 2020). A relative abundance of *Desulfovibrio* and *Sutterella* was associated with less sociable behaviour and autism (Wang et al. 2011; De Angelis et al. 2013).

3 Impact of MGB Axis in Neurological Disorders

The dysregulation of gut microbiome composition is attributed to the pathomechanism of several neurological disorders (Fig. 7.2). In the next section, we will discuss the role of gut microbiome in a few neurological conditions based on the current evidence obtained from animal and human studies.

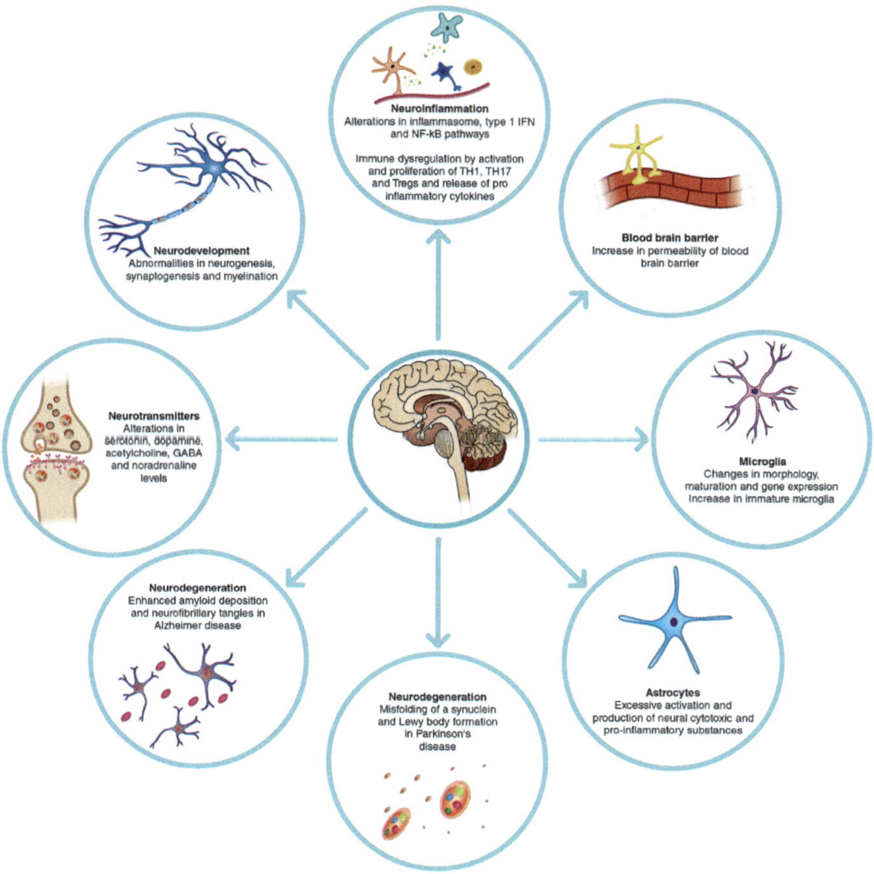

Fig. 7.2 Pathophysiological mechanisms in the central nervous system due to gut dysbiosis. *GABA* Gamma aminobutyric acid; *IFN* interferons; *NF-kB* nuclear factor kappa light chain enhancer of activated B cells; *Tregs* regulatory T cells; *TH1* T helper type 1 cells; *TH17* T helper 17 cells

3.1 Multiple Sclerosis

Multiple sclerosis (MS) is a chronic inflammatory and degenerative disease of the CNS characterised by the presence of widespread 'plaques' in the brain, spinal cord and optic nerve. MS plaques are pathologically heterogeneous revealing combinations of demyelination, inflammation, gliosis, axonal loss and remyelination (Reich et al. 2018). The commonest clinical phenotype of the disease at onset is the relapsing-remitting MS (RRMS) which occurs in more than 85%, while the remaining have progressive onset MS. The therapeutic approach is centred around immunomodulatory disease-modifying therapies (DMTs) which reduce the relapses but have unsatisfactory effect on the disease progression and long-term disability (Rae-Grant et al. 2018).

Immune dysregulation and pro-inflammatory immune shift are the hallmarks of MS, and active research is ongoing to identify the trigger for this. The pathogenesis is thought to be multifactorial with environmental influence prevailing over genetic factors. Evidence for the role of gut dysbiosis in MS susceptibility, pathogenesis, and disease course is compelling. Studies in animal models and persons with MS show alteration of gut microbiota which contribute to the change in immune homeostasis. CNS inflammation simulating MS can be induced in experimental animals by priming and activating the immune system with injection of myelin peptides and this is referred to as the EAE model. EAE can also be spontaneously generated in transgenic mice which express myelin-specific immune receptors (Glatigny and Bettelli 2018). Gut microbiota is essential for EAE induction as was evidenced by low frequency and reduced severity of EAE in GF mice. These mice had concomitant decreased pro-inflammatory TH17 differentiation, low levels of pro-inflammatory cytokines and enhanced Tregs. EAE could be re-established in them by allowing gut colonisation with segmented filamentous bacteria (Ivanov et al. 2009). Transfer of gut microbiota from MS patients could also induce EAE. In a study of 34 sets of monozygotic twins who were discordant for MS, transfer of gut microbiota from an affected twin induced spontaneous EAE more frequently in GF mice compared to the healthy twin (Berer et al. 2017). Conversely, administration of normal gut commensal bacteria, particularly those belonging to phyla Bacteroidetes and Firmicutes, was shown to reduce the incidence of EAE and the inflammatory response in susceptible models (Mowry and Glenn 2018).

In human studies, no difference was noted in the overall diversity and abundance of bacteria in MS and non-MS guts (Mowry and Glenn 2018; Mirza et al. 2020). The composition and relative richness of bacterial groups at sub-phylum level in MS were different from controls across multiple studies (Mirza et al. 2020). Members of phylum Bacteroidetes (*Bacteroidaceae, Bacteroides* and *Prevotella*) and phylum Firmicutes (*Fecalibacterium*) were reduced in MS gut (Cantarel et al. 2015; Miyake et al. 2015; Chen et al. 2016). These bacteria have key role in the generation of SCFAs which are anti-inflammatory. Studies also showed an increased abundance of methanogenic bacterial genera, *Methanobrevibacter* and *Akkermensia* (Tremlett et al. 2016; Jangi et al. 2016). They have pro-inflammatory effects related to respectively recruiting and activating immune cells and altering genes involved in immune responses. However, methanogenic bacteria may be increased secondary to constipation in MS and hence their abundance may result from MS than cause it. A recent study noted that the profile of gut microbiota in treatment-naïve MS varies across different ethnic groups with relative abundance of *Clostridia* being a common observation in all the groups (Ventura et al. 2019). Exposure to DMT can result in significant alterations in the beta diversity of gut microbiota (Tremlett et al. 2016; Mirza et al. 2020).

Studies in paediatric MS are of particular interest as children have fewer exposure to environmental stimuli and hence could shed light on the pathogenic role of the gut microbiota. Phylum Actinobacteria and genus *Desulfovibrio* were more abundant in MS gut. (Tremlett et al. 2016). These bacteria are also increased in other autoimmune inflammatory diseases including inflammatory bowel disease (Tremlett and Waubant 2018). A three-fold higher risk of an early relapse was noted among paediatric MS with Fusobacteria being absent in stool samples (Tremlett et al. 2016).

3.2 Neuromyelitis Optica Spectrum Disorder

Neuromyelitis optica spectrum disorder (NMSOD) is a severe multiphasic demyelinating disease driven by autoantibodies against aquaporin-4 water channels (AQP4). The B-cell responses in NMOSD are strongly influenced by T-cell-mediated cytotoxicity. Gut microbiota studies in NMOSD patients revealed overabundance of *Clostridium perfringens* (Cree et al. 2016). This is of particular interest in the pathogenesis of the disease as AQP4 displays high degree of homology to an adenosine triphosphate-binding cassette transporter permease of *C. perfringens* and can show cross-reaction (Varrin-Doyer et al. 2012). A recent study from Indian NMOSD demonstrated increase in *Clostridium boltea* which also shares sequence similarity with AQP4 (Pandit et al. 2021).

3.3 Parkinson's Disease

Parkinson's disease (PD), a progressive neurodegenerative disease, is characterised by degeneration of the dopaminergic neurons projecting from the substantia nigra. PD is an alpha-synucleinopathy and its pathological hallmark is the presence of alpha-synuclein (α-synuclein) immunoreactive inclusions called Lewy bodies within neurons. Mutations of the gene for α-synuclein can cause familial forms of PD (Kalia and Lang 2016). Though the dominant clinical presentations of PD are motor disability, a range of non-motor signs have been described in PD with olfactory, sleep, gastrointestinal, autonomic, neuropsychiatric and sensory dysfunction

(Poewe 2008). These non-motor symptoms often precede the neurological manifestations and have severe impact on the QOL (Pfeiffer 2016).

Gastrointestinal symptoms are frequent in PD among which constipation and delayed intestinal motility are the most common and distressing (Poewe 2008). Gut dysbiosis influences the gastrointestinal manifestations of PD as well as has role in the pathogenesis and severity of the disease. Mice transplanted with faecal microbiota from PD patients developed the classical motor symptoms of PD and neuroinflammation (Sampson et al. 2016). Neuroinflammation is being increasingly recognised as being integral to PD pathogenesis (Hirsch and Hunot 2009). The main mechanisms proposed for the influence of gut on PD include activation of the ENS through interactions with gut microbiota, activation of the local immune system and increase in gut permeability (Mulak 2015).

The contribution of gut in the pathogenesis was suggested by the detection of α-synuclein in the ENS in early PD (Hilton et al. 2014). Misfolding of α-synuclein could be induced by damage to enteric neurons and enteric glial cells leading to the accumulation of the inclusion bodies (Mulak 2015; Yang et al. 2019). Gut dysbiosis or an alternate environmental factor initiates the neuronal damage. Vagus nerve can transmit the α-synuclein pathology from the ENS to the dorsal vagal nucleus (Holmqvist et al. 2014). This hypothesis is supported by evidence from truncal vagotomy which prevented the spread of α-synuclein from gut to brain and the subsequent neurodegeneration in mouse models (Kim et al. 2019). Changes in gut flora and small intestinal bacterial overgrowth in PD contribute to inflammation and leaky gut which are hypothesised as substrates for the subsequent neurodegeneration (Tan et al. 2014; Yang et al. 2019).

In general, colonic biopsies and faecal analysis have revealed non-uniform data on gut microbiota in PD compared to controls. An increased diversity of bacterial species in PD gut was recorded in one study (Keshavarzian et al. 2015). Two studies showed a reduced abundance of Prevotellaceae species which reduced SCFAs implicated in PD pathogenesis. Other microbials noted to be decreased are Lachnospiraceae, Firmicutes, *Clostridium coccoides, Bacteroides fragilis, Faecalibacterium prausnitzii,* Enterococcaceae and Bacteroidetes, whereas Verrucomicrobiaceae species were increased in PD guts (Cryan et al. 2019). Enterobacteriaceae was associated with increased severity of postural instability and gait dysfunction in PD (Scheperjans et al. 2015).

Absorption and metabolism of levo-dopa which is the primary treatment for PD is influenced by the gut microbiota. Tyrosine decarboxylase from gut microbiota in mice reduced the plasma concentration of L-dopa (van Kessel et al. 2019). Conversely, PD medications have been linked to change in gut microbial status. Bacillaceae species was relatively abundant in patients who received levo-dopa compared to alternate therapy (Heintz-Buschart et al. 2018). A combination of monoamine oxidase inhibitors and amantadine also resulted in an increased microbial richness in PD (Bedarf et al. 2017).

3.4 Alzheimer's Disease

Alzheimer's disease (AD) is the most common neurodegenerative disease which manifests clinically as progressive loss of episodic memory and other neurocognitive functions. The typical pathological substrates in the brain are extracellular deposition of amyloid plaques and dystrophic neurites containing hyperphosphorylated tau protein (DeTure and Dickson 2019). More recent evidence has linked neuroinflammation and systemic inflammation with AD pathogenesis (Heneka et al. 2015). Genetic and environmental factors play key roles in AD pathogenesis, the best understood being increasing age and the presence of the high-risk ε4 allele of apolipoprotein E gene (Xu and Wang 2016). Studies in animal models and gut microbiota in AD patients suggest a strong influence of the latter in AD pathogenesis. GF mice had reduced neuroinflammation and lower amyloid burden in brain. Antibiotics were shown to alter the neuropathology in transgenic mice by reducing accumulation of amyloid plaques (Jiang et al. 2017; Bostanciklioğlu 2019).

In human studies, alpha (within sample) and beta (between sample) diversities were reduced in AD compared to controls. Bacterial flora was different in AD compared to healthy controls with reduced abundance of Firmicutes in AD. Pro-inflammatory Escherichia and Shigella were increased with increase in inflammatory markers NLRP3, CXCL2, IL-6, and IL-1β in amyloid positive and cognitively impaired patients compared to cognitively preserved patients. AD gut also demonstrated reduced anti-inflammatory strain *Bifidobacterium breve* strain A1 and *Escherichia rectale* (Vogt et al. 2017; Cattaneo et al. 2017).

3.5 Autism Spectrum Disorder

Autism spectrum disorder (ASD) manifests in early childhood and is defined by the clinical features of stereotyped behaviours and rituals and deficits in communication and social interaction. The cause of ASD is largely attributed to the genetic and environmental factors and their interplay (Lord et al. 2018). Neuro-inflammation and gut microbiota has also been implicated as a cause for ASD in the recent studies (Li et al. 2017). Maternal factors including vaginal infections, periodontitis, prescription drug usage (sodium valproate), diet and obesity, mode of delivery (caesarean section vs vaginal delivery) and postnatal factors such as antibiotics and breastfeeding, host genetics and environment shape the gut microbiota in early life (Tamburini et al. 2016). Various human and animal studies have shown that these factors can alter the composition of gut microbiota and their metabolic products thus predisposing a child for autism (Li et al. 2017).

Many clinical and physiological changes seen in ASD including diarrhoea, constipation, gaseous distension, gastroesophageal reflux disease, intolerance to certain food, abnormal metabolites and neuro-inflammation are attributed to the less diverse gut microbiota in the ASD children (Pulikkan et al. 2019). Aberrant behaviours of irritability, hyperactivity, social withdrawal and stereotypy were found to be more in ASD children with gastrointestinal problems (Chaidez et al. 2014; Moradi et al. 2021). Histopathology from ASD subjects with gastrointestinal symptoms was inconsistent with some showing lymphoid nodular hyperplasia, follicular lymphoid hyperplasia, ulcerative colitis and others with either non-specific inflammation or normal findings (Kang et al. 2014). From pyrosequencing study of faecal microflora, higher counts of microbial organisms belonging to Bacteroidetes and Proteobacterium phyla in children with ASD were detected as compared to neurotypical children in whom Firmicutes and Actinobacterium were predominant (Finegold et al. 2010). Similar findings were also observed in a recent study in ASD children between age group 2–4 years which showed higher colonisation by Bacteroidetes, Proteobacteria and butyrate producing *Faecalibacterium prausnitzii*, whereas a significant reduction of *Bifidobacterium longum* was observed (Coretti et al. 2018). Contradictory to the above findings, an increase in the ratio of Firmicutes/Bacteroidetes in ASD was reported in some studies (Strati et al. 2017; Moradi et al. 2021). In ASD subjects with constipation, high levels of Escherichia, Shigella and Clostridium cluster XVIII were found (Strati et al. 2017).

As mentioned in the initial part of this chapter, gut microbiota and their metabolites play a central role in brain behaviour modulated via immune, neural and endocrine pathways originating from leaky intestines. In ASD, metabolites from the transformed gut microbiome metabolism lead to increased production of SCFAs including acetate, propionate and butyrate (Li et al. 2017). A decrease in the gut colonisation by *E. coli* affects the catabolism of 3,3 phenylpropionate. SCFAs activate inflammasomes, innate immune system receptors and sensors thereby modulating downstream inflammatory pathways involving Tregs and TH17 cells contributing to aberrant behaviour in ASD. Levels of pro-inflammatory markers such as TNFα, IL-10, TGFβ, NT and SORT-1 were found to be significantly higher in ASD patients as compared to controls thus confirming that the immune pathways were altered in them (Carissimi et al. 2019). Desulfovibrio sp. and *B. fragilis*, a predominant gut coloniser in ASD, produce LPS which is an important antigenic component triggering immunological response (Finegold et al. 2010). Thus, intestinal dysbiosis has effects on immunological, biochemical and neuroendocrine system which ultimately influences the brain and behaviour.

3.6 Glioma

Table 7.1 provides a summary of the pattern in gut dysbiosis observed in major neurodegenerative diseases as per reliable research studies conducted in the recent past.

Tumours of CNS cause substantial morbidity and mortality, and the most common histologic type is glioma (Patel et al. 2019). The gut microbiota was recently implicated in the pathogenesis of glioma since the microbiota exert pleiotropic effects in tumour progression, growth, survival, transformation and response to

Table 7.1 Gut dysbiosis in neurological disorders

Study	Study population	Results
Multiple sclerosis (MS)		
Jangi et al. (2016)	60 patients with MS and 43 healthy controls	↑*Methanobrevibacter* and *Akkermansia* & ↓ *Butyricimonas* ↑*Prevotella* and *Sutterella* and ↓ *Sarcina* in MS patients treated with DMT compared with untreated patients
Chen et al. (2016)	31 RRMS patients and 36 healthy controls	MS patients showed ↑*Pseudomonas*, *Mycoplana*, *Haemophilus*, *Blautia*, and *Dorea* genera ↑*Parabacteroides*, *Adlercreutzia* and *Prevotella* genera in healthy controls
Tremlett et al. (2016)	18 paediatric RRMS patients and 17 controls	↑*Desulfovibrionaceae* (*Bilophila*, *Desulfovibrio* and *Christensenellaceae*) and ↓ *Lachnospiraceae* and *Ruminococcaceae* in MS patients
Parkinson's disease (PD)		
Hill-Burns et al. (2017)	197 PD cases and 130 controls	↑*Bifidobacteriaceae*, *Christensenellaceae*, *Lachnospiraceae*, *Lactobacillaceae*, *Pasteurellaceae* and *Verrucomicrobiaceae* families
Li et al. (2019)	51 PD patients and 48 healthy controls	↑*Akkermansia* and ↓ *Lactobacillus* in PD patients
Lin et al. (2019)	80 PD patients and 77 healthy controls	↑*Verrucomicrobia*, *Mucispirillum*, *Porphyromonas*, *Lactobacillus*, and *Parabacteroides* ↑*Prevotella* in controls
Autism spectrum disorder (ASD)		
Coretti et al. (2018)	11 ASD children and 14 healthy controls	↑*Faecalibacterium prausnitzii* and ↓ *Bifidobacterium longum* in ASD
Pulikkan et al. (2018)	30 ASD children and 24 healthy controls	↑*Lactobacillaceae*, *Bifidobacteraceae*, and *Veillonellaceae* ↑*Prevotellaceae* in controls
Ma et al. (2019)	45 ASD and 45 healthy children	↓in relative abundance of *Lachnoclostridium*, *Tyzzerella* subgroup 4 and *Flavonifractor* in ASD Firmicutes/Bacteroidetes ratio not significantly different between the ASD and healthy controls
Zurita et al. (2020)	25 ASD cases and 35 controls	↑*Bacteroides*, *Akkermansia*, *Coprococcus* and *Ruminococcus* in ASD

Note: This table includes only the recent and larger studies published after 2016
DMT disease-modifying therapy; *RRMS* relapsing-remitting multiple sclerosis.

therapeutic agents. A recent study of tumours involving breast, lung, brain, ovary, pancreas, melanoma and bone revealed presence of intracellular bacteria in both cancer and immune cells (Nejman et al. 2020). In the mouse glioma models, reduction in Bacteroidetes and Firmicutes were seen with an abundance of Verrucomicrobia especially of Akkermansia genus after tumour growth was observed. The metabolites such as dihydroxy phenylacetic acid (DOPAC), 5-hydroxy indole acetic acid (5-HIAA), adenosine, histamine, acetate, propionate, butyrate, norepinephrine, GABA, tryptophan, valerate and aspartic acid levels were found to be reduced in them (Dono et al. 2020). The circulating immune cells such as T and B cells and macrophages cross the BBB and play a role in gliomagenesis. Tumour survival and growth is promoted by an immunosuppressed microenvironment created by immune cells and cytokine profile in which Tregs play a crucial role (Mehrian-Shai et al. 2019).

4 Therapeutic Interventions

The gut microbiome composition can be manipulated by the application of probiotics, prebiotics, diet modifications and faecal microbial transplantation. Relying on such alternative interventions to augment the gut microbiome, post proper evaluation and trials will reduce the huge burden on complex drugs and associated side effects, and promote the QOL of patients suffering from neurological disorders.

Probiotics are lactic acid-producing bacteria (Lactococci, Lactobacilli, Bifidobacteria and Saccharomycetes) that augment the IEB and may prevent intestinal inflammatory diseases. Prebiotics are non-digestible oligosaccharides that encourage the growth of beneficial bacteria. A few studies have reported beneficial effects of probiotics/prebiotics in neuropsychiatric symptoms such as anxiety and depression (Li et al. 2017).

Two studies in MS compared outcomes of probiotic preparations with *Lactobacillus* species versus placebo and noted improvement in markers of inflammation; however, intestinal bacterial loads were not assessed in them (Kouchaki et al. 2017; Tamtaji et al. 2017). Another study showed an anti-inflammatory shift of the flora with an increased abundance of *Lactobacillus*, and decreased abundance of *Akkermansia, Dorea* and *Blautia* with probiotic therapy (Tankou et al. 2018). Probiotics/prebiotics can potentially alter the gut dysbiosis in PD and have been tried as therapeutic strategies. Increased complete bowel movements were reported in one study in PD patients with probiotics/prebiotics compared to placebo (Barichella et al. 2016). Treatment with probiotics (*Lactobacillus* and *Bifidobacterium* species) produced an increase in mini-mental status examination (MMSE) scores in a small cohort of patients with AD (Akbari et al. 2016). Larger studies looking into changes in the gut flora and cognitive scores with therapy are needed to assess therapeutic effects further. In a recent systemic review of eight clinical trials that involved the use of pre/probiotic supplementation in children with ASD, prebiotics combined with a gluten and casein-free diet was found to improve certain gastrointestinal symptoms along with significant reduction in anti-sociability scores. But probiotics have limited evidence in alleviating the gastrointestinal or neurobehavioral symptoms in ASD children (Ng et al. 2019).

There is currently no standardised pre/probiotics regimen with regard to type and concentration of different strains, duration of treatments and monitoring parameters in the management of neurological disorders.

Potential therapeutic approaches sought for modification of microbiota in MS also include vitamin D3 supplementation and diet modifications. Vitamin D3 supplementation in a small sample of MS patients improved immune tolerance and reduced inflammation with anti-inflammatory shift of gut microbiota in the subgroup who were not on immunotherapy (Cantarel et al. 2015). Dietary modifications have also been shown to impact the inflammatory profile. A diet rich in simple sugars, saturated fat, salt and high calorie content has been associated with increased risk for autoimmune demyelinating diseases. This has been shown to increase inflammatory cells in the gut and induce gut dysbiosis (Thorburn et al. 2014). A shift to normal gut flora can be induced by a calorie-restricted diet rich in fruits, vegetables and fish although robust studies for the utility in MS are lacking (Mowry et al. 2018).

Faecal microbiota transplantation (FMT) is a procedure wherein the faecal microbiota from a healthy person is transferred to a patient with gut-microbiome dysbiosis. FMT in patients with PD improved constipation and non-gastrointestinal manifestations (Yang et al. 2019). It was also tried in ASD; however, the safety profile and efficacy of this intervention needs to be established (Li et al. 2017). The principle of microbiota transfer therapy (MTT) is based on FMT protocol with minor modifications. Oral vancomycin treatment is given for 14 days followed by fasting (12–24 h) and bowel cleansing. Standardised Human Gut Microbiota (SHGM) either orally or rectally is then administered to repopulate the gut microbiota for a period of 7–8 weeks. Gastrointestinal and behavioural symptoms in ASD subjects significantly improved with MTT (Kang et al. 2017). There is insufficient evidence to translate FMT and MTT to clinical practice at present.

5 Conclusion

The bi-directional pathways involving gut microbiome and human brain connect host behaviour with the gut function thereby subserving a major role in health and disease. The marked differences in the composition of gut microbiota across various studies suggest a strong role for ethnicity, dietary patterns and other environmental influences (Cryan et al. 2019). Nevertheless, the perturbations in the gut microbiota have been linked to disease risk and progression in various neurological disorders including neurodegenerative, neurodevelopmental and autoimmune diseases, tumours and stroke. Gut microbiome dysregulation may not be directly causative, but when combined with other risk factors can be responsible for various neurological disorders (Tremlett et al. 2017). Many of these studies are limited by small sample size and the wide variability in the methods for collection, storage and analysis of samples (Cryan et al. 2019). Therefore, in-depth, standardised and multicentric studies at a global level are highly warranted to substantiate the relationship between gut microbiota and the numerous neurological diseases in which they have been implicated. Such studies possess the potential to identify biomarkers for disease onset, activity and co-morbid conditions. Simultaneously, the role of physiotherapy and non-strenuous, aerobic exercises in alleviating the symptoms in neurological disorders by strengthening the gut-brain axis can be further explored to gauge its potency as a promising intervention strategy in the future.

References

Akbari E, Asemi Z, Daneshvar Kakhaki R et al (2016) Effect of probiotic supplementation on cognitive function and metabolic status in Alzheimer's disease: a randomized, double-blind and controlled trial. Front Aging Neurosci 8:256. https://doi.org/10.3389/fnagi.2016.00256

Aresti Sanz J, El Aidy S (2019) Microbiota and gut neuropeptides: a dual action of antimicrobial activity and neuroimmune response. Psychopharmacology 236:1597–1609. https://doi.org/10.1007/s00213-019-05224-0

Arpaia N, Campbell C, Fan X et al (2013) Metabolites produced by commensal bacteria promote peripheral regulatory T-cell generation. Nature 504:451–455

Barichella M, Pacchetti C, Bolliri C et al (2016) Probiotics and prebiotic fiber for constipation associated with Parkinson disease: an RCT. Neurology 87:1274–1280. https://doi.org/10.1212/WNL.0000000000003127

Bedarf JR, Hildebrand F, Coelho LP et al (2017) Functional implications of microbial and viral gut metagenome changes in early stage L-DOPA-naïve Parkinson's disease patients. Genome Med 9:39. https://doi.org/10.1186/s13073-017-0428-y

Berer K, Gerdes LA, Cekanaviciute E et al (2017) Gut microbiota from multiple sclerosis patients enables spontaneous autoimmune encephalomyelitis in mice. Proc Natl Acad Sci U S A 114:10719–10724. https://doi.org/10.1073/pnas.1711233114

Blander JM, Longman RS, Iliev ID et al (2017) Regulation of inflammation by microbiota interactions with the host. Nat Immunol 18:851–860. https://doi.org/10.1038/ni.3780

Bostanciklioğlu M (2019) The role of gut microbiota in pathogenesis of Alzheimer's disease. J Appl Microbiol 127:954–967. https://doi.org/10.1111/jam.14264

Browning KN, Verheijden S, Boeckxstaens GE (2017) The Vagus nerve in appetite regulation, mood, and intestinal inflammation. Gastroenterology 152:730–744. https://doi.org/10.1053/j.gastro.2016.10.046

Cani PD (2018) Human gut microbiome: hopes, threats and promises. Gut 67:1716–1725. https://doi.org/10.1136/gutjnl-2018-316723

Cantarel BL, Waubant E, Chehoud C et al (2015) Gut microbiota in multiple sclerosis: possible influence of Immunomodulators. J Investig Med 63:729–734. https://doi.org/10.1097/JIM.0000000000000192

Carissimi C, Laudadio I, Palone F et al (2019) Functional analysis of gut microbiota and immunoinflammation in children with autism spectrum disorders. Dig Liver Dis 51:1366–1374. https://doi.org/10.1016/j.dld.2019.06.006

Cattaneo A, Cattane N, Galluzzi S et al (2017) Association of brain amyloidosis with proinflammatory gut bacterial taxa and peripheral inflammation markers in cognitively impaired elderly. Neurobiol Aging 49:60–68. https://doi.org/10.1016/j.neurobiolaging.2016.08.019

Cheung SG, Goldenthal AR, Uhlemann A-C et al (2019) Systematic review of gut microbiota and major depression. Front Psychiatry 10:34. https://doi.org/10.3389/fpsyt.2019.00034

Chaidez V, Hansen RL, Hertz-Picciotto I (2014) Gastrointestinal problems in children with autism, developmental delays or typical development. J Autism Dev Disord 44:1117–1127. https://doi.org/10.1007/s10803-013-1973-x

Chen J, Chia N, Kalari KR et al (2016) Multiple sclerosis patients have a distinct gut microbiota compared to healthy controls. Sci Rep 6:28484. https://doi.org/10.1038/srep28484

Coretti L, Paparo L, Riccio MP et al (2018) Gut microbiota features in young children with autism Spectrum disorders. Front Microbiol 9:3146. https://doi.org/10.3389/fmicb.2018.03146

Cree BAC, Spencer CM, Varrin-Doyer M et al (2016) Gut microbiome analysis in neuromyelitis optica reveals overabundance of Clostridium perfringens. Ann Neurol 80:443–447. https://doi.org/10.1002/ana.24718

Cryan JF, O'Riordan KJ, Cowan CSM et al (2019) The microbiota-gut-brain Axis. Physiol Rev 99:1877–2013. https://doi.org/10.1152/physrev.00018.2018

De Angelis M, Piccolo M, Vannini L et al (2013) Fecal microbiota and metabolome of children with autism and pervasive developmental disorder not otherwise specified. PLoS One 8:e76993. https://doi.org/10.1371/journal.pone.0076993

DeTure MA, Dickson DW (2019) The neuropathological diagnosis of Alzheimer's disease. Mol Neurodegener 14:32. https://doi.org/10.1186/s13024-019-0333-5

Dietz KC, Casaccia P (2010) HDAC inhibitors and neurodegeneration: at the edge between protection and damage. Pharmacol Res 62:11–17. https://doi.org/10.1016/j.phrs.2010.01.011

Dono A, Patrizz A, McCormack RM et al (2020) Glioma induced alterations in fecal short-chain fatty acids and neurotransmitters. CNS Oncology 9:CNS57. https://doi.org/10.2217/cns-2020-0007

Dzidic M, Boix-Amorós A, Selma-Royo M et al (2018) Gut microbiota and mucosal immunity in the neonate. Med Sci 6:56. https://doi.org/10.3390/medsci6030056

El Aidy S, Dinan TG, Cryan JF (2014) Immune modulation of the brain-gut-microbe axis. Front Microbiol 5:146. https://doi.org/10.3389/fmicb.2014.00146

Erny D, Hrabě de Angelis AL, Jaitin D et al (2015) Host microbiota constantly control maturation and function of microglia in the CNS. Nat Neurosci 18:965–977. https://doi.org/10.1038/nn.4030

Finegold SM, Dowd SE, Gontcharova V et al (2010) Pyrosequencing study of fecal microflora of autistic and control children. Anaerobe 16:444–453. https://doi.org/10.1016/j.anaerobe.2010.06.008

Fung TC, Olson CA, Hsiao EY (2017) Interactions between the microbiota, immune and nervous systems in health and disease. Nat Neurosci 20:145–155. https://doi.org/10.1038/nn.4476

Gill SR, Pop M, DeBoy RT et al (2006) Metagenomic analysis of the human distal gut microbiome. Science 312:1355–1359. https://doi.org/10.1126/science.1124234

Glatigny S, Bettelli E (2018) Experimental autoimmune encephalomyelitis (EAE) as animal models of multiple sclerosis (MS). Cold Spring Harb Perspect Med 8:a028977. https://doi.org/10.1101/cshperspect.a028977

Gubert C, Kong G, Renoir T et al (2018) Exercise, diet and stress as modulators of gut microbiota: implications for neurodegenerative diseases. Neurobiol Dis. https://doi.org/10.1016/j.nbd.2019.104621

Heintz-Buschart A, Pandey U, Wicke T et al (2018) The nasal and gut microbiome in Parkinson's disease and idiopathic rapid eye movement sleep behavior disorder: nose and gut microbiome in PD and iRBD. Mov Disord 33:88–98. https://doi.org/10.1002/mds.27105

Heneka MT, Carson MJ, Khoury JE et al (2015) Neuroinflammation in Alzheimer's disease. Lancet Neurol 14:388–405. https://doi.org/10.1016/S1474-4422(15)70016-5

Hill-Burns EM, Debelius JW, Morton JT et al (2017) Parkinson's disease and Parkinson's disease medications have distinct signatures of the gut microbiome. Mov Disord 32(5):739–749. https://doi.org/10.1002/mds.26942

Hilton D, Stephens M, Kirk L et al (2014) Accumulation of α-synuclein in the bowel of patients in the pre-clinical phase of Parkinson's disease. Acta Neuropathol 127:235–241. https://doi.org/10.1007/s00401-013-1214-6

Hirsch EC, Hunot S (2009) Neuroinflammation in Parkinson's disease: a target for neuroprotection? Lancet Neurol 8:382–397. https://doi.org/10.1016/S1474-4422(09)70062-6

Holmqvist S, Chutna O, Bousset L et al (2014) Direct evidence of Parkinson pathology spread from the gastrointestinal tract to the brain in rats. Acta Neuropathol 128:805–820. https://doi.org/10.1007/s00401-014-1343-6

Holzer P, Farzi A (2014) Neuropeptides and the microbiota-gut-brain Axis. In: Lyte M, Cryan JF (eds) Microbial endocrinology: the microbiota-gut-brain Axis in health and disease. Springer, New York, NY, pp 195–219

Ivanov II, Atarashi K, Manel N et al (2009) Induction of intestinal Th17 cells by segmented filamentous bacteria. Cell 139:485–498. https://doi.org/10.1016/j.cell.2009.09.033

Jangi S, Gandhi R, Cox LM et al (2016) Alterations of the human gut microbiome in multiple sclerosis. Nat Commun 7:12015. https://doi.org/10.1038/ncomms12015
Jiang C, Li G, Huang P et al (2017) The gut microbiota and Alzheimer's disease. JAD 58:1–15. https://doi.org/10.3233/JAD-161141
Johnson KV-A (2020) Gut microbiome composition and diversity are related to human personality traits. Human Microbiome J 15:100069. https://doi.org/10.1016/j.humic.2019.100069
Kalia LV, Lang AE (2016) Evolving basic, pathological and clinical concepts in PD. Nat Rev Neurol 12:65–66. https://doi.org/10.1038/nrneurol.2015.249
Kang D-W, Adams JB, Gregory AC et al (2017) Microbiota transfer therapy alters gut ecosystem and improves gastrointestinal and autism symptoms: an open-label study. Microbiome 5:10. https://doi.org/10.1186/s40168-016-0225-7
Kang V, Wagner GC, Ming X (2014) Gastrointestinal dysfunction in children with autism spectrum disorders. Autism Res 7:501–506. https://doi.org/10.1002/aur.1386
Keshavarzian A, Green SJ, Engen PA et al (2015) Colonic bacterial composition in Parkinson's disease. Mov Disord 30:1351–1360. https://doi.org/10.1002/mds.26307
Kim S, Kwon S-H, Kam T-I et al (2019) Transneuronal propagation of pathologic α-Synuclein from the gut to the brain models Parkinson's disease. Neuron 103:627–641.e7. https://doi.org/10.1016/j.neuron.2019.05.035
Kouchaki E, Tamtaji OR, Salami M et al (2017) Clinical and metabolic response to probiotic supplementation in patients with multiple sclerosis: a randomized, double-blind, placebo-controlled trial. Clin Nutr 36:1245–1249. https://doi.org/10.1016/j.clnu.2016.08.015
Li C, Cui L, Yang Y et al (2019) Gut microbiota differs between Parkinson's disease patients and healthy controls in Northeast China. Front Mol Neurosci 12:171. https://doi.org/10.3389/fnmol.2019.00171
Li Q, Han Y, Dy ABC, Hagerman RJ (2017) The gut microbiota and autism Spectrum disorders. Front Cell Neurosci 11:120. https://doi.org/10.3389/fncel.2017.00120
Lin CH, Chen CC, Chiang HL et al (2019) Altered gut microbiota and inflammatory cytokine responses in patients with Parkinson's disease. J Neuroinflammation 16(1):129. https://doi.org/10.1186/s12974-019-1528-y
Lord C, Elsabbagh M, Baird G, Veenstra-Vanderweele J (2018) Autism spectrum disorder. Lancet 392:508–520. https://doi.org/10.1016/S0140-6736(18)31129-2
Lowry CA, Hollis JH, de Vries A et al (2007) Identification of an immune-responsive mesolimbocortical serotonergic system: potential role in regulation of emotional behavior. Neuroscience 146:756–772. https://doi.org/10.1016/j.neuroscience.2007.01.067
Lyte M, Cryan JF (eds) (2014) Microbial endocrinology: the microbiota-gut-brain Axis in health and disease. Springer, New York, NY
Ma B, Liang J, Dai M et al (2019) Altered gut microbiota in Chinese children with autism spectrum disorders. Front Cell Infect Microbiol 9:40. https://doi.org/10.3389/fcimb.2019.00040
Malan-Muller S, Valles-Colomer M, Raes J et al (2018) The gut microbiome and mental health: implications for anxiety–and trauma-related disorders. OMICS: J Integr Biol 22:90–107. https://doi.org/10.1089/omi.2017.0077
Mehrian-Shai R, Reichardt JKV, Harris CC, Toren A (2019) The is, paving the way to brain cancer. Trends in Cancer 5:200–207. https://doi.org/10.1016/j.trecan.2019.02.008
MetaHIT Consortium (additional members), Arumugam M, Raes J et al (2011) Enterotypes of the human gut microbiome. Nature 473:174–180. https://doi.org/10.1038/nature09944
Mirza A, Forbes JD, Zhu F et al (2020) The multiple sclerosis gut microbiota: a systematic review. Mult Scler Relat Disord 37:101427. https://doi.org/10.1016/j.msard.2019.101427
Miyake S, Kim S, Suda W et al (2015) Dysbiosis in the gut microbiota of patients with multiple sclerosis, with a striking depletion of species belonging to clostridia XIVa and IV clusters. PLoS One 10:e0137429. https://doi.org/10.1371/journal.pone.0137429
Moos WH, Faller DV, Harpp DN et al (2016) Microbiota and neurological disorders: a gut feeling. BioResearch Open Access 5:137–145. https://doi.org/10.1089/biores.2016.0010

Moradi K, Ashraf-Ganjouei A, Tavolinejad H et al (2021) The interplay between gut microbiota and autism spectrum disorders: a focus on immunological pathways. Prog Neuro-Psychopharmacol Biol Psychiatry 106:110091. https://doi.org/10.1016/j.pnpbp.2020.110091

Mowry EM, Azevedo CJ, McCulloch CE et al (2018) Body mass index, but not vitamin D status, is associated with brain volume change in MS. Neurology 91:e2256–e2264. https://doi.org/10.1212/WNL.0000000000006644

Mowry EM, Glenn JD (2018) The dynamics of the gut microbiome in multiple sclerosis in relation to disease. Neurol Clin 36:185–196. https://doi.org/10.1016/j.ncl.2017.08.008

Mulak A (2015) Brain-gut-microbiota axis in Parkinson's disease. WJG 21:10609. https://doi.org/10.3748/wjg.v21.i37.10609

Nayak D, Roth TL, McGavern DB (2014) Microglia development and function. Annu Rev Immunol 32:367–402. https://doi.org/10.1146/annurev-immunol-032713-120240

Nejman D, Livyatan I, Fuks G et al (2020) The human tumor microbiome is composed of tumor type–specific intracellular bacteria. Science 368:973–980. https://doi.org/10.1126/science.aay9189

Ng QX, Loke W, Venkatanarayanan N et al (2019) A systematic review of the role of prebiotics and probiotics in autism Spectrum disorders. Medicina (Kaunas) 55:129. https://doi.org/10.3390/medicina55050129

Pan W, Stone KP, Hsuchou H et al (2011) Cytokine signaling modulates blood-brain barrier function. Curr Pharm Des 17:3729–3740. https://doi.org/10.2174/138161211798220918

Pandit L, Cox LM, Malli C et al (2021) Clostridium bolteae is elevated in neuromyelitis optica spectrum disorder in India and shares sequence similarity with AQP4. Neurol Neuroimmunol Neuroinflamm 8:e907. https://doi.org/10.1212/NXI.0000000000000907

Pascale A, Marchesi N, Marelli C et al (2018) Microbiota and metabolic diseases. Endocrine 61:357–371. https://doi.org/10.1007/s12020-018-1605-5

Patel AP, Fisher JL, Nichols E et al (2019) Global, regional, and national burden of brain and other CNS cancer, 1990–2016: a systematic analysis for the global burden of disease study 2016. Lancet Neurol 18:376–393. https://doi.org/10.1016/S1474-4422(18)30468-X

Pfeiffer RF (2016) Non-motor symptoms in Parkinson's disease. Parkinsonism Relat Disord 22:S119–S122. https://doi.org/10.1016/j.parkreldis.2015.09.004

Poewe W (2008) Non-motor symptoms in Parkinson's disease. Eur J Neurol 15(Suppl 1):14–20. https://doi.org/10.1111/j.1468-1331.2008.02056.x

Pulikkan J, Maji A, Dhakan DB et al (2018) Gut microbial dysbiosis in Indian children with autism spectrum disorders. Microb Ecol 76(4):1102–1114. https://doi.org/10.1007/s00248-018-1176-2

Pulikkan J, Mazumder A, Grace T (2019) Role of the gut microbiome in autism Spectrum disorders. Adv Exp Med Biol 1118:253–269. https://doi.org/10.1007/978-3-030-05542-4_13

Rae-Grant A, Day GS, Marrie RA et al (2018) Practice guideline recommendations summary: disease-modifying therapies for adults with multiple sclerosis: report of the guideline development, dissemination, and implementation subcommittee of the American academy of neurology. Neurology 90:777–788. https://doi.org/10.1212/WNL.0000000000005347

Reich DS, Lucchinetti CF, Calabresi PA (2018) Multiple sclerosis. N Engl J Med 378:169–180. https://doi.org/10.1056/NEJMra1401483

Rothhammer V, Mascanfroni ID, Bunse L et al (2016) Type I interferons and microbial metabolites of tryptophan modulate astrocyte activity and central nervous system inflammation via the aryl hydrocarbon receptor. Nat Med 22:586–597. https://doi.org/10.1038/nm.4106

Sampson TR, Debelius JW, Thron T et al (2016) Gut microbiota regulate motor deficits and Neuroinflammation in a model of Parkinson's disease. Cell 167:1469–1480.e12. https://doi.org/10.1016/j.cell.2016.11.018

Scheperjans F, Aho V, Pereira PAB et al (2015) Gut microbiota are related to Parkinson's disease and clinical phenotype. Mov Disord 30:350–358. https://doi.org/10.1002/mds.26069

Sender R, Fuchs S, Milo R (2016) Revised estimates for the number of human and bacteria cells in the body. PLoS Biol 14:e1002533. https://doi.org/10.1371/journal.pbio.1002533
Silva YP, Bernardi A, Frozza RL (2020) The role of short-chain fatty acids from gut microbiota in gut-brain communication. Front Endocrinol 11:25. https://doi.org/10.3389/fendo.2020.00025
Smith PM, Howitt MR, Panikov N et al (2013) The microbial metabolites, short-chain fatty acids, regulate colonic Treg cell homeostasis. Science 341:569–573. https://doi.org/10.1126/science.1241165
Strandwitz P (2018) Neurotransmitter modulation by the gut microbiota. Brain Res 1693:128–133. https://doi.org/10.1016/j.brainres.2018.03.015
Strati F, Cavalieri D, Albanese D et al (2017) New evidences on the altered gut microbiota in autism spectrum disorders. Microbiome 5:24. https://doi.org/10.1186/s40168-017-0242-1
Tamburini S, Shen N, Wu HC, Clemente JC (2016) The microbiome in early life: implications for health outcomes. Nat Med 22:713–722. https://doi.org/10.1038/nm.4142
Tamtaji OR, Kouchaki E, Salami M et al (2017) The effects of probiotic supplementation on gene expression related to inflammation, insulin, and lipids in patients with multiple sclerosis: a randomized, double-blind, placebo-controlled trial. J Am Coll Nutr 36:660–665. https://doi.org/10.1080/07315724.2017.1347074
Tan AH, Mahadeva S, Thalha AM et al (2014) Small intestinal bacterial overgrowth in Parkinson's disease. Parkinsonism Relat Disord 20:535–540. https://doi.org/10.1016/j.parkreldis.2014.02.019
Tankou SK, Regev K, Healy BC et al (2018) A probiotic modulates the microbiome and immunity in multiple sclerosis. Ann Neurol 83:1147–1161. https://doi.org/10.1002/ana.25244
Thorburn AN, Macia L, Mackay CR (2014) Diet, metabolites, and "western-lifestyle" inflammatory diseases. Immunity 40:833–842. https://doi.org/10.1016/j.immuni.2014.05.014
Thursby E, Juge N (2017) Introduction to the human gut microbiota. Biochem J 474:1823–1836. https://doi.org/10.1042/BCJ20160510
Tremlett H, Bauer KC, Appel-Cresswell S et al (2017) The gut microbiome in human neurological disease: a review: gut microbiome. Ann Neurol 81:369–382. https://doi.org/10.1002/ana.24901
Tremlett H, Fadrosh DW, Faruqi AA et al (2016) Gut microbiota in early pediatric multiple sclerosis: a case−control study. Eur J Neurol 23:1308–1321. https://doi.org/10.1111/ene.13026
Tremlett H, Waubant E (2018) The gut microbiota and pediatric multiple sclerosis: recent findings. Neurotherapeutics 15:102–108. https://doi.org/10.1007/s13311-017-0574-3
van Kessel SP, Frye AK, El-Gendy AO et al (2019) Gut bacterial tyrosine decarboxylases restrict levels of levodopa in the treatment of Parkinson's disease. Nat Commun 10:310. https://doi.org/10.1038/s41467-019-08294-y
Varrin-Doyer M, Spencer CM, Schulze-Topphoff U et al (2012) Aquaporin 4-specific T cells in neuromyelitis optica exhibit a Th17 bias and recognize clostridium ABC transporter. Ann Neurol 72:53–64. https://doi.org/10.1002/ana.23651
Ventura RE, Iizumi T, Battaglia T et al (2019) Gut microbiome of treatment-naïve MS patients of different ethnicities early in disease course. Sci Rep 9:16396. https://doi.org/10.1038/s41598-019-52894-z
Vogt NM, Kerby RL, Dill-McFarland KA et al (2017) Gut microbiome alterations in Alzheimer's disease. Sci Rep 7:13537. https://doi.org/10.1038/s41598-017-13601-y
Xu R, Wang Q (2016) Towards understanding brain-gut-microbiome connections in Alzheimer's disease. BMC Syst Biol 10:63. https://doi.org/10.1186/s12918-016-0307-y
Yang D, Zhao D, Ali Shah SZ et al (2019) The role of the gut microbiota in the pathogenesis of Parkinson's disease. Front Neurol 10:1155. https://doi.org/10.3389/fneur.2019.01155
Vijay N, Morris M (2014) Role of Monocarboxylate transporters in drug delivery to the brain. CPD 20:1487–1498
Wall R, Cryan JF, Ross RP et al (2014) Bacterial neuroactive compounds produced by Psychobiotics. In: Lyte M, Cryan JF (eds) Microbial endocrinology: the microbiota-gut-brain Axis in health and disease. Springer, New York, New York, NY, pp 221–239

Wang L, Christophersen CT, Sorich MJ et al (2011) Low relative abundances of the mucolytic bacterium Akkermansia muciniphila and Bifidobacterium spp. in feces of children with autism. Appl Environ Microbiol 77:6718–6721. https://doi.org/10.1128/AEM.05212-11

Wu H-J, Wu E (2012) The role of gut microbiota in immune homeostasis and autoimmunity. Gut Microbes 3:4–14. https://doi.org/10.4161/gmic.19320

Zurita MF, Cárdenas PA, Sandoval ME et al (2020) Analysis of gut microbiome, nutrition and immune status in autism spectrum disorder: a case-control study in Ecuador. Gut Microbes 11(3):453–464. https://doi.org/10.1080/19490976.2019.1662260

Chapter 8
Modern Perspectives in Controlling Human Diseases through Probiotic Intervention

J. Devika Das, V. S. Akhila, Merin Paul, Aparna Shankar, Iype Joseph, and Sabu Thomas

1 Introduction

The human gut ecosystem is a diverse mélange of microbial community entangled in a delicate symbiotic relationship with the host body. The past decades of microbiome-associated studies and advancements in molecular biology have unraveled many constituent organisms of this complex community particularly those colonizing gastrointestinal tract (GI) and their importance in regulating different aspects of host physiology. A major proportion of these beneficial microbes are regarded as probiotics and perform a multitude of actions viz., competitive inhibition of pathogens, production of active metabolites, regulating host mucosal immunity, and activation of different inter compartmental organ axes. Probiotics are primarily therapeutic agents that modulate the gut microbial ecology in order to attain healthy functional eubiosis state.

The use of probiotics as prophylactic agents against infectious and metabolic conditions has gained considerable acceptance among the scientific community. Fascinatingly, the probiotic viability and the mechanisms underlying their presumed beneficial functionalities are still dubious and seem to be strain and disease-specific. The studies on classical probiotics encompassing strains of *Lactobacillus* generated limited effects on the gut microbiota and associated traits. Overall, the development of targeted next-generation probiotics as putative biodrugs holds promise for better

J. Devika Das · V. S. Akhila · M. Paul · A. Shankar · I. Joseph · S. Thomas (✉)
Pathogen Biology, Rajiv Gandhi Centre for Biotechnology,
Thiruvananthapuram, Kerala, India
e-mail: sabu@rgcb.res.in

S. Thomas (ed.), *Human Microbiome*,
https://doi.org/10.1007/978-981-16-7672-7_8

innovation in the therapeutic industry (Cordaillat-Simmons et al. 2020). Moreover, the emergence of omics technologies has been very useful in the selection of probiotic strains of interest and in elucidating its possible therapeutic effects on various gastrointestinal diseases and metabolic disorders. Despite its tremendous therapeutic potential, the accelerated long-term usage of probiotics exacerbates the concerns in spreading antibiotic-resistant genes in physiological environment (Suez et al. 2019). Emerging concepts of postbiotics/paraprobiotics may circumvent the risk associated with the use of live intact cells and elicit ample advantages over probiotics. Indeed, the increased consumer awareness toward natural therapeutic interventions has accelerated the use of probiotics worldwide which began with the consumption of non-specific probiotics in common food sources including yogurt, tofu, pickles, etc. before 1930. The realization of the benefits of such food sources led to the development of Yakult, one of the first specific commercial probiotic products comprising *Lactobacillus casei* Shirota and *Bifidobacterium breve* strains in 1935 and has been followed by the manufacturing of multitude probiotic products in different countries which cater to definite health requirements (Fig. 8.1). Such a scenario demands more comprehensive studies to elucidate the molecular mechanisms behind the beneficial attributes of probiotics and their interactions with other bacterial communities in the gut, posing minimal risk to human health thereby opening doors to "personalized probiotics."

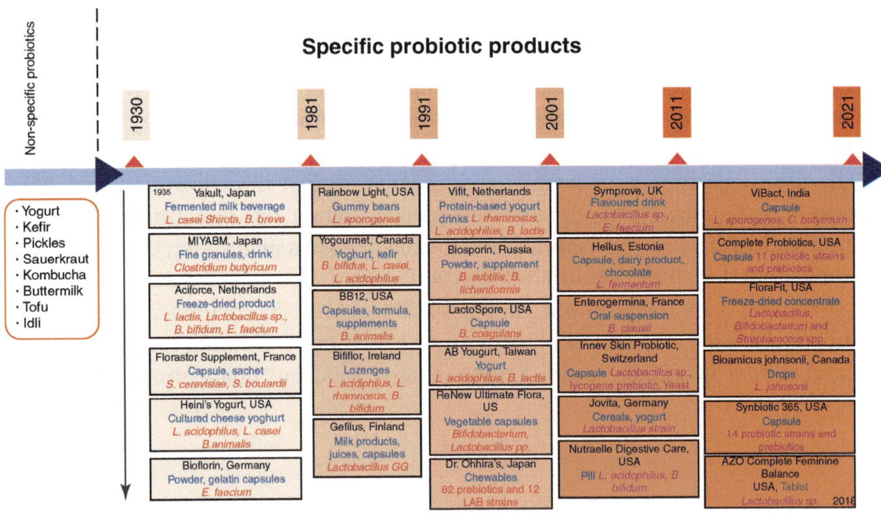

Fig. 8.1 Timeline of the development of probiotics as dietary supplements. A snapshot of non-specific and specific probiotic products represented in a timeline. The selected specific products include the brand name, country of origin, formulation of the product, and details of the corresponding active probiotic component

2 Streamlining the Paradigms of Probiotic Therapy

Probiotics, as mentioned in the classical definition, are live microorganisms that benefit health in numerous ways. Probiotic therapy majorly targets the disturbed intestinal flora and exerts their beneficial effects by maintenance of gut integrity, preventing bacterial translocation, modulating the gut microbiota, and immunomodulation of the host system (Kotzampassi and Giamarellos-Bourboulis 2012). Since the mechanism of probiotic action is mostly strain-dependent, properties exhibited by one strain cannot be blindly extrapolated to other strains of the same species (Oelschlaeger 2010). Probiotics are also termed as immunobiotics due to their beneficial effects on the health by activation of mucosal immune apparatus at different mucosal sites (Clancy 2003). Most of the probiotics used nowadays comprise yeasts (*Saccharomyces boulardii* and *Saccharomyces cerevisiae*), bacteria (*Bifidobacterium, Lactobacillus*), other non-pathogenic strains of *Bacillus* sp. and *Escherichia coli* (only Gram-negative bacteria administered as probiotics) (de Vrese and Schrezenmeir 2008). More recently, FDA has coined the term Live Bio therapeutic Product (LBP) to define "a biological product containing live organisms that is applicable for the prevention, treatment or cure of a disease of human beings and is not a vaccine" (FDA 2016). Although probiotics cannot replace antibiotics as a therapeutic strategy, it can be used as a complement to reduce infections, thus help to reduce the unwanted use of antibiotics. However, several techno-functional limitations, viability, and the risk of transmission of acquired antibiotic resistance determinants hamper the application of live organisms as prophylactic and therapeutic agents against clinical disorders (Das et al. 2020).

Prebiotics and synbiotics are closely related to probiotics wherein the former are mainly non-digestible fibers which positively influence the host's health by enhancing the growth of friendly microorganisms in the intestine (de Vrese and Schrezenmeir 2008) and moreover the latter refers to a combination of prebiotics and probiotics given together. Synbiotics imply synergism in which a prebiotic compound particularly favors a probiotic as in the case of oligofructose and *Bifidobacteria* (Schrezenmeir and de Vrese 2001). It is also important to note that many of the proposed beneficial traits of probiotics are facilitated by the production of secondary metabolites such as short-chain fatty acids, functional proteins, polysaccharides, extracellular polysaccharides, and peptidoglycan-derived muropeptides. Therefore, apart from live probiotics, dead forms of these bacteria or their secretory metabolites can also be used for improving health. Paraprobiotics and postbiotics are two relatively new terms introduced in the field of functional foods which impart numerous beneficial effects. Postbiotics comprise the non-viable products such as metabolic byproducts secreted by probiotics in cell-free supernatants such as secreted proteins, enzymes, short-chain fatty acids, biosurfactants, peptides, and organic acids (Nataraj et al. 2020). On the other hand, paraprobiotics are inactivated probiotics or ghost probiotics (Tsilingiri and Rescigno 2013). Paraprobiotics indicates the use of inactivated or non-viable microbial cells or cell fractions that may be intact or ruptured and contain teichoic acids, peptidoglycans, or surface proteins which are immunologically active and thereby impart immune activation of the host system (Ananta and Knorr 2009; Adams 2010; Taverniti and Guglielmetti 2011). Apart from ameliorating the general wellbeing, several clinical trials and experimental studies emphasize the beneficial effects of strain-specific probiotic interventions in improving symptoms associated with precise health conditions. Figure 8.2 depicts how prophylactic and therapeutic consumption of probiotics affects the human system.

3 Interaction of Enteric Microbiota with Brain, Liver, and Lung Functions

Gut dysbiosis has been proved in both gastrointestinal as well as non-gastrointestinal diseases. Recent evidence of breast cancer and microbiota variation postulates the interplay of gut microbes in the progression of breast cancer (Laborda-Illanes et al. 2020). Such cross-talk between gut microbiota and different parts of human body established the concept of the gut-lung axis, gut-liver axis, and gut-brain axis.

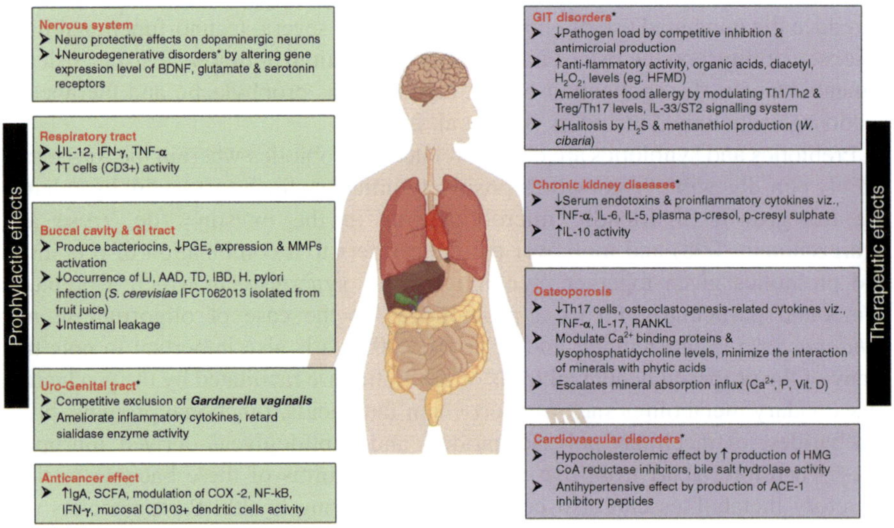

Fig. 8.2 Prophylactic and therapeutic effects of probiotics in human. The aforementioned effects noted are a combined data acquired from different preclinical and clinical studies. *Denotes prophylactic/therapeutic effects of probiotics proven by human clinical trials. *AAD* antibiotic-associated diarrhea, *IBD* inflammatory bowel diseases, *TD* traveler's diarrhea, *LI* lactose intolerance, *MMP* matrix metallo proteinases, *BDNF* brain-derived neurotrophic factor, *SCFA* short-chain fatty acids, *COX-2* cyclooxygenase-2, *HFMD* hand-foot-mouth disease, *RANKL* receptor activator of nuclear factor kappa-B ligand, *IL* interleukin, *TNF* tumor necrosis factor, *IFN* interferon, *PGE* prostaglandin E, *Ig* immunoglobulin, *NF-κB* nuclear factor kappa-light-chain-enhancer of activated B cells, *ACE-1* - angiotensin-converting Eenzyme-1, ↑ - increase, ↓ - decrease

The gut-lung axis concept postulates that alterations of intestinal microbiota may have a profound effect on lung health. Microbes or its bioactive compounds absorbed into the circulation may alter lung function by host immune cells, resulting in systemic cytokine release. Gut microbiota influences the lung health through its metabolites or endotoxins (Zuo et al. 2020). On the other hand, gut-brain axis is a two-way communication between central nervous system (CNS), enteric nervous system, and gastrointestinal tract. The interplay between gut-brain axis and microbiota is by means of endocrine, immune, neural, and humoral routes (Carabotti et al. 2015). One theory explaining the gut-brain axis is the "leaky gut theory." Leaky gut is a syndrome which is influenced by many factors like stress, unhealthy diet, alcohol use which in turn increases the intestinal permeability. Such factors result in the passage of microorganisms to different parts of the body. This theory also postulated that the disruption of the gut barrier causes the macromolecules to pass through the blood-brain barrier resulting in neuroinflammation (Slykerman et al. 2017; Garcia Bueno et al. 2016). Thus, restoring the leaky gut barrier is suggestive of one of the effective therapies for these types of neurodisorders (Ait-Belgnaoui et al. 2012). Moreover, probiotic strains like *L. helveticus* NS8 reduced post-restraint anxiety in stress-induced experimental mice by lowering adrenocorticotrophic and corticosterone hormones (Liang 2015). Such interventions lead to the emergence of "psychobiotics," an advanced branch of probiotic research wherein beneficial microbes are administered for improving mental health by altering commensal gut microbes (Sarkar et al. 2016). Altogether, psychobiotics, via the bacteria-brain communication, strengthen both the immune system as well as the enteric nervous system.

Numerous interesting studies are available on the effects of probiotics on enterohepatic system. The liver is anatomically linked to gut microbiota by enterohepatic circulation particularly through the portal duct. However, the role of gut dysbiosis on liver disease severity was only recently explored. Leaky gut further paves the way for translocation of bacteria from the gut, migration of immune cells to the liver which activate the adaptive and innate immune system thereby leading to tumor progression and subsequent liver injury (Konturek et al. 2018). In a recent study from Ponziani et al. (2018) the profiling of gut microbiota of hepatocellular carcinoma patients showed a decrease in anti-inflammatory bacteria like *Bifidobacteria* which also correlates with enhanced intestinal and liver inflammation with severe hepatocarcinogenesis.

4 Probiotic Interventions in Infectious Diseases

4.1 Antibiotic-associated Diarrhea

Antibiotic-associated diarrhea (AAD) occurs as a result of disruption of the normal gut flora during the exposure of antibiotics. Antibiotics action on anaerobes plays a crucial role in AAD and the risk is higher in case of treatment with aminopenicillin-clavulanate combination, clindamycin, and cephalosporins (Wiström et al. 2001). Major pathogens associated with AAD include *Clostridium difficle, Staphylococcus aureus*, *Salmonella* sp., *C. perfringens*, *Klebsiella oxytoca*, and *Candida* sp. followed by some unidentified pathogens also (Barbut and Meynard 2002). *C. difficle* infections fall under the category "difficult to cure" due to the increased incidence of antibiotic resistance acquisitions in this pathogen. Reduction in the acidic environment of the gut increases *C. difficle* growth and infection subsequent to germination of the clostridial acid-resistant spores. Though these infections are considered as a distinct hurdle resulting from rampant antibiotic use, the restricted therapeutic options further worsen the situation.

Probiotics are now considered as satisfactory alternative to treat the adverse symptoms of AAD. Reestablishing the dysbiosed gut is the prime target for AAD treatment (D'Souza et al. 2002). A study conducted by Cremonini et al. (2002) reports the effectiveness of probiotics in AAD owing to lowering the colonic pH and escalating the production of short-chain fatty acids (SCFA) such as acetic acid, propionic acid, butyric acid, and formic acid (Ríos-Covián et al. 2016). The beneficial microbes of paramount importance in this milieu include *L. casei, L. acidophilus, L. bulgaricus, Streptococcus thermophilus, B. bifidum, B. longum,* and *S. boulardii* (Guo et al. 2019; Liao et al. 2021).

4.2 Infectious Diarrhea

Gastrointestinal disturbance manifested by diarrhea is one of the leading culprits of hospitalization and mortality globally irrespective of ages. Rotavirus is the major cause of acute infantile diarrhea with 600,000 child deaths per year, under the age of 5 years (Parashar et al. 2006). The infant formula containing *L. reuteri, L. rhamnosus* GG, *B. animalis* Bb12, and *L. casei* Shirota generated better prophylactic effect on the recurrence of Rota viral diarrhea (Shah 2007; Szajewska et al. 2001). Studies on *Lactobacillus* sp., *Bifidobacterium* sp., and *S. boulardii* were found to be effective in controlling and treating the occurrence of viral diarrhea in young children (Grandy et al. 2010; Dalgic et al. 2011; Azagra-Boronat et al. 2020).

The second major type of diarrheal disease is the traveler's diarrhea, which mostly affects people who embark on long-distance travels (Riddle et al. 2017). 80% of the diarrheal diseases were caused by bacterial pathogens and administration of antibiotics is the most effective treatment strategy to treat these infections (Gascón 2006). The study conducted by Fourniat et al. observed that *E. coli* adhesion to HeLa cells was inhibited by heat-killed forms of *L. acidophilus*; however, the phenomenon was lost post lysis of the same (1992). To this date, numerous preclinical, clinical, and meta-analysis studies have been conducted and are still being conducted to further elucidate the significance of beneficial microbes on prophylactic and therapeutic arm of diarrheal disease management.

4.3 Peptic Ulcer due to **Helicobacter pylori** *Infection*

Helicobacter pylori are one of the major bacteria causing peptic ulcer which leads to gastric cancer. Triple combination therapy using antibiotics like amoxicillin and clarithromycin along with a proton pump inhibitor is the commonly used treatment strategy. But isolates showing more than 20% resistance to these antibiotics were reported from different geographical areas. The resistance of *H. pylori* towards different antibiotics makes them difficult to eradicate and recurrent infections by *H. pylori* leads to gastric cancer is also a concern. Hence the use of probiotics with the standard therapeutic regimen has emerged as an auxiliary treatment to control *H. pylori* infection in humans. Probiotics limit the *H. pylori* growth by decreasing the acidity of the stomach and improving gut mucosal barrier by producing SCFAs like acetic acid, lactic acid, and propionic acid (Goderska et al. 2018). The use of heat-inactivated and lyophilized culture of *L. acidophilus* showed improved eradication of *H. pylori* when administered along with standard anti-*H. pylori* treatment. This provides a more guarded method of treatment when compared to live probiotic intervention (Canducci et al. 2000). In addition to *L. acidophilus*, recent studies showed that *L. gasseri, L. reuteri*, and *Pediococcus* strains were effective in *H. pylori* eradication. Although probiotic treatment alone cannot be used as a treatment for *H. pylori* infection, its co-administration with antibiotics increases the speed of eradication and reduces antibiotic-associated side effects (Eslami et al. 2019).

4.4 Necrotizing Enterocolitis

Necrotizing enterocolitis (NEC) is the common cause of neonatal mortality that affects 20–25% newborns worldwide who are under 32 weeks of gestation (Deshpande et al. 2015). The major reason for this is the excessive pro-inflammatory response subsequent to stimulation with pathogenic bacteria (Hsueh et al. 2003). Many studies suggested a positive correlation with an altered gut microbiome and necrotizing enterocolitis development (Claud and Walker 2008). Current treatment regimens for these conditions are scanty. Multiple studies have reported the efficacy of probiotics as a prophylactic intervention in necrotizing enterocolitis (Claud and Walker 2008; Deshpande et al. 2015; Martin and Walker 2008). A recent network meta-analysis of randomized controlled trials (RCTs) by Beghetti et al. emphasized the benefits of *L. acidophilus* LB and *B. lactis* Bb-12/B94 in reducing NEC-associated risks (2021).

4.5 Respiratory Infections

The efficacy of host defense against pulmonary pathogens is mainly via activation of neutrophils and alveolar macrophages. For example, in the case of Pneumococcal infection, the systemic protection is offered by anticapsular antibodies and cytokines (Racedo et al. 2006). The study using *L. rhamnosus* strains against *S. pneumoniae* showed a significant reduction in the pathogen load and its dissemination in the lungs and other organs. The immune cells play a major role in alleviating infection by increasing cytokine levels by Th2 (IL-4, IL-10, and IL-6) and Th1 (IFN- γ) responses (Salva et al. 2010). *K. pneumoniae* is another clinical pathogen which causes high mortality on account of its drug-resistant characteristics. It is a common inhabitant of the gut, but can exert an inflammatory response in the respiratory epithelial cells while on infection. A recent study evaluates the effects of oral administration of *L. plantarum* CRM653 in mice infected with *K. pneumoniae* showed diminished lung inflammatory response with reduced numbers of neutrophils, macrophages, and suppression of TNF-α, IL-6, and T_{reg} responses (Vareille-Delarbre et al. 2019).

Seasonal viral attacks are another major respiratory tract-associated infections in children and elder ones. Evidence suggests that gut dysbiosis plays a role in viral respiratory infections. The major cause of influenza-associated mortality is the excessive production of inflammatory cytokines. Oral administration of *L. rhamnosus* in influenza-infected mice showed a diminished infiltration of mononuclear cells, reduced inflammatory response, and helper T cell-mediated antiviral activity in lungs (Song et al. 2016). In addition to the live probiotics, heat-killed forms also aid in controlling respiratory infections. Oral administration of heat-killed *L. plantarum* stimulated the production of interferon-beta (IFN-β) in the serum of intranasally challenged mice with influenza virus compared with control groups (Maeda et al. 2009). Cases of patients with severe gastrointestinal symptoms were reported in association with SARS CoV-2 infection. As angiotensin-converting enzyme 2 (ACE2) receptors (receptors used by SARS-CoV-2 to enter host cell) are abundant in epithelial surfaces of small intestinal mucosa, it may help in the viral entry and multiplication. Gut microbiota perturbations are closely linked to lung microbiota dysbiosis that increases the severity of the disease (Mohandas and Vairappan 2020). Since gut dysbiosis can be rectified by probiotics, consumption of such natural products is gaining importance in the pandemic scenario due to its immunomodulatory properties (Sundararaman et al. 2020).

5 Probiotic Interventions in Non-communicable Diseases

The rapid transition in socio-demographic factors and environment, followed by alteration in behavioral risk factors such as adaptation to modern lifestyle, physical inactivity, alcohol consumption, and tobacco use have splurged the occurrence of noncommunicable diseases (NCD) worldwide. In general, cardiovascular diseases, cancer, diabetes, and chronic respiratory diseases are considered to be the major contributors to morbidity and mortality stemming from NCD. The accumulating evidences from numerous population-based studies, animal models, and cell line studies emphasize the putative correlation between microbiome composition and the insurgence in the development of non-communicable diseases (Xu et al. 2020;

Sommariva et al. 2020). Furthermore, the GI microbial ecosystem generates considerable amounts of diverse metabolites that produce protective effects on the maintenance of host immune, digestive, and neuroendocrine system. Thus, eubiosis of gut microenvironment is crucial for the prevention and development of metabolic disorders in host body. Emerging findings propound the relevance of probiotics as therapeutic/dietary supplements with paramount health benefits, thereby positively affecting the diversity and functions of commensal gut microbiota.

An expanding number of clinical trials support the probiotic intervention on cardiovascular health. Cardiovascular diseases are generally associated with hypercholesterolemia and dyslipidemia. Hypocholesterolemic effect of probiotics is imparted by hampering cholesterol absorption from intestine (Tomaro-Duchesneau et al. 2014), assimilating cholesterol on its cell surface (Liong and Shah 2005), redirecting cholesterol for bile acid synthesis by increasing the breakdown of bile acids (Ettinger et al. 2014) and by producing short-chain fatty acids that reduce hepatogenic cholesterol production by inactivating hydroxyl-methyl-glutaryl coenzyme A reductase (HMG CoA reductase) (De Preter et al. 2007). Eventually, it also produces a significant influence on lipid profile by reducing low-density lipoprotein and circulating triglycerides along with an increase in high-density lipoprotein (Tomaro-Duchesneau et al. 2014). Subsequently, the hypotensive effect is mediated by a reduction in nitrogen oxide production in macrophages, reactive oxygen species production, and augmented production of angiotensin-converting enzyme inhibitor peptides like isoleucyl–prolyl–proline (Ile-Pro-Pro) and valyl–prolyl–proline (Val-Pro-Pro) (Khalesi et al. 2014).

Diabetes mellitus (Type 2) and obesity are contemplated as growing epidemics and their presence are associated with multiple gastrointestinal complications (Azpiroz and Malagelada 2016). Data gathered from meta-analysis and animal studies using probiotics indicates that habitual intake of these beneficial microbes can enhance insulin sensitivity, glucose turnover rate, and antioxidant status. It aids in regulating glucose metabolism and thereby reduces HbA1c, fasting blood glucose levels, the inflammatory markers (TNF-α and IL-6), and C-reactive protein levels in host body (Kobyliak et al. 2016; Firouzi et al. 2017). Probiotic intervention can also modulate the up/downregulation of gut hormones linked with glucose homeostasis such as glucagon-like peptide-1 and gastric inhibitory polypeptide (Parnell and Reimer 2009). Gut dysbiosis also influences host metabolic pathways through its effect on nutrient uptake, energy harvest, fat accumulation, and appetite. The short-chain fatty acids, primarily acetate, butyrate, and propionate derived from microbial metabolism, impart a protective effect against certain NCDs such as diabetes, colorectal cancer, and ulcerative colitis (Bolognini et al. 2016). In obese patients, the expression of tight junction proteins like zonulin and occludin is highly compromised and this alters intestinal permeability favoring entry of pathogens and their metabolites/toxins into bloodstream. Studies have also emphasized that the gut microbiota together with diet could reduce colonic permeability thereby halting the pathogen entry and consequent inflammation on body (Sivamaruthi et al. 2019). Obese and obese-diabetic subjects are characterized by chronic low-grade inflammation with an elevated basal concentration of pro-inflammatory cytokines, especially IL-6, TNF, and MCP-1(Fernandez and Manuel 2017). Contrastingly, malnutrition also fuels up immune dysfunction. The beneficial microbiota orchestrates the temporal balance between pro-inflammatory, inflammatory, and immunoregulatory cytokines there by facilitating effective pathogen clearance with minimal immune response-related disorders to the body.

Furthermore, probiotics have been gaining much attention due to their anticancer properties. The oncobiotic mechanisms were assumed by binding and degradation of potential dietary carcinogens (Orrhage et al. 2002), regulation of tumor cell proliferation and apoptosis, modulation of bacterial enzymes implicated in oncogenesis (Raman et al. 2013), and generation of metabolites having tumor-suppressing characteristics (Requena et al. 2018). Moreover, inclusion of beneficial microbes with antioxidant property in specific enteral diet of patients suffering with acute respiratory distress syndrome (ARDS) may facilitate better oxygenation during the ventilation period (Singer et al. 2019).

The risk factors related to the development of NCDs are thus strongly interrelated and are conducive to each other. Aging is closely linked with decline in microbial diversity which can be correlated with defects in T- and B-cell functions, escalated cholesterol assimilation, and inflammation rates, thus paving the way for the availability of more assembled pathogen entry points at the cellular level. Several animal model studies substantiated the role of probiotics in controlling the development of different NCDs and certain findings were taken to next level by demonstrating its effect on human trials (Kocsis et al. 2020).

6 Role of Probiotics in Anticancer Therapy

Growing evidence suggests that an imbalance in the normal flora of the gastrointestinal (GI) system might associate with a number of diseases such as inflammation, impaired metabolism, dysregulation of the immune system, and even cancer. It has already been proven that the diverse human gut microbiota participates in a complex interaction with the GI tract and the immune system (Vivarelli et al. 2019). In addition to the involvement of immune system in cancer progression and metastasis, various studies showed that a patient's microbiome also influences both disease progression and response to therapy in cancer. The results from these studies validated the findings of earlier preclinical studies and support the idea of rational modulation of the microbiome as a valid therapeutic strategy in cancer patients (Gopalakrishnan et al. 2018; Routy et al. 2018).

Recent studies have shown that certain gut bacteria have pro and anti-effects on cancer treatments. It is not yet known as to how and which bacterial species specifically contribute to the anti-tumor response induced by chemotherapy. Chemotherapy has secondary effects that include increased permeability of the intestinal barrier and thus facilitates the entry of the gut bacteria into the bloodstream. In response to the leaky gut, body elicits an immune response and is believed to be beneficial for patients, since it can also destroy cancer cells. Researchers showed that two species of bacteria, *Enterococcus hirae* and *Barnesiella intestinihominis*, together potentiate the therapeutic effects of cyclophosphamide, a chemotherapeutic agent used to treat many cancers. Several preclinical models demonstrate that the anti-tumor immune response induced by cyclophosphamide is optimized after oral administration of *E. hirae*. Treatment by oral administration of *B. intestinihominis* also enabled a similar effect. The tumor is therefore attacked directly by cyclophosphamide and indirectly by this "booster" effect of the bacteria (Daillère et al. 2016).

The effect of various microbes related to cancer is gaining much attention in cancer research, particularly in colorectal cancer. Several studies have shown the relationship of microbiota/microbial-derived factors in either the promotion or suppression of tumor development and progression, thus eliciting an effect on the anti-cancer treatments. It has been demonstrated that certain probiotic strains suppress *H. pylori*-related gastric cancer (Kuo et al. 2013) and cervical cancer induced by human Papilloma virus in the host body (Verhoeven et al. 2013). Besides, the administration of probiotics has helped to replenish the patient's gut flora which had been affected by chemotherapy (Gao et al. 2015). The possibility of cancer prevention by administering probiotics is outweighed by the concept of probiotics as a supportive care for patients undergoing anticancer treatment. A recent report of a randomized controlled trial (RCT) has demonstrated that probiotics can reduce the toxicity elucidated by chemotherapy and radiation therapy (Hassan et al. 2018).

Even with the emergence of new treatments such as targeted therapies and immunotherapy, advanced and persistent cancers require systemic chemotherapy (Bradbury et al. 2017). However, systemic chemotherapy has certain side effects including nausea, vomiting, and diarrhea which may worsen the nutritional status, reduce immune function, delay the treatment cycle, and increase the treatment costs in many cases (Andreyev et al. 2014; Marx et al. 2016; Tong et al. 2009). Many of the studies reported that chemotherapy-induced diarrhea may be caused by intestinal epithelial cell apoptosis, intestinal barrier dysfunction, alterations to the intestinal flora, and pro-inflammatory cytokine production (Tian et al. 2019).

Most studies suggest that probiotics are effective against chemotherapy-induced diarrhea. Two independent meta-analyses showed that probiotic administration in cancer patients significantly decreased the incidence of radiotherapy-induced diarrhea (Qiu et al. 2019; Wei et al. 2018). Tian et al. (2019) reported a positive correlation of probiotic consumption and reduced episodes of diarrhea in lung cancer patients. However, there was no difference in the rates of nausea or vomiting between the test and control groups. Another study highlighted the results of clinical trials which evaluated the preventive effect of probiotics on infections associated with colorectal cancer resection. It was shown that probiotics significantly reduced the post-infection and pneumonia in patients who undergo colorectal cancer surgery compared with placebo (Ouyang et al. 2019).

Mucositis is another major problem associated with strong chemotherapy in cancer patients. Intestinal mucositis is a cytotoxic effect in the chemotherapy of cancer mainly due to the drug 5-fluorouracil (5-FU). Because of its high therapeutic efficacy, it is commonly used in different types of cancers like breast, colorectal, stomach, and head and neck cancers (Longley et al. 2003). But its non-specific action in all types of cells produces many side effects including intestinal mucositis (Duncan and Grant 2003). Many studies evidenced that 5-FU treatment modifies the relative abundance of microbiota such as *Lactobacillus, Bacteroidetes, Clostridium, Staphylococcus, E. coli*, and *Streptococcus* (Stringer et al. 2009).

In a study published by Jiang et al. (2019), the usage of probiotics significantly reduced the rate of grade 3 mucositis in patients with nasopharyngeal cancer undergoing chemo radiotherapy, when compared with placebo (16% vs 46%). Probiotics also helped to increase the levels of CD4+, CD8+, and CD3+ T cells. However, there was no difference in the suppression of bone marrow, levels of lymphocytes, leukocytes, hemoglobin, albumin, or weight loss.

The growing evidence using animal models suggests that the efficacy of anticancer therapies is directly linked to the composition of the microbiota. It was found that animals devoid of any microbiota failed to respond to chemotherapy compared to normal mice during tumor challenge. Also, upon addition of certain bacterial species, the efficacy of immune checkpoint inhibitors in mouse models improved. In addition, several studies suggest that the difference in microbiota was clearly distinguishable in patients who respond to immune checkpoint inhibition and those who do not respond. Therefore, the theory of treating patients with antibiotics during anticancer therapy is currently a topic of ongoing research (Vivarelli et al. 2019).

7 Major Concerns Associated with Probiotic Intervention

Bacteria naturally present in food supplements and probiotics constitute a potential source of antibiotic resistance determinants. Owing to human consumption, safety of these organisms is of paramount importance as their resistance toward antibiotics can be one of the possible threats. The possibility of these bacteria to transfer resistance determinants horizontally to pathogens and commensal gut microbiota intensifies the issue. The main probiotic bacteria commercially available are of the genera *Lactobacillus* and *Bifidobacterium*, along with *Bacillus* or *E. coli* strains. Some of the aforementioned species are used as starters and adjunct cultures in the food industry. With some exceptions, antibiotic resistance in these beneficial microbes does not constitute a safety concern in itself, when mutations or intrinsic resistance mechanisms are responsible for the resistance phenotype. In fact, some probiotic strains with intrinsic antibiotic resistance could be useful for restoring the gut microbiota after antibiotic treatment. However, specific antibiotic resistance determinants carried on mobile genetic elements, such as tetracycline resistance genes, are often detected in the typical probiotic genera and constitute a reservoir of resistance for potential food or gut pathogens, thus representing a serious safety issue (Gueimonde et al. 2013). Plasmid-associated antibiotic resistance, which occasionally occurs, is a matter of concern as it can be detrimental to the use of probiotics owing to the possibility of the resistance spreading to harmful microorganisms inhabiting the same niche. Further, the presence of transferable antibiotic resistance genes even to a less represented member of the gut microbial community poses a safety hazard and needs to be taken into account. Probiotic safety is beleaguered with the scarcity of well-designed and targeted studies resulting from diverse collection methodologies, clinical endpoints, and analytical rigor and probiotic strains of interest, which need to be dealt in the right perspective (Sharma et al. 2014; Suez et al. 2019). The administration of paraprobiotics and postbiotics slows down the pace of antibiotic resistance transmission since the bacteria are not consumed in their live form, thereby eliminating the presence of resistance determinants.

8 Delineation of Microbial Profiling Using Multi-omic Technology

The breakthrough of different omics technologies provides a broad range of high throughput methods that enable an in-depth understanding of probiotic physiology. Probio-genomics is a new area fostered by the consistent progression of these techniques and has significantly contributed to gathering more information about the evolution, genetic diversity and also assists in unfolding the molecular mechanism behind the purported health-promoting activities of beneficial microbes.

Insights into genetic motifs, patterns, and elucidation of generic pathways decipher the functional dimensions of a particular isolate of interest. Armed with this knowledge, the so-called difficult to detect novel secondary metabolites and other antimicrobial substances can be easily determined by adaptation of these advanced tools (Corr et al. 2007; Stoyanova et al. 2012). This was further exemplified by the detection of various gene clusters of mucus binding pili responsible for intestinal adherence (Douillard et al. 2013), detection of epigenetic alterations (Casadesus and Low 2006), and motifs responsible for coding resistance to crucial antibiotics (Proença et al. 2017) of different *Lactobacillus* species. Of note, the rapid progress in single-cell genomic techniques can be employed to identify the least abundant bacterial species within a heterogenous bacterial community sample (Yao et al. 2017). In contrast, metagenomics discerns an overview of species abundance in the microbial ecosystem and highlights the common metabolic pathways available in the particular microbiome (Huttenhower et al. 2012). It also helps in addressing the alteration in functional activity of gut microbiota in response to dietary and xenobiotic perturbations (Maurice et al. 2013). Furthermore, it is also imperative in envisaging amino acid metabolism (den Hengst et al. 2005) and in discovering strains with unique traits like increased thermo resistance and enhanced threshold for heightened stress state as reported in certain strains of *Lactococcus* (Chen et al. 2015) and *Lactobacillus* species respectively (Liu et al. 2016).

Proteomics confers the proteome map of the bacteria thereby helping in determining the biological function of the cell. Notably, the advent of proteomics dispenses knowledge regarding the adaptation of gut microbiota under harsh gastrointestinal conditions like low pH, presence of bile salts (Wu et al. 2012), and osmotic stress inside the body (Zhang et al. 2010). Combinatorial holistic approach of proteomics with metagenomics and transcriptomics imparts critical information regarding robust biomarkers involved in the functional process of bacteria. Metabolomics/metabolic profiling of probiotic bacteria unwinds the target of their actions. This technique is also efficacious in determining the distinct metabolic changes that occur under different meticulous conditions and also add insights into rate-limiting steps which are crucial for the generation of useful metabolites (Ohtake et al. 2017). Besides this, it also helps to underpin the relation between gastrointestinal diseases and metabolites generated by the commensal microbes (Nicholson et al. 2005). Nonetheless, the volume of information gathered from metabolomic approach facilitates better optimization of strains by understanding their functional traits in therapeutic/food industry (Khangwal and Shukla 2019). The emergence of culturomics, a fluxomic analysis, furnishes detailed description of the beneficial microbiota at the cellular level and it is widely used for identifying new strains that are generally not detected via other techniques (Kambouris et al. 2018). It is worth emphasizing that an inclusive approach using integrated multi-omics techniques yields a more complete picture of the microbial composition, its diversity, and in determining the molecular mechanisms by which probiotics enact their beneficial attributes. The integrated multi-omic analysis has revealed different human enterotypes (Arumugam et al. 2011) and microbial divergences, in response to different human diseases (Sekirov et al. 2010) as well as environmental factors (Candela et al. 2012). As an example, Bianchi et al. (2020) used an integrative approach to elucidate the changes in microbial metabolism, amino acid production, and health-promoting attributes of *L. paracasei, S. thermophilus,* and *Bifidobacteria* strains in response to various technological challenges and formulations respectively.

9 Bioengineered Probiotics

Even though the proven effect of probiotic strains in the prevention and treatment of intestinal diseases is well studied, the non-specificity and ineffectiveness of the same in certain recipients is worrisome. The factors determining the efficiency of the probiotic treatment include the strain, dose, route of administration, and the formulation of probiotic preparation. Moreover, the manufacturing process and probiotic delivery system also affects the overall activity (Grzeskowiak et al. 2011). Recent studies showed that the gut microbiome diversity/indigenous flora among different ethnic groups may also affect probiotic efficacy (Barzegari and Saei 2012). Furthermore, non-specific nature of traditional probiotics often fails to address the entry and multiplication of enteric pathogens. Wide acceptability and usage of genome sequencing and genetic engineering have enabled researchers to develop bioengineered probiotics which gain attention as a therapeutic strategy. Genetic modification via bioengineering enables the development of new putative strains expressing foreign genes geared toward more specificity and desired functionalities. The use of bioengineered probiotics led to the concept of biodrug for the prevention/

treatment of specific entities and also in arming the host defense system by acting as adjuvant for vaccine delivery (Amalaradjou and Bhunia 2013).

The new advancement in the field of probiotics is the use of recombinant technology to create bioengineered probiotics. The recombinant probiotics were originally generated for the mucosal delivery of therapeutic and prophylactic molecules including DNA, peptides, single-chain variable fragments, cytokines, enzymes, and allergens. The colonization capacity, gastric acid and bile salts tolerance, and sustained colonization in the mucosal surface make probiotic bacteria an excellent delivery system in vivo (Wells 2011; Sleator and Hill 2008).

During the past two decades, the field of probiotic genetic engineering has advanced as a leading area of research for applications in humans and animals (LeBlanc et al. 2013; Amalaradjou and Bhunia 2013). Designer probiotics have been used for various purposes like to improve nutritional and health, as therapeutic agents, in cancer treatment, to prevent metabolic diseases, etc. Many species of human intestinal bacteria are used to produce recombinant probiotics. Some of the promising candidates are *L. lactis* (LL-Thy12) to induce suppressor T cells and treating the inflammatory diseases, *Salmonella typhimurium* A1-R and non-pathogenic *E. coli* Nissle 1917 to target tumor and treating metastatic cancer, *L. lactis* sAGX0085 carrying an *htff1* cassette producing hTFF1for reducing the severity and the course of radiation-induced oral mucositosis, *L. casei* pSW501for treating Type I diabetes, acylphosphatidylethanolamines (NAPE)-expressing *E. coli* Nissle 1917 to reduce the levels of obesity, *L. rhamnosus* JB-1 to improve the enteric nerve function and behavioral aspects, etc. Although many of the aforementioned strains showed promising activity in in-vivo systems, thorough studies and clinical trials are warranted before its human application (Kumar et al. 2016).

Taken together, adjuvant therapy using probioceuticals having direct or indirect impact on multi-organ system will preferably tune the host defense system against multitude of clinical indications and the subsequent outcomes in the body. However, it is necessary to gather more information to assess the risks associated with their biosafety and their ability to cause allergy due to prolonged habitual consumption.

10 Next-generation Probiotics: Metamorphosis from Commensal Bacteria to Novel Biodrugs

Today, the probiotic industry is mainly occupied by the so-called conventional probiotics encompassing strains of *Lactobacillus* and *Bifidobacterium*. Although they possess GRAS (Generally Regarded as Safe)/QPS (Qualified Presumption of Safety) status, its rampant regular usage doesn't aim against specific diseases. Indeed, evolving evidences suggest that the routine administration of traditional probiotics to hospitalized patients with potential indications has been involved with adverse effects or worsen the existing comorbidities (Yelin et al. 2019). Hence, probiotic formulations have to be tailored against the specific target prior to its endorsement as prophylactic/therapeutic usage.

Advanced studies on gut microbiota and its possible interactions with host body followed by availability of advanced genetic sequencing tools and the bioinformatics platforms paved the way for identifying novel candidates as next-generation probiotics (NGP). NGPs are developed to target-specific diseases and it belongs to diverse genera with majority being identified on the basis of comparative analysis on healthy and unhealthy individuals. *Prevotella copri* of phylum Bacteroidetes is a promising NGP candidate that was proved to regulate glucose homeostasis via modulating intestinal gluconeogenesis (De Vadder et al. 2016). In parallel, *A. muciniphila,* a mucin utilizing organism of phylum Verrucomicrobia, synchronies the endocannabinoid system, thereby playing a pivotal role in the regulation of inflammation and type-2 diabetes-associated obesity (Plovier et al. 2017). Importantly, its utility is further extended as an oncobiotic (Ansaldo et al. 2019) and is known to produce an immunomodulatory protein termed "Amuc-1100" (Plovier et al. 2017).

Nontoxigenic *Bacteroides fragilis* strains possess a unique capsular polysaccharide with zwitterionic motifs termed "Polysacharride A," which regulates neuroinflammations (Wang et al. 2014) and prevent viral encephalitis (Ramakrishna et al. 2019). Even though studies are limited, *Christensenella minuta* of phylum Firmicutes and *Parabacteroides goldsteinii* of phylum Bacteroidetes demonstrated potent probiotic effects against metabolic disorders such as obesity (Goodrich et al. 2016). On top of these, *Pediococcus pentosaceus* categorized as a lactic acid bacteria was widely studied for its hypocholesterolemic effect and production of bacteriocins (Syakila et al. 2019). *Gordonibacter* species of phylum *Actinobacteria* isolated from human gut were found to be capable of converting ellagic acid found in nuts and pomegranate to urolithin metabolites such as urolithin A, B, and isourolithin A. The anti-inflammatory, anti-carcinogenic, cardioprotective, and neuroprotective properties of these metabolites categorized *Gordonibacter* as a propitious NGP candidate for further studies (Selma et al. 2017). Intensive research on bacterial species such as *Faecalibacterium prausnitzii, Eubacterium limosum, E. faecium, Bifidobacterium* sp., *Collinsella aerofaciens,* and *Burkholderia cepacia* deciphered promising effects on anticancer immunotherapies (Chaput et al. 2017, 2019). The specific beneficial characteristics of these NGP candidates demand more longitudinal studies before it is commercialized either for targeted therapy or for the amelioration of GI diseases.

11 Conclusion and Future Perspectives

The impact of gut microbiota on the maintenance and restoration of host wellbeing by alleviating the symptoms of infectious diseases, non-communicable diseases and improving mental health has boosted the administration of both probiotics and paraprobiotics. Application of probiotics along with antibiotics and other drugs like chemotherapeutic agents might reduce the drug's ill effects and improve its efficiency in a great extent. In addition to conventional probiotics and its bioengineered counterparts, new advancements in the area have led to the identification of novel species of probiotics which broadens the scope of probiotics in applied research. However, concerns regarding the development of drug resistance and its transfer along with some isolated reports of sepsis have limited the use of live probiotics in immunocompromised individuals. The advancement of new technologies, deep sequencing of the strains for human consumption, and detailed studies of the altered microbial pattern related to specific diseases is greatly warranted to establish the extended use of probiotics in the health sector.

References

Adams CA (2010) The probiotic paradox: live and dead cells are biological response modifiers. Nutr Res Rev 23(1):37–46. https://doi.org/10.1017/S0954422410000090

Ait-Belgnaoui A, Durand H, Cartier C, Chaumaz G, Eutamene H, Ferrier L et al (2012) Prevention of gut leakiness by a probiotic treatment leads to attenuated HPA response to an acute psychological stress in rats. Psychoneuroendocrinology 37:1885–1895. https://doi.org/10.1016/j.psyneuen.2012.03.024

Amalaradjou MA, Bhunia AK (2013) Bioengineered probiotics, a strategic approach to control enteric infections. Bioengineered 4:379–387

Ananta E, Knorr D (2009) Comparison of inactivation pathways of thermal or high pressure inactivated Lactobacillus rhamnosus ATCC 53103 by flow cytometry analysis. Food Microbiol 26(5):542–546. https://doi.org/10.1016/j.fm.2009.01.008

Andreyev J, Ross P, Donnellan C, Lennan E, Leonard P, Waters C, Wedlake L, Bridgewater J, Glynne-Jones R, Allum W et al (2014) Guidance on the management of diarrhoea during cancer chemotherapy. Lancet Oncol 15:e447–e460

Ansaldo D, Slayden LC, Ching KL, Koch MA, Wolf NK, Plichta DR, Brown EB, Graham DB, Xavier RJ, Moon JJ, Barton GM (2019) Akkermansia muciniphila induces intestinal adaptive immune responses during homeostasis. Science 21:1179–1184. https://doi.org/10.1126/science.aaw7479

Arumugam M, Raes J, Pelletier E et al (2011) Enterotypes of the human gut microbiome. Nature 473:174–180. https://doi.org/10.1038/nature09944

Azpiroz F, Malagelada C (2016) Diabetic neuropathy in the gut: pathogenesis and diagnosis. Diabetologia 59:404–408. https://doi.org/10.1007/s00125-015-3831-1

Barbut F, Meynard JL (2002) Managing antibiotic associated diarrhoea. BMJ 324(7350):1345–1346. https://doi.org/10.1136/bmj.324.7350.1345

Barzegari A, Saei AA (2012) Designing probiotics with respect to the native microbiome. Future Microbiol 7:571–575

Beghetti I, Panizza D, Lenzi J, Gori D, Martini S, Corvaglia L, Aceti A (2021) Probiotics for preventing necrotizing enterocolitis in preterm infants: a network meta-analysis. Nutrients 13(1):192. https://doi.org/10.3390/nu13010192

Bianchi L, Laghi L, Correani V, Schifano E, Landi C, Uccelletti D, Mattei B (2020) A combined proteomics, metabolomics and in vivo analysis approach for the characterization of probiotics in large-scale production. Biomol Ther 10(1):157. https://doi.org/10.3390/biom10010157

Bolognini D, Tobin AB, Milligan G, Moss CE (2016) The pharmacology and function of receptors for short-chain fatty acids. Mol Pharmacol 89(3):388–398. https://doi.org/10.1124/mol.115.102301

Bradbury P, Sivajohanathan D, Chan A, Kulkarni S, Ung Y, Ellis PM (2017) Postoperative adjuvant systemic therapy in completely resected non-small cell lung cancer: a systematic review. Clin Lung Cancer 18:259–273

Candela M, Biagi E, Maccaferri S, Turroni S, Brigidi P (2012) Intestinal microbiota is a plastic factor responding to environmental changes. Trends Microbiol 20:385–391. https://doi.org/10.1016/j.tim.2012.05.003

Canducci F, Armuzzi A, Cremonini F, Cammarota G, Bartolozzi F, Pola P, Gasbarrini G, Gasbarrini A (2000) A lyophilized and inactivated culture of Lactobacillus acidophilus increases Helicobacter pylori eradication rates. Aliment Pharmacol Ther 14(12):1625–1629. https://doi.org/10.1046/j.1365-2036.2000.00885.x

Carabotti M, Scirocco A, Maselli MA, Severi C (2015) The gut-brain axis: interactions between enteric microbiota, central and enteric nervous systems. Ann Gastroenterol 28(2):203–209

Casadesus J, Low D (2006) Epigenetic gene regulation in the bacterial world. Microbiol Mol Biol Rev 70(3):830–856. https://doi.org/10.1128/mmbr.00016-06

Chang CJ, Lin TL, Tsai YL, Wu TR, Lai WF, Lu CC, Lai HC (2019) Next generation probiotics in disease amelioration. J Food Drug Anal 27(3):615–622. https://doi.org/10.1016/j.jfda.2018.12.011

Chaput N, Lepage P, Coutzac C, Soularue E, Le Roux K, Monot C, Boselli L, Routier E, Cassard L, Collins M, Vaysse T, Marthey L, Eggermont A, Asvatourian V, Lanoy E, Mateus C, Robert C, Carbonnel F (2017) Baseline gut microbiota predicts clinical response and colitis in metastatic melanoma patients treated with ipilimumab. Ann Oncol 28(6):1368–1379. https://doi.org/10.1093/annonc/mdx108

Chen J, Shen J, Hellgren LI, Jensen RP, Solem C (2015) Adaptation of Lactococcus lactis to high growth temperature leads to a dramatic increase in acidification rate. Sci Rep 5:14199. https://doi.org/10.1038/srep14199

Clancy R (2003) Immunobiotics and the probiotic evolution. FEMS Immunol Med Microbiol 38(1):9–12. https://doi.org/10.1016/S0928-8244(03)00147-0

Claud EC, Walker WA (2008) Bacterial colonization, probiotics, and necrotizing enterocolitis. J Clin Gastroenterol 42(2):46–52. https://doi.org/10.1097/MCG.0b013e31815a57a8

Cordaillat-Simmons M, Rouanet A, Pot B (2020) Live biotherapeutic products: the importance of a defined regulatory framework. Exp Mol Med 52:1397–1406. https://doi.org/10.1038/s12276-020-0437-6

Corr SC, Li Y, Riedel CU, O'Toole PW, Hill C, Gahan CG (2007) Bacteriocin production as a mechanism for the antiinfective activity of Lactobacillus salivarius UCC118. Proc Natl Acad Sci U S A 104(18):7617–7621. https://doi.org/10.1073/pnas.0700440104

Cremonini F, Di Caro S, Santarelli L, Gabrielli M, Candelli M, Nista EC, Lupascu A, Gasbarrini G, Gasbarrini A (2002) Probiotics in antibiotic-associated diarrhoea d igestive and liver disease. Dig Liv Dis 34(2):78–80

Daillère R, Vétizou M, Waldschmitt N, Yamazaki T, Isnard C, Poirier-Colame V, Duong CPM, Flament C, Lepage P, Roberti MP, Routy B, Jacquelot N, Apetoh L, Becharef S, Rusakiewicz S, Langella P, Sokol H, Kroemer G, Enot D, Roux A, Eggermont A, Tartour E, Johannes L, Woerther PL, Chachaty E, Soria JC, Golden E, Formenti S, Plebanski M, Madondo M, Rosenstiel P, Raoult D, Cattoir V, Boneca IG, Chamaillard M, Zitvogel L (2016) Enterococcus hirae and Barnesiella intestinihominis facilitate cyclophosphamide-induced therapeutic immunomodulatory effects. Immunity 45(4):931–943. https://doi.org/10.1016/j.immuni.2016.09.009

Dalgic N, Sancar M, Bayraktar B, Pullu M, Hasim O (2011) Probiotic, zinc and lactose free formula in children with rotavirus diarrhoea: Are they effective? Pediatr Int 53:677–682

Das DJ, Shankar A, Johnson JB, Thomas S (2020) Critical insights into antibiotic resistance transferability in probiotic Lactobacillus. Nutrition 69:110567. https://doi.org/10.1016/j.nut.2019.110567

De Preter V, Vanhoutte T, Huys G, Swings J, De Vuyst L, Rutgeerts P, Verbeke K (2007) Effects of Lactobacillus casei Shirota, Bifidobacterium breve and oligofructose-enriched inulin on colonic nitrogen-protein metabolism in healthy humans. Am J Physiol Gastrointest Liver Physiol 292(1):358–368. https://doi.org/10.1152/ajpgi.00052.2006

de Vrese M, Schrezenmeir J (2008) Probiotics, prebiotics, and synbiotics. Adv Biochem Eng Biotechnol 111:1–66. https://doi.org/10.1007/10_2008_097

De Vadder F, Kovatcheva-Datchary P, Zitoun C, Duchampt A, Bäckhed F, Mithieux G (2016) Microbiota-produced succinate improves glucose homeostasis via intestinal gluconeogenesis. Cell Metabol 24:151–157. https://doi.org/10.1016/j.cmet.2016.06.013

den Hengst CD, Van Hijum SA, Geurts JM, Nauta A, Kok J, Kuipers OP (2005) The Lactococcus lactis CodY regulon: identification of a conserved cis-regulatory element. J Biol Chem 280(40):34332–34342. https://doi.org/10.1074/jbc.m502349200

Deshpande G, Rao S, Patole S (2015) Probiotics in neonatal intensive care–back to the future.Aust N Z. J Obstet Gynaecol 55(3):210–217. https://doi.org/10.1111/ajo.12328

Douillard FP, Ribbera A, Järvinen HM, Kant R, Pietilä TE, Randazzo C, Paulin L, Laine PK, Caggia C, von Ossowski I, Reunanen J, Satokari R, Salminen S, Palva A, de Vos WM (2013) Comparative genomic and functional analysis of Lactobacillus casei and Lactobacillus rhamnosus strains marketed as probiotics. Appl Environ Microbiol 79(6):1923–1933. https://doi.org/10.1128/aem.03467-12

D'Souza AL, Rajkumar C, Cooke J, Bulpitt CJ (2002) Probiotics in prevention of antibiotic associated diarrhoea: meta-analysis. BMJ 324:1361–1364

Duncan M, Grant G (2003) Mucositis-Causes and possible treatments. Alim Pharmacol Ther 18:853–874. https://doi.org/10.1046/j.0269-2813.2003.01784.x

Eslami M, Yousefi B, Kokhaei P, Jazayeri Moghadas A, Sadighi Moghadam B, Arabkari V, Niazi Z (2019) Are probiotics useful for therapy of Helicobacter pylori diseases? Comp Immunol Microbiol Infect Dis 64:99–108. https://doi.org/10.1016/j.cimid.2019.02.010

Ettinger G, MacDonald K, Reid G, Burton JP (2014) The influence of the human microbiome and probiotics on cardiovascular health. Gut Microbes 5(6):719–728. https://doi.org/10.4161/19490976.2014.983775

FDA (2016) Early clinical trials with live biotherapeutic products: chemistry, manufacturing, and control information

Fernandez C, Manuel A (2017) Obesity, respiratory disease and pulmonary infections. Ann Res Hosp. https://doi.org/10.21037/arh.2017.08.06

Firouzi S, Majid HA, Ismail A, Kamaruddin NA, Barakatun-Nisak MY (2017) Effect of multi-strain probiotics (multi-strain microbial cell preparation) on glycemic control and other diabetes-related outcomes in people with type 2 diabetes: a randomized controlled trial. Eur J Nutr 56:1535–1550

Fourniat J, Colomban C, Linxe C, Karam D (1992) Heat-killed Lactobacillus acidophilus inhibits adhesion of Escherichia coli B41 to HeLa cells. Ann Rech Vet 23(4):361–370

Gao Z, Guo B, Gao R, Zhu Q, Qu W, Qin H (2015) Probiotics modify human intestinal mucosa-associated microbiota in patients with colorectal cancer. Mol Med Rep 12(4):6119–6127

García Bueno B, Caso JR, Madrigal JL, Leza JC (2016) Innate immune receptor toll-like receptor 4 signalling in neuropsychiatric diseases. Neurosci Biobehav Rev 64:134–147. https://doi.org/10.1016/j.neubiorev.2016.02.013

Gascón J (2006) Epidemiology, etiology and pathophysiology of traveler's diarrhoea. Digestion 73(1):102–108

Goderska K, Agudo Pena S, Alarcon T (2018) Helicobacter pylori treatment: antibiotics or probiotics. Appl Microbiol Biotechnol 102(1):1–7. https://doi.org/10.1007/s00253-017-8535-7

Goodrich JK, Davenport ER, Beaumont M, Jackson MA, Knight R, Ober C, Spector TD, Bell JT, Clark AG, Ley RE (2016) Genetic determinants of the gut microbiome in UK twins. Cell Host Microbe 19(5):731–743. https://doi.org/10.1016/j.chom.2016.04.017

Gopalakrishnan V, Spencer CN, Nezi L, Reuben A, Andrews MC, Karpinets TV et al (2018) Gut microbiome modulates response to anti-PD-1 immunotherapy in melanoma patients. Science 359:97–103. https://doi.org/10.1126/science.aan4236

Grandy G, Medina M, Soria R, Teran CG, Araya M (2010) Probiotics in the treatment of acute rotavirus diarrhoea. A randomized, double-blind, controlled trial using two different probiotic preparations in Bolivian children. BMC Infect Dis 10:253–259

Grzeskowiak L, Isolauri E, Salminen S, Gueimonde M (2011) Manufacturing process influences properties of probiotic bacteria. Br J Nutr 105:887–894. https://doi.org/10.1017/S0007114510004496

Gueimonde M, Borja S, de Los Reyes-Gavilán CG, Abelardo M (2013) Antibiotic resistance in probiotic bacteria. Front Microbiol 4:202. https://doi.org/10.3389/fmicb.2013.00202

Guo Q, Goldenberg JZ, Humphrey C, El Dib R, Johnston BC (2019) Probiotics for the prevention of pediatric antibiotic-associated diarrhoea. Cochrane Database of Syst Rev 4(4):CD004827. https://doi.org/10.1002/14651858.CD004827.pub5

Hassan H, Rompola M, Glaser AW, Kinsey SE, Philips RS (2018) Systematic review and meta-analysis investigating the efficacy and safety of probiotics in people with cancer. Support Care Cancer 26(8):2503–2509

Hsueh W, Caplan MS, Qu XW et al (2003) Neonatal necrotizing enterocolitis: clinical considerations and pathogenetic concepts. Pediatr Dev Pathol 6(1):6–23

Huttenhower C, Gevers D, Knight R, Abubucker S, Badger JH, Chinwalla AT et al (2012) Structure, function and diversity of the healthy human microbiome. Nature 486:207–214. https://doi.org/10.1038/nature11234

Azagra-Boronat I, Massot-Cladera M, Knipping K, Garssen J, Amor KB, Knol J, Franch À, Castell M, Rodríguez-Lagunas MJ, Pérez-Cano FJ (2020) Strain-specific probiotic properties of Bifidobacteria and Lactobacilli for the prevention of diarrhoea caused by rotavirus in a preclinical model. Nutrients 12(2):498. https://doi.org/10.3390/nu12020498

Jiang C, Wang H, Xia C et al (2019) A randomized, double-blind, placebo-controlled trial of probiotics to reduce the severity of oral mucositis induced by chemoradiotherapy for patients with nasopharyngeal carcinoma. Cancer 125(7):1081–1090

Kambouris ME, Pavlidis C, Skoufas E, Arabatzis M, Kantzanou M, Velegraki A, Patrinos GP (2018) Culturomics: a new kid on the block of OMICS to enable personalized medicine. OMICS 22(2):108–118. https://doi.org/10.1089/omi.2017.0017

Khalesi S, Sun J, Buys N, Jayasinghe R (2014) Effect of probiotics on blood pressure: a systematic review and meta-analysis of randomized, controlled trials. J Hypertens 64(4):897–903. https://doi.org/10.1161/hypertensionaha.114.03469

Khangwal I, Shukla P (2019) Combinatory biotechnological intervention for gut microbiota. Appl Microbiol Biotechnol 103(9):3615–3625. https://doi.org/10.1007/s00253-019-09727

Kobyliak N, Virchenko O, Falalyeyeva T (2016) Pathophysiological role of host microbiota in the development of obesity. Nutr J 23:15–43. https://doi.org/10.1186/s12937-016-0166-9

Kocsis T, Molnárv B, Németh D et al (2020) Probiotics have beneficial metabolic effects in patients with type 2 diabetes mellitus: a meta-analysis of randomized clinical trials. Sci Rep 10:11787. https://doi.org/10.1038/s41598-020-68440-1

Konturek PC, Harsch IA, Konturek K, Schink M, Zopf Y (2018) Gut-liver axis: How intestinal bacteria affect the liver. Fortschr Med 160(5):11–15. https://doi.org/10.1007/s15006-018-1051-6

Kotzampassi K, Giamarellos-Bourboulis EJ (2012) Probiotics for infectious diseases: more drugs, less dietary supplementation. Int J Antimicrob Agents 40(4):288–296. https://doi.org/10.1016/j.ijantimicag.2012.06.006

Kumar M, Yadav AK, Verma V, Singh B, Mal G, Nagpal R, Hemalatha RR (2016) Bioengineered probiotics as a new hope for health and diseases: an overview of potential and prospects. Future Microbiol 11:4. https://doi.org/10.2217/fmb.16.4

Kuo CH, Wang SS, Lu CY, Hu HM, Kuo FC, Weng BC, Wu CC, Liu CJ, Tsai PY, Lee TC, Chen LW, Cheng KH, Chang LL, Wu DC (2013) Long-term use of probiotic-containing yogurts is a safe way to prevent Helicobacter pylori: based on a Mongolian gerbil's model.

Laborda-Illanes A, Sanchez-Alcoholado L, Dominguez-Recio ME, Jimenez-Rodriguez B, Lavado R, Comino-Méndez I, Alba E, Queipo-Ortuño MI (2020) Breast and gut microbiota action mechanisms in breast cancer pathogenesis and treatment. Cancers (Basel) 12(9):2465. https://doi.org/10.3390/cancers12092465

LeBlanc JG, Aubry C, Cortes-Perez NG et al (2013) Mucosal targeting of therapeutic molecules using genetically modified lactic acid bacteria: an update. FEMS Microbiol Lett 344:1–9

Liang S (2015) Administration of Lactobacillus helveticus NS8 improves behavioral,cognitive, and biochemical aberrations caused by chronic restraint stress. Neuroscience 310:561–577
Liao W, Chen C, Wen T, Zhao Q (2021) Probiotics for the prevention of antibiotic-associated diarrhoea in adults: a meta-analysis of randomized placebo-controlled trials. J Clin Gastroenterol 55(6):469–480. https://doi.org/10.1097/MCG.0000000000001464
Liong MT, Shah NP (2005) Acid and bile tolerance and cholesterol removal ability of Lactobacilli strains. J Dairy Sci 88(1):55–66
Liu J, Deng Y, Peters BM, Li L, Li B, Chen L, Xu Z, Shirtliff ME (2016) Transcriptomic analysis on the formation of the viable putative non-culturable state of beer-spoilage Lactobacillus acetotolerans. Sci Rep 6:36753. https://doi.org/10.1038/srep36753
Longley DB, Harkin DP, Johnston PG (2003) 5-Fluorouracil: Mechanisms of action and clinical strategies. Nat Rev Cancer 3:330–338. https://doi.org/10.1038/nrc1074
Maeda N, Nakamura R, Hirose Y, Murosaki S, Yamamoto Y, Kase T, Yoshikai Y (2009) Oral administration of heat-killed Lactobacillus plantarum L-137 enhances protection against influenza virus infection by stimulation of type I interferon production in mice. Int Immuno pharmacol 9(9):1122–1125. https://doi.org/10.1016/j.intimp.2009.04.015
Martin CR, Walker WA (2008) Probiotics: role in pathophysiology and prevention in necrotizing enterocolitis. Semin Perinatol 32:127–137
Marx W, Kiss N, McCarthy AL, McKavanagh D, Isenring L (2016) Chemotherapy-induced nausea and vomiting: A narrative review to inform dietetics practice. J Acad Nutr Diet 116:819–827
Maurice CF, Haiser HJ, Turnbaugh PJ (2013) Xenobiotics shape the physiology and gene expression of the active human gut microbiome. Cell 152:39–50. https://doi.org/10.1016/j.cell.2012.10.052
Mohandas S, Vairappan B (2020) SARS-CoV-2 infection and the gut-liver axis. J Dig Dis 21(12):687–695. https://doi.org/10.1111/1751-2980.12951
Nataraj BH, Ali SA, Behare PV, Yadav H (2020) Postbiotics-paraprobiotics: the new horizons in microbial biotherapy and functional foods. Microb Cell Factories 19(1):168. https://doi.org/10.1186/s12934-020-01426-w
Nicholson JK, Holmes E, Wilson ID (2005) Gut microorganisms, mammalian metabolism and personalized health care. Nat Rev Microbiol 3(5):431–438. https://doi.org/10.1038/nrmicro1152
Oelschlaeger TA (2010) Mechanisms of probiotic actions–a review. Int J Med Microbiol 00(1):57–62. https://doi.org/10.1016/j.ijmm.2009.08.005
Ohtake T, Pontrelli S, Laviña WA, Liao JC, Putri SP, Fukusaki E (2017) Metabolomics-driven approach to solving a CoA imbalance for improved 1-butanol production in Escherichia coli. Metab Eng 41:135–143. https://doi.org/10.1016/j.ymben.2017.04.003
Orrhage KM, Annas A, Nord CE, Brittebo EB, Rafter JJ (2002) Effects of lactic acid bacteria on the uptake and distribution of the food mutagen Trp-P-2 in mice. Scand J Gastroenterol 37(2):215–221. https://doi.org/10.1080/003655202753416902
Ouyang X, Li Q, Shi M et al (2019) Probiotics for preventing postoperative infection in colorectal cancer patients: a systematic review and meta-analysis. Int J Color Dis 34:459–469
Parashar UD, Gibson CJ, Bresee JS, Glass RI (2006) Rotavirus and severe childhood diarrhoea. Emerg Infect Dis 12:304–306
Parnell JA, Reimer RA (2009) Weight loss during oligofructose supplementation is associated with decreased ghrelin and increased peptide YY in overweight and obese adults. Am J Clin Nutr 89:1751–1759. https://doi.org/10.3945/ajcn.2009.27465
Plovier H, Everard A, Druart C et al (2017) A purified membrane protein from Akkermansia muciniphila or the pasteurized bacterium improves metabolism in obese and diabetic mice. Nat Med 23:107–113. https://doi.org/10.1038/nm.4236
Ponziani FR, Bhoori S, Castelli C, Putignani L, Rivoltini L, Del Chierico F, Sanguinetti M, Morelli D, Paroni Sterbini F, Petito V et al (2018) Hepatocellular carcinoma is associated with gut microbiota profile and inflammation in nonalcoholic fatty liver disease. Hepatology 69(1):107–120. https://doi.org/10.1002/hep.30036

Proença JT, Barral DC, Gordo I (2017) Commensal-to-pathogen transition: One-single transposon insertion results in two pathoadaptive traits in Escherichia coli–macrophage interaction. Sci Rep 7(1):4504. https://doi.org/10.1038/s41598-017-04081-1

Qiu G, Yu Y, Wang Y, Wang X (2019) The significance of probiotics in preventing radiotherapy-induced diarrhoea in patients with cervical cancer: a systematic review and meta-analysis. Int J Surg 65:61–69

Racedo S, Villena J, Medina M, Agüero G, Rodríguez V, Alvarez S (2006) Lacto bacillus casei administration reduces lung injuries ina Streptococcus pneumoniae infection in mice. Microbes Infect 8(9–10):2359–2366. https://doi.org/10.1016/j.micinf.2006.04.022

Ramakrishna C, Kujawski M, Chu H, Li L, Mazmanian SK, Cantin EM (2019) Bacteroides fragilis polysaccharide A induces IL-10 secreting B and T cells that prevent viral encephalitis. Nat Commun 10:2153. https://doi.org/10.1038/s41467-019-09884-6

Raman M, Ambalam P, Kondepudi KK, Pithva S, Kothari C, Patel AT, Purama RK, Dave JM, BRM V (2013) Potential of probiotics, prebiotics and synbiotics for management of colorectal cancer. Gut Microbes 4:181–192. https://doi.org/10.4161/gmic.23919

Requena T, Martínez-Cuesta MC, Peláez C (2018) Diet and microbiota linked in health and disease. Food Funct 9(2):688–704. https://doi.org/10.1039/c7fo01820g

Riddle MS, Connor BA, Beeching NJ, DuPont HL, Hamer DH, Kozarsky P, Libman M, Steffen R, Taylor D, Tribble DR, Vila J, Zanger P, Ericsson CD (2017) Guidelines for the prevention and treatment of travelers' diarrhoea: a graded expert panel report. J Travel Med 1(24):57–74. https://doi.org/10.1093/jtm/tax026

Ríos-Covián D, Ruas-Madiedo P, Margolles A, Gueimonde M, de los Reyes-Gavilán CG, Salazar N (2016) Intestinal short chain fatty acids and their link with diet and human health. Front Microbiol 7:1–9. https://doi.org/10.3389/fmicb.2016.00185

Routy B, Le Chatelier E, Derosa L, Duong CPM, Alou MT, Daillère R et al (2018) Gut microbiome influences efficacy of PD-1-based immunotherapy against epithelial tumors. Science 5:91–97. https://doi.org/10.1126/science.aan3706

Salva S, Villena J, Alvarez S (2010) Immunomodulatory activity of Lactobacillus rhamnosus strains isolated from goat milk: impact on intestinal and respiratory infections. Int J Food Microbiol 141(1–2):82–89. https://doi.org/10.1016/j.ijfoodmicro.2010.03.013

Sarkar A, Lehto SM, Harty S, Dinan TG, Cryan JF, Burnet PWJ (2016) Psychobiotics and the manipulation of bacteria-gut-brain signals. Trends Neurosci 39(11):763–781. https://doi.org/10.1016/j.tins.2016.09.002

Schrezenmeir J, de Vrese M (2001) Probiotics, prebiotics and synbiotics–approaching a definition. Am J Clin Nutr 73(2):361–364. https://doi.org/10.1093/ajcn/73.2.361s

Sekirov I, Russell SL, Antunes LC, Finlay BB (2010) Gut microbiota in health and disease. Physiol Rev 90:859–904

Selma MV, David B, María LC, María RV, Rocío GV, Alex M, Juan EC, Francisco TBA (2017) Isolation of human intestinal bacteria capable of producing the bioactive metabolite isourolithin a from ellagic acid. Front Microbiol 8:152. https://www.frontiersin.org/article/10.3389/fmicb.2017.01521

Shah NP (2007) Functional cultures and health benefits. Int Dairy J 17(11):1262–1277. https://doi.org/10.1016/j.idairyj.2007.01.014

Sharma P, Tomar SK, Goswami P, Sangwan V, Singh R (2014) Antibiotic resistance among commercially available probiotics. Int Food Res J 57:176–195

Singer P, Blaser AR, Berger MM, Alhazzani W, Calder PC, Casaer MP, Hiesmayr M, Mayer K, Montejo JC, Pichard C, Preiser JC, van Zanten ARH, Oczkowski S, Szczeklik W, Bischoff SC (2019) ESPEN guideline on clinical nutrition in the intensive care unit. Clin Nutr 38(1):48–79. https://doi.org/10.1016/j.clnu.2018.08.037

Sivamaruthi BS, Kesika P, Suganthy N, Chaiyasut C (2019) A review on role of microbiome in obesity and antiobesity properties of probiotic supplements. Biomed Res Int 2019:3291367. https://doi.org/10.1155/2019/3291367

Sleator RD, Hill C (2008) Battle of the bugs. Science 321:1294–1295. https://doi.org/10.1126/science.321.5894.1294b

Slykerman RF, Thompson J, Waldie KE, Murphy R, Wall C, Mitchell EA (2017) Antibiotics in the first year of life and subsequent neurocognitive outcomes. Acta Paediatr 106:87–94. https://doi.org/10.1111/apa.13613

Sommariva M, LeNoci V, Bianchi F, Camelliti S, Balsari A, Tagliabue E, Sfondrini L (2020) The lung microbiota: role in maintaining pulmonary immune homeostasis and its implications in cancer development and therapy. Cell Mol Life Sci 77:2739–2749. https://doi.org/10.1007/s00018-020-03452-8

Song JA, Kim HJ, Hong SK, Lee DH, Lee SW, Song CS, Kim KT, Choi IS, Lee JB, Park SY (2016) Oral intake of Lactobacillus rhamnosus M21 enhances the survival rate of mice lethally infected with influenza virus. J Microbiol Immunol Infect 49(1):16–23. https://doi.org/10.1016/j.jmii.2014.07.011

Stoyanova LG, Ustyugova EA, Netrusov AI (2012) Antibacterial metabolites of lactic acid bacteria: their diversity and properties. Appl Biochem Microbiol 48:229–243. https://doi.org/10.1134/s0003683812030143

Stringer AM, Gibson RJ, Logan RM, Bowen JM, Yeoh ASJ, Hamilton J et al (2009) Gastrointestinal microflora and mucins may play a critical role in the development of 5-fluorouracil-induced gastrointestinal mucositis. Exp Biol Med 234:430–441. https://doi.org/10.3181/0810-RM-301

Suez J, Zmora N, Segal E, Elinav E (2019) The pros, cons, and many unknowns of probiotics. Nat Med 25:716–729

Sundararaman A, Ray M, Ravindra PV, Halami PM (2020) Role of probiotics to combat viral infections with emphasis on COVID-19. Appl Microbiol Biotechnol 104(19):8089–8104. https://doi.org/10.1007/s00253-020-10832-4

Syakila RN, Lim SM, Agatonovic-Kustrin S, Lim FT, Ramasamy K (2019) In vitro assessment of Pediococci and Lactobacilli induced cholesterol-lowering effect using digitally enhanced high-performance thin-layer chromatography and confocal microscopy. Anal Bioanal Chem 411(6):1181–1192. https://doi.org/10.1007/s00216-018-1544-2

Szajewska H, Kotowska M, Mrukowicz JZ, Armańska M, Mikołajczyk W (2001) Efficacy of Lactobacillus GG in prevention of nosocomial diarrhoea in infants. J Pediatr 138(3):361–365. https://doi.org/10.1067/mpd.2001.111321

Taverniti V, Guglielmetti S (2011) The immunomodulatory properties of probiotic microorganisms beyond their viability (ghost probiotics: proposal of paraprobiotic concept). Genes Nutr 6(3):261–274. https://doi.org/10.1007/s12263-011-0218-x

Tian Y, Li M, Song W, Jiang R, Li YQ (2019) Effects of probiotics on chemotherapy in patients with lung cancer. Oncol Lett 17(3):2836–2848

Tomaro-Duchesneau C, Jones ML, Shah D, Jain P, Saha S, Prakash S (2014) Cholesterol assimilation by Lactobacillus probiotic bacteria: An in vitro investigation. Biomed Res Int 2014:380316

Tong H, Isenring E, Yates P (2009) The prevalence of nutrition impact symptoms and their relationship to quality of life and clinical outcomes in medical oncology patients. Support Care Cancer 17:83–90

Tsilingiri K, Rescigno M (2013) Postbiotics: what else? Benef Microbes 4(1):101–107. https://doi.org/10.3920/BM2012.0046

Vareille-Delarbre M, Miquel S, Garcin S, Bertran T, Balestrino D, Evrard B, Forestier C (2019) Immunomodulatory effects of lactobacillus plantarum on inflammatory response induced by klebsiella pneumoniae. Infect Immun 87(11):e00570–e00519. https://doi.org/10.1128/IAI.00570-19

Verhoeven V, Renard N, Makar A, Van Royen P, Bogers JP, Lardon F, Peeters M, Baay M (2013) Probiotics enhance the clearance of human Papillomavirus-related cervical lesions: a prospective controlled pilot study. Eur J Cancer Prev 22(1):46–51. https://doi.org/10.1097/CEJ.0b013e328355ed23

Vivarelli S, Salemi R, Candido S et al (2019) Gut microbiota and cancer: from pathogenesis to therapy. Cancers (Basel) 11(1):38

Wang Y, Telesford KM, Ochoa-Repáraz J, Haque-Begum S, Christy M, Kasper EJ, Wang L, Wu Y, Robson SC, Kasper DL, Kasper LH (2014) An intestinal commensal symbiosis factor controls neuroinflammation via TLR2-mediated CD39 signalling. Nat Commun 5:4432. https://doi.org/10.1038/ncomms5432

Wei D, Heus P, van de Wetering FT, van Tienhoven G, Verleye L, Scholten RJ (2018) Probiotics for the prevention or treatment of chemotherapy or radiotherapy related diarrhoea in people with cancer. Cochrane Database Syst Rev 8(CD008831)

Wells J (2011) Mucosal vaccination and therapy with genetically modified lactic acid bacteria. Annu Rev Food Sci Technol 2:423–445. https://doi.org/10.1146/annurev-food-022510-133640

Wiström J, Norrby SR, Myhre EB, Eriksson S, Granström G, Lagergren L, Englund G, Nord CE, Svenungsson B (2001) Frequency of antibiotic-associated diarrhoea in 2462 antibiotic-treated hospitalized patients: a prospective study. J Antimicrob Chemother 47(1):43–50. https://doi.org/10.1093/jac/47.1.43

Wu C, Zhang J, Chen W, Wang M, Du G, Chen J (2012) A combined physiological and proteomic approach to reveal lactic-acid-induced alterations in Lactobacillus casei Zhang and its mutant with enhanced lactic acid tolerance. Appl Microbiol Biotechnol 93:707–722. https://doi.org/10.1007/s00253-011-3757-6

Xu H, Wang X, Feng W, Liu Q, Zhou S, Liu Q, Cai L (2020) The gut microbiota and its interactions with cardiovascular disease. Microb Biotechnol 13(3):637–656. https://doi.org/10.1111/1751-7915.13524

Yao G, Yu J, Hou Q, Hui W, Liu W, Kwok LY, Menghe B, Sun T, Zhang H, Zhang W (2017) A perspective study of koumiss microbiome by metagenomics analysis based on single-cell amplification technique. Front Microbiol 8:165. https://doi.org/10.3389/fmicb.2017.00165

Yelin I, Flett KB, Merakou C, Mehrotra P, Stam J, Snesurd E, Hinkle M, Lesho E, McGann P, McAdam AJ, Sandora TJ, Kishnoy R, Priebe GP (2019) Genomic and epidemiological evidence of bacterial transmission from probiotic capsule to blood in ICU patients. Nat Med 25:1728–1732. https://doi.org/10.1038/s41591-019-0626-9

Zhang Y, Zhang Y, Zhu Y, Mao S, Li Y (2010) Proteomic analyses to reveal the protective role of glutathione in resistance of Lactococcus lactis to osmotic stress. Appl Environ Microbiol 76:3177. https://doi.org/10.1128/aem.02942-09

Zuo T, Zhang F, Lui GCY et al (2020) Alterations in gut microbiota of patients with COVID19 during time of hospitalization. Gastroenterology 159(3):944955

Chapter 9
Microbiome Association of Polypharmacy in Geriatric Population

Gururaja Perumal Pazhani, Murugan Veerapandian, Rajkumar Malayandi, and Thandavarayan Ramamurthy

1 Introduction

The increase of aged population in the twenty-first century is not only due to the advancement in medical science, but also several other factors such as education, industrialization, nutritive foods, hygienic practices, and stable income. Global population growth rate is rapid at one billion per decade and it is expected that the world population count will surpass 10 billion by 2040 (Figdor and Gulabivala 2008). An increase in geriatric population is the major trend accounted for the global population. However, growing socioeconomic, emotional, and health issues are greatly affecting the quality of the geriatric population (Allaband et al. 2019). Other factors, including nature of the disease, race, income, and health-related beliefs can contribute to the use of multiple drugs (Rossi et al. 2007). Several government agencies and research institutes are working constantly to improve the quality and wellbeing of the geriatric population. Healthcare is the primary and foremost important aspect that improves quality of life, in which drug therapies

G. P. Pazhani (✉)
Chettinad School of Pharmaceutical Sciences, Chettinad Academy of Education and Research, Kelambakkam, Tamil Nadu, India

M. Veerapandian
CSIR-Central Electrochemical Research Institute, Karaikudi, Tamil Nadu, India

Academy of Scientific & Innovative Research, Ghaziabad, Uttar Pradesh, India

R. Malayandi
USV Private Limited, Mumbai, Maharashtra, India

T. Ramamurthy
National Institute of Cholera and Enteric Diseases, Kolkata, West Bengal, India

S. Thomas (ed.), *Human Microbiome*,
https://doi.org/10.1007/978-981-16-7672-7_9

play an important role. On the other hand, modern sedentary lifestyle habits result in various metabolic, cardiovascular, and psychological disorders that lead to consumption of numerous medicines.

The word polypharmacy is derived from the Greek word, meaning 'more than one Pharmacy'. Researchers use this term to describe the use of several medications that are mostly not specific, less effective, and with or without similar biological activities (Maher et al. 2014). An important factor that needs to be considered is the use of the herbal supplements, nutraceutical, and over-the-counter medications. Polypharmacy is a major cause of concern in the elderly than young adults because they are at risk for more disease conditions and many drugs are prescribed at times. Polypharmacy administration in older hospitalized patients has shown to be associated with gut microbiota imbalance. Proton pump inhibitors and drugs used to treat mental and psychotic disorders are prone to induce the gut microbiota dysbiosis. However, mechanisms of drug-driven alteration in the gut microbiota are yet to be established (Ticinesi et al. 2017). Inappropriate treatment and irregular use of drugs might lead to adverse effects such as gastrointestinal pH change, mucosal permeability alteration and could promote microbial diversity in the GIT. Recently, it has been reported globally that 30% to 50% of the elderly people take more than 5 drugs needlessly and the highest number of drugs is being consumed by those who are in nursing care units (Maher et al. 2014; Kim and Parish 2017). Figure 9.1 depicts the general factors governing the geriatric healthcare and possible complementary advancement required from modern analytical techniques.

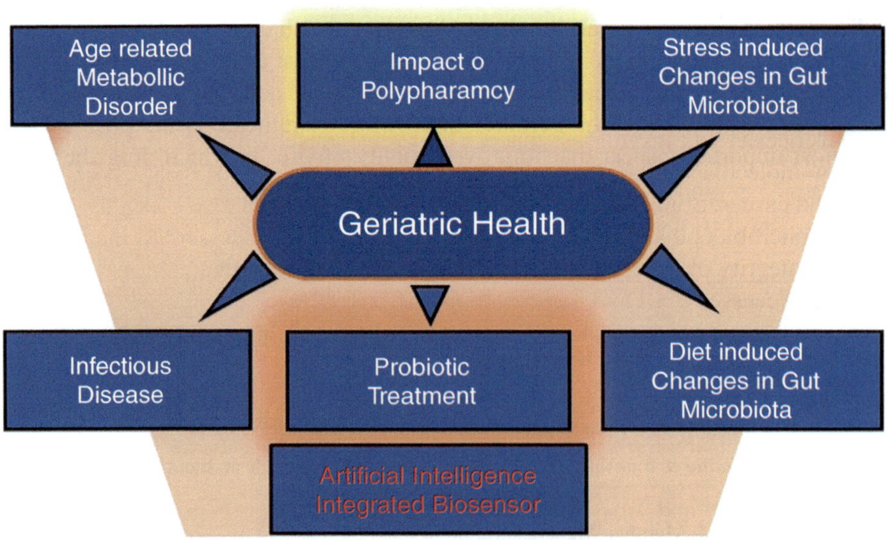

Fig. 9.1 Schematic illustration on the geriatric health and its influencing factors with need of possible intervention from artificial intelligence-enabled biosensor for better geriatrics

2 Chronic Conditions in Elderly Population and Their Treatment

Chronic pain is a very common health problem in elderly people. Poor management of pain causes disability, social distancing, depression, and anxiety. With age, there are gradual changes in the immune system, particularly the mast cell, which leads to high sensitivity to any external stimuli, neuroinflammation and neuralgia and the conditions may further be exaggerated due to polypharmacy. Due to central nervous system illnesses, chronic cough is common in elderly with multifactorial syndromes (Song et al. 2019). Protracted itch is another challenge associated with elderly people, which is caused by multifactorial illnesses like anxiety, contact dermatitis, neural degeneration, diabetes, and allergic reaction associated with drugs taken orally and also those that are topically applied to cure various skin related diseases. Diuretics like thiazides and antihypertensive drugs like calcium channel blockers have been reported to cause itchiness in the elderly with or without rashes (Valdes-Rodriguez et al. 2015). Another major chronic illness in the elderly is heart failure with no definite cure. The exact cause of heart failure in the elderly may not be with a single aetiology, but it is generally associated with comorbid conditions. The management of heart failure with polypharmacy in the elderly is a great challenge (Alghamdi and Chan 2017). Comorbidity was considered the cause of poor quality of life, depression, and increased chances of death. Population-based studies reported from a developed country showed that 67% of the elderly people had multimorbidity combined with the increase in age among the study population (Salive 2013). The elderly people who were above 85 years of age had 81% of multimorbidity.

Worldwide, chronic inflammatory rheumatic (CIR) arthritis is increasing year on year among the elderly population. Patients who have CIR have more chances of developing cardiovascular disease (CVD), difficulties in breathing, cancer, and irritable bowel syndromes characterized with inflammatory markers like interleukins and reaction proteins, and serum levels higher than the normal range of 0.2 to 3 mg/L. However, the exact mechanism of how the inflammatory markers induce comorbidities are yet to be established and clinicians are advocated to conduct prospective trials to access the pathophysiological involvement of inflammatory markers in elderly patients. In addition, standardized procedures are advocated for clinicians to assess and prevent comorbidities (Salive 2013). An integrated therapeutic approach with polypharmacy is preferred for the effective management of CIR (Lahaye et al. 2019).

Persistent constipation is a common problem associated with elderly people. The cause of constipation is associated with gut microbial composition, decreased intake of high fibre food and water, lack of physical activity, and various gastrointestinal, metabolic, neurogenic, and psychogenic disorders. Medications like iron supplements, non-steroidal anti-inflammatory agents, opiates, calcium channel blockers, diuretics (non-potassium sparing), antihistamine, antidepressants, antipsychotics, antacids, and anticholinergics could also be the reason for constipation

(Gras-Miralles and Cremonini 2013). Constipation leads to a poor quality of life and demands modifications in the lifestyle, eating habits, and effective medication with fewer side-effects or nutraceutical intake. Kidney disorder is another most common problem in elderly people and its prevalence is high among the geriatric population from developed countries. The epidemiological data have shown that renal failure in elderly patients was diagnosed with nephrotoxins. In patients with kidney failure, drugs may not work to improve kidney function; hence, treatment like dialysis or a kidney transplant is generally recommended (Olyaei and Bennett 2009).

A recent study has suggested that polyneuropathy has increased over the years in developed countries. The status of polyneuropathy in the elderly has been about 1% to 7%. The reason for such variation could be due to difference in assessment protocol, study populations, and also due to variation in polyneuropathy in developed and developing countries. Communicable diseases such as leprosy, diabetes, alcoholic addiction, cardiovascular disease, and polypharmacy are the mainstay for the cause of polyneuropathy (Hanewinckel et al. 2016). Rhinitis and sinusitis are other important health-related conditions in the elderly population that affect the quality of their life. Nearly 15% of the elderly population are affected in developed countries and clinical management between elderly and adult populations is challenging due to comorbidity and drug interactions resulting from polypharmacy (Hsu and Suh 2018).

Malnutrition is common among the geriatric population, especially those admitted in hospitals and long-term care centres. The reason for malnutrition may be due to lonesomeness, impaired metabolic activities induced by the production of cytokines, ignorance, inappropriate consumption of a balanced diet, and stress due to unaccomplished desire, environment, and psychological factors. The malnourished elderly population is vulnerable to opportunistic infections, which increases hospital stay and weakens muscle strength. Thorough examination is required to understand the cause of malnutrition, including the possibilities of interaction and adverse reactions of multiple medications (Feldblum et al. 2007). The best intervention strategies such as vitamins, balanced diet, and nutritional supplements are required to treat malnutrition in elderly people.

3 Age-related Metabolic and Functional Changes in Geriatrics

Blood biomarkers are employed to understand the disease progression and metabolic changes in elderly people. Currently, nineteen biomarkers have been identified to detect haematological, and metabolic disorders, lipid profile, and inflammatory biosignatures. The presence of single or multiple biomarkers confirms the dysfunction of metabolic activities, but does not indicate the diseased state (Sebastiani et al. 2017). Studies on the mechanism of control of the intestinal barrier have found a close association of digestive and non-digestive disorders. The elderly people are susceptible to malnutrition and therefore, there is an improper maintenance of nutrient and electrolyte balance in the gastrointestinal tract (GIT). The intestinal barrier is weak among malnourished elderly thereby altering the structure-function of the gut and also allowing intestinal permeability. As a result of this, mucosal inflammation and neural activation are often reported (Farré and Vicario 2017). Difficulties in swallowing, known as dysphagia, is more common in elderly people. Lower oesophageal sphincter function was seen among asymptomatic older individuals than healthy young adults. However, the exact cause of dysphagia in the elderly is not yet known (Besanko et al. 2014). Gastroesophageal reflux disease (GERD) has been reported among 20% of the US population, mostly among the elderly group. Physiological change, lifestyle diseases, comorbidities, and polypharmacy are found to aggravate GERD (Commisso and Lim 2019). Older age and comorbidities are responsible for loss in muscle mass and strength in obese and non-obese elderly people. This defect is due to metabolic dysfunction and functional deterioration (Buch et al. 2016). The microbes residing in the GIT have an important role in regulating age-related muscle weakness in bioenergetic pathways. Gut microbiota are complexly involved in performing the normal physiological functions of the host, nutrient absorption, and synthesis of amino acids, small and long-chain fatty acids. These chemicals contribute to human health and disease and are also involved in 'gut-muscle axis'. The microbial diversity in the gut affects muscle mass and metabolic functions that could lead to insulin resistance and glucose tolerance. The skeletal muscle tissues are involved in the disposal of glucose which is indirectly controlled by the microbiota (Grosicki et al. 2018). The optimal intestinal microbiota is vital for glycogen storage, muscle protein synthesis, and mitochondrial biogenesis and function (Przewłócka et al. 2020).

Platelet count and its complex activation process have been found to decrease in older people, which alter the hematopoietic tissue and vascular health condition. Antiplatelet treatment and oxidative stress lower the count and affects the function. The function of platelet has been linked with mRNA and microRNA expression of cancer-causing genes and also with thrombotic disease (Jones 2016). *Bifidobacteria* are considered to maintain a healthy gut and colonization of this species differs among elderly people (Kato et al. 2017). However, a molecular-based study has identified an increase in the diversity of bacterial population with ageing. The gut microbial populations are directly influenced by nutrition, disease condition, and use of medicines (Tiihonen et al. 2009). In addition to this, culture and non-culture-based studies have found that the composition of microbial populations is different in the elderly and young adults due to several confounding factors. Alteration in the microbial composition causes various biological changes irrespective of age. It has been correlated with strong changes in immune responses, host metabolism, and muscle weakness (O'Toole and Jeffery 2018). Considerable diversity of gut microbiome has been correlated with Parkinson's disease. This diversity affects the intestinal barrier function and immune system of the elderly that leads to gastrointestinal dysfunction and motor symptoms of Parkinson's disease (Bedarf et al. 2017). The SARS-CoV-2 virus has been implicated in cardiopulmonary and varying vascular complications in the elderly population. Viruses indirectly control the bacterial populations in the nose, mouth, respiratory tract, and intestinal milieu since virus particles may target the surfactin of the bacterial cells, which lead to microbial imbalance or maladaptation in the respective niches (Zeppa et al. 2020). There are several metabolic and functional changes that occur in the elderly under residential care, but the most common among them is injury (Gustavsson et al. 2018).

4 Normal Life Expectancy of People in Different Countries

Human life expectancy has been greatly improved at varying degrees in different countries. Nutritious foods, good medical care, pollution-free environment, stress-free lifestyle, and cultural and biological differences have been reported to influence life expectancy. The life span has increased by 2.2 years per decade since 1960 in the UK. However, this increase has not directly reflected on the quality of life in elderly people. An increase in the life of elderly people has been crippled with a disability and vulnerability to several diseases (Brown 2015). A survey conducted among member countries of the Organisation for Economic Co-operation and Development during 2011 has shown that Switzerland recorded the highest life expectancy at 82.5 years. However, there is a substantial geographical variation and influences of the social standing of neighbourhoods that determine the life expectancy rates (Moser et al. 2014). Recently, detailed surveys on life expectancy in developed and developing countries have been reviewed and are available online (https://www.infoplease.com/world/health-and-social-statistics/life-expectancy-countries.Accessed on January 1, 2021).

5 Infectious Disease and Treatment Options in Geriatrics

Infectious diseases are increasing globally on a yearly basis. The vector and zoonotic disease are quite common and are transmitted through food, water, and direct contact (McArthur 2019). Elderly people are vulnerable to infectious diseases due to weak immune systems, changes in physiological conditions, and declining anatomical structure. The spectrum of infectious diseases in elderly people is very wide. It is apparent that the ageing immune system directs to decreased actions of macrophages, natural killer cells, and function of dendritic cells. The weaker immune system and comorbidities cause diminished defence to infectious agents. The diagnosis of infectious disease in the elderly is very challenging due to the atypical clinical presentation of symptoms and is worsened due to polypharmacy. Both urinary tract infection (UTI) and asymptomatic bacteriuria cause major health burden in elderly people. The treatment options for these conditions are recommended based on the comorbidities, severity of illness, living conditions, antimicrobial resistance pattern prevailing in the locality and compatibility with other drugs consumed by patients. The case-control study among the children had shown that the gut environment is responsible for UTI infection (Paalanne et al. 2018). However, the same is yet to be established in elderly people. Sepsis is the most commonly reported complication due to infections of the respiratory tract, bloodstream, and urinary tract. Bloodstream and respiratory infections often cause febrile in the elderly (Liang 2016). *Haemophilus influenza* type *B* (HiB) and *Streptococcus pneumonia* are the main causative agents of meningitis in the elderly and is less commonly caused by other pathogens like *Listeria monocytogenes*, group *B Streptococcus, E. coli,* and *K. pneumoniae.* The mortality rate associated with meningitis was more than 20% among the age group of 65 years in the USA. A study conducted in North India has shown that meningitis caused the highest mortality (20–30%) (Madhumita and Gupta 2011). However, the incidence of bacterial meningitis decreased in developed countries after the introduction of Meningococcal conjugate vaccine (Thigpen et al. 2011). In India, Pneumococcal conjugate vaccine was introduced in 2017 for children in North India and later it has been extended into other parts of India (Sachdeva 2017). Generally, for adults, vaccination is given only under special conditions.

The number of infectious diseases is expanding constantly among elderly people, particularly among the immunocompromised. Smoking habits increase the risk of bacterial and viral infections by changing the structural integrity of the respiratory tract and by weakening the immune system (Arcavi and Benowitz 2004). Antimicrobial therapy is the mainstay in the control of infectious diseases in the elderly. In addition to this, adjuvant corticosteroid therapy is also administered to control the inflammation associated with bacterial and viral infections. Prolonged hospitalization and consumption of antibiotics may destabilize the gastrointestinal microbiota and also cause resistance to the pathogens, particularly *Clostridium difficile*, which cause antibiotic-associated diarrhoea. Probiotics are highly recommended to control antibiotic-associated diarrhoea.

6 Impact of Polypharmacy in Geriatric Patients

Elderly people are mostly under multiple drug therapy to address various diseases. These drugs interact and influence the pharmacokinetics and pharmacodynamics. In addition, the anatomical, physiological, and pathological changes also influence the pharmacokinetic and pharmacodynamic properties of the drugs (Templier et al. 2016). The physiological changes, especially the absorption, accumulation, and metabolism, result in altered pharmacokinetics (Browne et al. 2016). The significant reduction in metabolism and elimination in elderly patients is associated with high plasma drug concentration and dose-related toxicities. Also, the gastrointestinal absorption significantly changes on account of physiological factors such as reduction in the small intestinal surface area, delayed gastric emptying, and elevated gastrointestinal pH. Increased absorption of biopharmaceutical classification system class III/IV compounds require sufficient residential time. However, elevated gastric pH impairs the absorption profile of basic drugs such as calcium salts. Albeit the delayed-release formulations are premeditated to release the drug in the small intestine, elevated gastric pH results in an earlier release of such drugs. Ageing is associated with increase in the total body fat content and decrease in water content (Stewart 2012). Hence, this altered composition significantly influences the distribution kinetics of drugs and dosage levels. Lipophilic compounds significantly reduce plasma drug concentration to levels lower than the therapeutic drug concentration. This may lead to therapeutic failure, especially with narrow therapeutic drugs. Plasma proteins such as albumin and alpha 1 acid glycoprotein play an important role in the distribution kinetics. Decreased serum protein levels are associated with elderly patients.

There is a drastic reduction (30–40%) in hepatic metabolism in the case of elderly subjects when compared with normal young adults (Takahasi et al. 2018). The reduction in hepatic metabolic clearance results in the accumulation of not only drugs but also metabolites. Hence, customized optimization of the dose is essential in order to achieve the desired plasma drug concentration for an optimal therapy. Impaired renal elimination of drugs is another important pharmacokinetic change related to ageing. The age-dependent reduction in creatinine clearance is observed above 40 years. Though the glomerular filtration rate is poor in aged patients, their serum creatinine levels often remain normal. Existence of normal serum creatinine levels can mislead clinicians to overlook the kidney function. Less muscle mass and lack of physical activities result in less creatinine secretion. Reductions in tubular function with age are parallel to those in glomerular function. These renal-associated factors significantly impact the elimination of water-soluble drugs. Concomitant administration of multiple drugs also exaggerates the pharmacokinetic phenomenon. Therefore, the rational synthesis of prescription is essential for elderly subjects.

The age-related physiological changes and concomitant disease states in the elderly significantly impact the pharmacodynamics of the drugs. The age-related pharmacodynamic effect is clinically important due to the adverse drug reaction and side effects. For example, idiosyncratic age-related kidney dysfunction is the most common side effect of non-steroidal anti-inflammatory drugs. Similarly, bisphosphonates-induced nephrotoxicity is more common with elderly population when compared with adults. The other common pharmacodynamic effects with elderly patients are associated with central, sympathetic and parasympathetic nervous systems, and cardiac system which are well described in the literature.

The gut microbiota has an impact on drug metabolism and affects its activity and increases the toxic effect. Prodrugs like prontosil and neoprontosil are converted into sulfanilamide by the action of microbes. The microbiota plays an important role in bacteria-driven drug metabolism by secreting secondary metabolites that cause drug degradation by many chemical processes like oxidation, conjugation, deacylation, decarboxylation, hydrolysis, etc. (Wilson and Nicholson 2017). Many drugs introduced in clinical practice are withdrawn after a few years of use due to side effects and toxicity which may arise due to the improper study of drug-gut microbial interaction and its short-term and long-term outcomes in drug pharmacokinetics, thus influencing its toxic effects. It is very important to understand the interaction of a new drug with gut microbiome in subjects belonging to different age groups in order to propose the use of drugs and personalized medicine in the future.

The concept of pharmacomicrobiomics was coined in 2010 to relate how the microbiome influences the xenobiotic actions, particularly drug absorption, distribution, metabolism, and elimination (ADME) in individuals. The genetic response of individuals to drugs has been analysed by pharmacogenomics. According to this discipline, the gut microbiota has been recognized as the organ and second genome of individuals with respect to drug action. The microbiome has been an important target for improving drug efficacy and safety due to the opportunities to manipulate its composition. It is believed that the genetic factors could explain 20–95% drug action variability on individuals. Therefore, research in the last few decades has focused on pharmacogenetics and pharmacogenomics to understand the individual genetic variability on drug response (Doestzada et al. 2018). Considering the importance of future personalized medicine, the Human Microbiome Project (HMP) was initiated, which studies the interplay between drugs and microorganisms, involving ADME, microbial metabolites, immune modulation, ectopic and translocation of drug metabolites, etc. The pharmacogenomics and pharmacomicrobiomics studies have shed light on the pathogenesis and treatment options for rheumatoid arthritis and spondyloarthritis (Scher et al. 2020). Based on the pharmacomicrobiomics approach, the pathogenic mechanisms and treatment options of many other chronic diseases are yet to be established.

7 Stress and Its Impact on Gut Microbiota in Geriatric Population

Stress is a kind of emotional symptom expressed in different indications like anger, anxiety, irritability, dementia, and mental distress, which are collectively called psychopathological symptoms. The bliss of life is varied according to socio-demographic factors such as income, marital status, desire, and social activity. Often, elderly people are not able to accomplish the socio-demographic factors that lead to psycho-emotional tension. The stressful events in the elderly make impacts on the immune system, inflammatory responses, and also on secretory glands. Medically, elderly people are not able to cope with stressors due to several external and internal health-related factors (Fali et al. 2018; Nikolakakis et al. 2019). In animal studies, psychological stress was found to affect the gut microbiota. However, this is not yet confirmed in humans.

Based on the animal experiments, it is believed that the possible bacterial population in stress-induced condition would comprise toxin producers like *Fusobacterium, Porphyromonas, Bacteroides, Ruminococcus* and is less likely to include *Lactobacillus* and *Bifidobacterium* (Carson et al. 2018). The chronic psychological stresses are shown to increase oxidative stress and aggravate biological ageing by creating an imbalance between oxidative stress and antioxidant status.

There is increasing evidence for the upregulation of reactive oxygen species (ROS)-producing enzymes and down-regulation of an antioxidant enzyme with ageing (Carson et al. 2018). Recently, trimethylamine N-oxide (TMAO) has been recognized as an important biomarker in blood for the diagnosis of atherosclerotic risk. The consumption of high-fat substances like lecithin and carnitine increases the concentration of TMAO in the gut by bacteria belonging to *Anaeroplasmataceae*, *Deferribacteraceae*, *Enterobacteriaceae*, and *Prevotellaceae* families. The bacterial population of these families is shown to vary by ageing and is also responsible for the progress of other metabolic and chronic kidney diseases (Velasquez et al. 2016; Violi et al. 2017). Antibody (IgA and IgM)-mediated inflammatory responses, interference with neurotransmitters, and signalling have been reported in depressed elderly due to lipopolysaccharides of the gut microbiota (Naseribafrouei et al. 2014). Depression affects the function of the central nervous system by complex pathophysiological mechanisms, which influence the quality of life (FarioliVecchioli et al. 2018). The *Prevotella* group and *Bacteroides* cluster have been identified for brain-gut microbiota interaction that attributes for good human health. A transcriptome analysis of faecal sample of healthy volunteers showed the presence of active pathways for the production of γ-aminobutyric acid (GABA) by *Bacteroides* species (Strandwitz et al. 2019).

In adult women, difference in faecal microbial population has been observed with respect to the composition of *Bacteroides* and *Parabacteroides,* which are shown to affect behaviour and cause mood swings (Tillisch et al. 2017). Among the healthy Korean adults, *Prevotella* and *Lachnospiraceae* are responsible for healthy emotions and promoting good mental health (Lee et al. 2020a, b). The age-matched analysis revealed that the composition of gut microbiota was found to be responsible for Alzheimer's disease (AD). Several microbial toxin levels are higher among patients with AD than the healthy adults (Zhuang et al. 2018). The persistence of depression in the elderly is due to the complex network communication between bacteria and parasites present in the gut milieu. Bacteria are thought to play an important role in the population of parasites in the gut and vice versa. Parasites significantly change the nutritional status in the elderly, mainly, *Ascaris lumbricoides* has shown to interplay between the gut and brain via the gut-brain axis by complex network and cause depression and behavioural changes (Ramírez-Carrillo et al. 2020).

8 Diet and Associated Gut Microbial Changes in Geriatrics

The gut microbiomes are generally stable in the adult age group. However, the composition may be altered by various factors like diet, gastric acidity, use of drugs, consumption of alcohol, nutritional status, geography, and environmental changes. The gut microbiota is a very complex ecosystem and may vary with person and times. Studies on gut microbiota are promising, as they play a key factor for maintaining good health in the old age group (Maynard and Weinkove 2018). Age-related alterations of microbiome stimulate diseases causing the environment, mostly directed to colorectal cancer in elderly people. The short-chain fatty acids (SCAs) like acetate, propionate, and butyrate are produced from carbohydrate and fat metabolism by bacteria. These SCAs play a vital role in protecting the gut and other organs. With ageing, decrease in the concentrations of SCAs in the gut have been reported in adult mice model (Lee et al. 2020a, b). This hypothesis is yet to be explained in aged humans.

Diet has a major role in governing healthy and unhealthy adults. A cohort study conducted in the United States and Turkey among the cirrhosis and healthy individuals revealed that fermented drinks, coffee, milk, vegetables, and carbonated drinks have modified microbial diversity. Higher microbial diversity has been predicated on increased hospitalization cases and on the cause of liver-associated diseases (Bajaj et al. 2018). Rich protein and animal food have been shown to support *Bacteroides* species and carbohydrates promote *Prevotella* species in the gut. The long-term consumption of these foods in varying quantities would modulate the microbiome composition in the gut. Dietary fibre increases the *Clostridia* and *Faecalibacterium prausnitzii* populations (Wu et al. 2011; Lin et al. 2018). With ageing, underfeeding and malnutrition cause microbial shifts in the gut. A study among the Thai healthy adults has shown to increase the populations of *Bacteroides* with increased age and lower that of *Bifidobacterium* species. However, rice consumption seems to promote the *Bacteroides* species in Thai healthy adults (La-Ongkham et al. 2020).

The people who consume gluten-rich food or medication can acquire celiac disease at any age and the prevalence of this disease is common among elderly people. Elderly people with celiac disease are predisposed to chronic diseases, malignancy, osteoporosis, and short life span. Hence, gluten-free foods are recommended for symptomatic celiac patients to reduce the complications (Cappello et al. 2016).

Malnutrition and adverse effects of multiple medications cause frailty cascade and increase vulnerability to various syndromes like urinary infections, susceptibility to various infectious diseases, chronic conditions, electrolyte imbalance, dry mouth, high body mass index, loss of appetite, and gastrointestinal problems (Little 2018). The highest risk of polypharmacy and malnutrition interaction was observed among women than men; however, the exact reason for this is unknown (Fávaro-Moreira et al. 2016). The exact linkage between multi-drug consumption and nutritional status is highly complex and more research is warranted for understanding to a better level the emerging diseases and the abrupt usage of medications with repurposing approach. In this line, several classes of drugs have been identified to understand the interaction with malnutrition status, particularly, proton pump inhibitors, CVS drugs, antidiabetic drugs, and blood thinners like statins (Little 2018). Several drugs are in use to prevent the loss of vitamins and minerals in elderly patients (Little 2018). Regular intake of more than three drugs per day was shown to decrease the status of vitamins of D, K, pyridoxine, and folic acid. The status of other micro and macronutrient depletions with medicine intake is yet to be established (Fabian et al. 2011). The nutritional status of elderly people is negatively correlated with an increased intake of medicines. Lifestyle disorders like cardiovascular diseases, diabetes, an increase in sodium concentration, and a decrease in the consumption of vital ingredients have been observed in polypharmacy.

Medications indirectly promote the excessive consumption of non-protein diets and less protein intake leads to poor health condition (Heuberger and Caudell 2011). The pharmacokinetic studies of various drugs are carried out based on the metabolizing enzymes involved in the oxidative and conjugation processes. The diet and nutritional status of individuals affect the drug-metabolizing enzyme activities, resulting in altered drug action. Patients with acute myeloid leukaemia undergoing chemotherapy with good nutritional support have been found to have reduced hospital stays and better legibility of survival (Deluche et al. 2017). Vitamin B6 and nutritional diet supplement in elderly and hospitalized patients proved to be beneficial for recovering from an abnormal immune response than the healthy people (Huang et al. 2005).

9 Biosensors for the Detection of Gut Microbes

Gut microbial detections are initiated with culture-based approaches, which are known for its high specificity and sensitivity on appropriate media (Figdor and Gulabivala 2008). Nevertheless, the traditional method of bacterial culture can bias its growth (Allaband et al. 2019). Further, culturing strategies are time-consuming, labour-intensive, and demand laboratory environment (Templier et al. 2016), which further cause disrupted colony interactions (Figdor and Gulabivala 2008). However, several others still believe that the physiology and phenotypes of bacteria cannot be revealed without traditional cultivation (Stewart 2012; Browne et al. 2016). Although metagenomics is an alternative approach to study gut microbiota, achieving desired culture concentration of the detected bacteria is challenging. Amrane et al. (2019) integrated metagenomics and culturomic analysis to study the gut microbiota of patients with *Clostridium difficile* infection. Using culturomic approach, they increased the gut bacterial population and found that a consortium of 37 cultivable bacteria is potential against *C. difficile*.

Biosensor platform is one of the modern healthcare approaches capable of clinical diagnosis across the patient bedside, leading to point-of-care diagnosis. Biosensors rely on optical and electrochemical approaches and are rapid, specific, and highly sensitive against target analytes. Biorecognition elements such as antibodies, oligo-nucleotides/peptides, phages, and enzymes associated with specific biomacromolecules are important components of the biosensor platform. Although there are several optical and electrochemical-based probing platforms for the detection of pathogen or its associated extra/intracellular components (Sosnowski et al. 2020), to the best of the author's knowledge, methods for direct and rapid analysis of disrupted gut microbiota or composition in clinical set-up have not yet been demonstrated.

Takahasi et al. (2018) demonstrated that a synthetic biology platform can provide an affordable approach for on-demand microbiome sample analysis with simplified protocols useful for large-scale patient cohorts. Figure 9.2 shows the paper-based diagnostic platform as a modern analytical tool for quantification of microbial markers and host RNAs from stool samples wherein the authors have developed a toehold switch sensor to trace the V3 hypervariable region of the 16S ribosomal RNA, mimicking the standard analysis of bacterial species via 16S ribosomal DNA sequencing. To improve the specificity, biorecognition elements on toehold switch sensors were designed based on bacterial species-associated mRNAs. Integration of nucleic acid sequence-based amplification and quantitative reverse transcription (RT-qPCR) enable sensitive microbiome analysis of the clinical stool samples. To construct a compact point-of-care biosensor platform, the entire process starting from sample preparation to target testing to readout needs to be field-deployable and have a rapid analytical response time of <30 min. For instance, sample preparation steps, including centrifugation, liquefaction, and homogenization of raw samples, and hydrolysis should be devised into portable platforms in addition to signal transduction units. Another major challenge in biosensor device integration is a seamless interpretation of results via user interface without the need for a sophisticated instrument. Artificial intelligence (AI) in healthcare may play a vital role in data

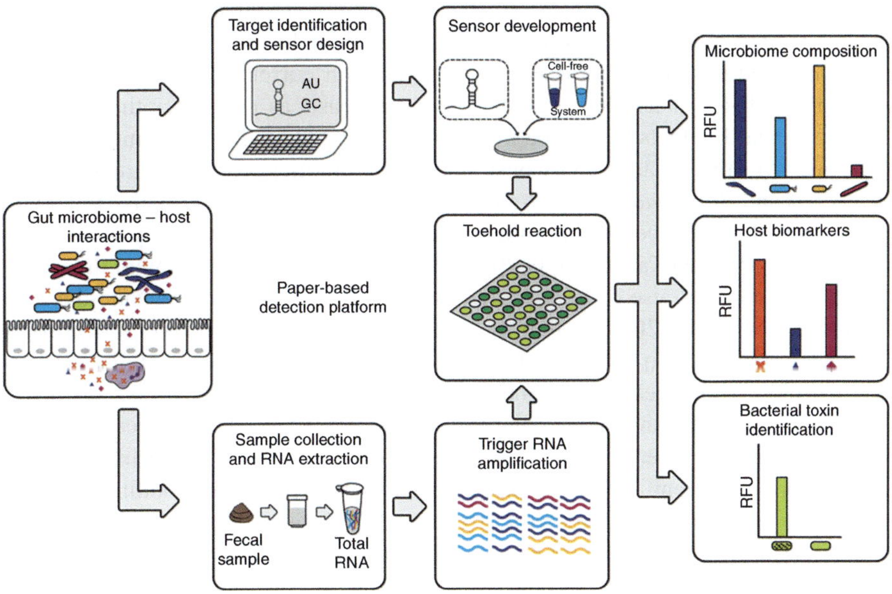

Fig. 9.2 Illustration on paper-based biosensor platform for gut microbiome analysis. In silico design of RNA toehold switch transducer and primers for RNA amplification to identify specific microbes or mRNA targets. Sensor components and primers are then integrated for paper-based biosensing. For the consequent study, total RNA can be extracted from faecal samples of patients through a commercial kit. Nucleic acid sequence-based amplification (NASBA) method is deployed for specific RNAs amplification and quantified using arrays of toehold switch sensors in paper-based reactions. From the calibration curve, specific microbial and host biomarker RNA concentrations of the analyte samples are detected. (Adapted from Takahasi et al. 2018, Nature Communications (Open access))

interpretation and continuous monitoring of clinical abnormalities. AI-enabled biosensors are known for complementing clinicians and patients and supporting care provision for chronic disease management by profiling real-time data. A recent review from Ladanza et al. (2020) highlighted that machine learning and deep learning can assist clinicians in processing and interpretation of massive data collected from gut microbiota. Integration of AI in modern devices can analyse the composition, structure, and complex interactions of microscopic organisms and can further aid in understanding their role in polydrug and food interactions, especially in geriatric population, thereby enabling precision medicine which may be more preventive, predictive and selective for better healthcare. To reach this stage, rigorous researches on the overall understanding of microbiota in various physiological locations or bodily fluids are absolutely essential.

10 Microbiome-Based Therapeutic Interventions in Geriatrics

Constipation is one of the gastrointestinal disorders frequently associated with adults. Its prevalence increases with age, particularly in hospitalized people who are under medications. The World Health Organization currently recommends using appropriate doses of live harmless bacteria and yeasts as probiotics that provides beneficial health effects by competitive antagonism and modulates immune system (Martínez-Martínez et al. 2017). Consumption of probiotics containing *Lactobacillus helveticus* and *Bifidobacterium longum* by depressed people was effective to correct the microbial imbalance and alter the pathophysiological conditions by modulating the gut microbiome (Kazemi et al. 2019). Short-term probiotic treatment in healthy adults was found to raise cellular innate immunity and to prevent oral candidiasis in the elderly (Ai et al. 2017; Miller et al. 2019). Several randomized clinical trials have been conducted to understand the role of probiotics with different disease conditions like Alzheimer, oral candidiasis, severe acute pancreatitis, psoriasis, Parkinson's disease, and type 2 diabetes (Besselink et al. 2008; Severance et al. 2017; Navarro-López et al. 2019; Tamtaji et al. 2019a, b). In all these trials, probiotics have proven potential benefits for reverting respective disease conditions.

The susceptibility to inflammatory rheumatic diseases in the elderly is a growing concern. The clinical presentations of this disease are diverse in the elderly as compared with young adults. In developed countries, biologics containing specific antibodies with immunomodulating properties have been successfully used for more than a decade as a therapeutic supplement for inflammatory rheumatic diseases. Tumour necrosis factor inhibitors have been used in rheumatoid arthritis for treatment and were found to be effective and well-tolerated by elderly people. This biological has been used as a second-line choice along with glucocorticoids and non-steroidal anti-inflammatory drugs. The benefits of biologicals in the elderly are that they promote the rational use of other drugs. However, in developing countries use of biologicals is minimal (Lahaye et al. 2015) and more studies are needed to prove their efficacy and safety in the elderly. The clinical trials on the use of vitamins had shown benefits for health in the elderly. Recently, the role of probiotics, prebiotics, and synbiotics on the physical health condition of elderly people has been reviewed, but there is no clear evidence on confounding factors and associated medications (Coutts et al. 2020).

The recurrent bacterial infections are common in the elderly like *Clostridium difficile* infection (CDI) due to the constant use of antimicrobials. Faecal microbiota transplant (FMT) is used to cure certain gastrointestinal disorders by collection of stool from a healthy individual and introduction during endoscopy (colonoscopy or enteroscopy) into the gut of the sick person. FMT has been successfully used to cure CDI and the success rate was about 90% (Liubakka and Vaughn 2016). The treatment for ulcerative colitis (UC) is complicated and many of the standard drugs used in the clinical practice may not respond. Several clinical trials on FMT for UC were found effective without any side effects (Moayyedi et al. 2015; Costello et al. 2019). However, the nature of faecal donor and time of UC infection is the confounding factors in the FMT. A trail on FMT by capsules and enriched FMT with *Lactobacillus* spp. showed similar outcomes (Garza-González et al. 2019).

11 Summary and Future Perspectives

Ageing is an inevitable process causing basic and functional characteristics in geriatrics making them vulnerable to communicable and non-communicable diseases. The normal life expectancy of people living in the developed countries is higher than in developing countries. There are clusters of correlations connecting nature of diseases, threats to health, affordability/accessibility to a quality health service, medication, lifestyle, and care. Chronic diseases and use of multiple drugs are very common in geriatrics. Multiple drugs and their metabolites could cause beneficial as well as harmful effects in the elderly. For instance, drug-drug interactions are common and reported to have unpredictable effects in geriatrics thereby influencing the gut microbiome. Antimicrobials dislocate symbiotic bacteria that support digestion, vitamin production, and detoxification. Also, commensal bacteria become resistant to commonly used drugs in clinical practice. Due to this reason, probiotics, prebiotics, and synbiotics are used for certain diseases that are not curable by antimicrobial agents particularly antibiotics-associated diarrhoea in the elderly.

Biologicals are found to be effective second-line drugs in geriatrics which help in reducing the usage of frontline drugs. The physiology of the gut microbiome has uncovered enormous beneficial effects on geriatric healthcare, suggesting the co-administration of probiotics to evade the harmful effect on gut microbiota. Nevertheless, more clinical trials are advocated to translate this knowledge into universal practice. Moreover, novel drug discovery and clinical trials are necessary for better understanding of the polydrug action and for achieving effective clinical practice. From the beginning of the twenty-first century, lifestyle has been correlated to individual health and quality of life. The desire of individuals to succeed in professional life could put personal wellbeing aside. Therefore, creating harmonious integration of work and lifestyle would yield better health conditions and drug-free life. There is tremendous support of growing evidence for the benefits of yoga to improve mental and emotional stability and lead a peaceful life. Practising breathing exercises and yogic lifestyle could possibly influence microbiome diversity in a favourable manner. Furthermore, modern analytical technologies are emerging as efficient non-invasive tools to detect vital biomarkers at minimum sample volume/concentration. With the advent of biomedical nanotechnology and bioengineering, a variety of biosensors have been recently devised for the rapid detection of clinical abnormalities. It is believed that emerging research in the above-mentioned transdisciplinary fields, in addition to the gut microbiome, would eventually provide new avenues on polydrug related therapeutics, biomacromolecules, and metabolites at high specificity/selectivity.

References

Ai R, Wei J, Ma D, Jiang L, Dan H, Zhou Y et al (2017) A meta-analysis of randomized trials assessing the effects of probiotic preparations on oral candidiasis in the elderly. Arch Oral Biol 83:187–192. https://doi.org/10.1016/j.archoralbio.2017.04.030

Alghamdi F, Chan M (2017) Management of heart failure in the elderly. Curr Opin Cardiol 32:217–223. https://doi.org/10.1097/HCO.0000000000000375

Allaband C, McDonald D, Vázquez-Baeza Y, Minich JJ, Tripathi A, Brenner DA, Knight R (2019) Microbiome 101: studying, analyzing, and interpreting gut microbiome data for clinicians. Clin Gastroenterol Hepatol 17:218–230. https://doi.org/10.1016/j.cgh.2018.09.017

Amrane S, Hocquart M, Afouda P, Kuete E, Pham TP, Dione N et al (2019) Metagenomic and culturomic analysis of gut microbiota dysbiosis during Clostridium difficile infection. Sci Rep 9:12807. https://doi.org/10.1038/s41598-019-49189-8

Arcavi L, Benowitz NL (2004) Cigarette smoking and infection. Arch Intern Med 164:2206–2216. https://doi.org/10.1001/archinte.164.20.2206

Bajaj JS, Idilman R, Mabudian L, Hood M, Fagan A, Turan D et al (2018) Diet affects gut microbiota and modulates hospitalization risk differentially in an international cirrhosis cohort. Hepatology 68:234–247. https://doi.org/10.1002/hep.29791

Bedarf JR, Hildebrand F, Coelho LP, Sunagawa S, Bahram M, Goeser F et al (2017) Functional implications of microbial and viral gut metagenome changes in early stage L-DOPA-naïve Parkinson's disease patients. Genome Med 9:39. https://doi.org/10.1186/s13073-017-0428-y

Besanko LK, Burgstad CM, Cock C, Heddle R, Fraser A, Fraser RJ (2014) Changes in esophageal and lower esophageal sphincter motility with healthy ageing. J Gastrointestin Liver Dis 23:243–248. https://doi.org/10.15403/jgld.2014.1121.233.lkb

Besselink MG, van Santvoort HC, Buskens E, Boermeester MA, van Goor H, Timmerman HM et al (2008) Dutch acute pancreatitis study group. Probiotic prophylaxis in predicted severe acute pancreatitis: a randomised, double-blind, placebo-controlled trial. Lancet 371:651–659. https://doi.org/10.1016/S0140-6736(08)60207-X

Brown GC (2015) Living too long: the current focus of medical research on increasing the quantity, rather than the quality, of life is damaging our health and harming the economy. EMBO Rep 16:137–141. https://doi.org/10.15252/embr.201439518

Browne HP, Forster SC, Anonye BO, Kumar N, Neville BA, Stares MD, Lawley TD (2016) Culturing of 'unculturable' human microbiota reveals novel taxa and extensive sporulation. Nature 533:543–546. https://doi.org/10.1038/nature1764
Buch A, Carmeli E, Boker LK, Marcus Y, Shefer G, Kis O et al (2016) Muscle function and fat content in relation to sarcopenia, obesity and frailty of old age–An overview. Exp Gerontol 76:25–32. https://doi.org/10.1016/j.exger.2016.01.008
Cappello M, Morreale GC, Licata A (2016) Elderly onset celiac disease: a narrative review. Clin Med Insights Gastroenterol 9:41–49. https://doi.org/10.4137/CGast.S38454
Carson TL, Wang F, Cui X, Jackson BE, Van Der Pol WJ, Lefkowitz EJ et al (2018) Associations between race, perceived psychological stress, and the gut microbiota in a sample of generally healthy black and white women: a pilot study on the role of race and perceived psychological stress. Psychosom Med 80:640–648. https://doi.org/10.1097/PSY.0000000000000614
Commisso A, Lim F (2019) Lifestyle modifications in adults and older adults with chronic gastroesophageal reflux disease (GERD). Crit Care Nurs Q 42:64–74. https://doi.org/10.1097/CNQ.0000000000000239
Costello SP, Hughes PA, Waters O, Bryant RV, Vincent AD, Blatchford P et al (2019) Effect of fecal microbiota transplantation on 8-week remission in patients with ulcerative colitis: a randomized clinical trial. JAMA 321:156–164. https://doi.org/10.1001/jama.2018.20046
Coutts L, Ibrahim K, Tan QY, Lim SER, Cox NJ, Roberts HC (2020) Can probiotics, prebiotics and synbiotics improve functional outcomes for older people: a systematic review. Eur Geriatr Med 11:975–993. https://doi.org/10.1007/s41999-020-00396-x
Deluche E, Girault S, Jesus P, Monzat S, Turlure P, Leobon S et al (2017) Assessment of the nutritional status of adult patients with acute myeloid leukemia during induction chemotherapy. Nutrition 41:120–125. https://doi.org/10.1016/j.nut.2017.04.011
Doestzada M, Vila AV, Zhernakova A, Koonen DPY, Weersma RK, Touw DJ et al (2018) Pharmacomicrobiomics: a novel route towards personalized medicine? Protein Cell 9:432–445. https://doi.org/10.1007/s13238-018-0547-2
Fabian E, Bogner M, Kickinger A, Wagner KH, Elmadfa I (2011) Intake of medication and vitamin status in the elderly. Ann Nutr Metab 58(2):118–125. https://doi.org/10.1159/000327351
Fali T, Vallet H, Sauce D (2018) Impact of stress on aged immune system compartments: overview from fundamental to clinical data. Exp Gerontol 105:19–26. https://doi.org/10.1016/j.exger.2018.02.007
FarioliVecchioli S, Sacchetti S, Nicolis di Robilant V, Cutuli D (2018) The role of physical exercise and Omega-3 fatty acids in depressive illness in the elderly. Curr Neuropharmacol 16:308–326. https://doi.org/10.2174/1570159X15666170912113852
Farré R, Vicario M (2017) Abnormal barrier function in gastrointestinal disorders. Handb Exp Pharmacol 239:193–217. https://doi.org/10.1007/164_2016_107
Fávaro-Moreira NC, Krausch-Hofmann S, Matthys C, Vereecken C, Vanhauwaert E, Declercq A et al (2016) Risk factors for malnutrition in older adults: a systematic review of the literature based on longitudinal data. Adv Nutr 7:507–522. https://doi.org/10.3945/an.115.011254
Feldblum I, German L, Castel H, Harman-Boehm I, Bilenko N, Eisinger M, Fraser D, Shahar DR (2007) Characteristics of undernourished older medical patients and the identification of predictors for undernutrition status. Nutr J 2(6):37. https://doi.org/10.1186/1475-2891-6-37
Figdor D, Gulabivala K (2008) Survival against the odds: microbiology of root canals associated with post-treatment disease: microbiology of post-treatment disease. Endod Top 18:62–77. https://doi.org/10.1111/j.1601-1546.2011.00259.x
Garza-González E, Mendoza-Olazarán S, Morfin-Otero R, Ramírez-Fontes A, Rodríguez-Zulueta P, Flores-Treviño S, Bocanegra-Ibarias P, Maldonado-Garza H, Camacho-Ortiz A (2019) Intestinal microbiome changes in fecal microbiota transplant (FMT) *vs.* FMT enriched with *lactobacillus* in the treatment of recurrent *Clostridioides difficile* infection. Can. J Gastroenterol Hepatol 4549298. https://doi.org/10.1155/2019/4549298

Grosicki GJ, Fielding RA, Lustgarten MS (2018) Gut microbiota contribute to age-related changes in skeletal muscle size, composition, and function: biological basis for a gut-muscle Axis. Calcif Tissue Int 102:433–442. https://doi.org/10.1007/s00223-017-0345-5

Gras-Miralles B, Cremonini F (2013) A critical appraisal of lubiprostone in the treatment of chronic constipation in the elderly. Clin Interv Ageing 8:191–200. https://doi.org/10.2147/CIA.S30729

Gustavsson J, Jernbro C, Nilson F (2018) There is more to life than risk avoidance–elderly people's experiences of falls, fall-injuries and compliant flooring. Int J Qual Stud Health Well-being 13:1479586. https://doi.org/10.1080/17482631.2018.1479586

Hanewinckel R, van Oijen M, Ikram MA, van Doorn PA (2016) The epidemiology and risk factors of chronic polyneuropathy. Eur J Epidemiol 31:5–20. https://doi.org/10.1007/s10654-015-0094-6

Heuberger RA, Caudell K (2011) Polypharmacy and nutritional status in older adults. Drugs Ageing 28:315–323. https://doi.org/10.2165/11587670-000000000-00000

Hsu DW, Suh JD (2018) Rhinitis and sinusitis in the geriatric population. Otolaryngol Clin North Am 51:803–813. doi: https://doi.org/10.1016/j.otc.2018.03.008. https://www.infoplease.com/world/health-and-social-statistics/life-expectancy-countries. Accessed on January 01, 2021

Huang YC, Chang HH, Huang SC, Cheng CH, Lee BJ, Cheng SY, Su KH (2005) Plasma pyridoxal 5′-phosphate is a significant indicator of immune responses in the mechanically ventilated critically ill. Nutrition 21:779–785. https://doi.org/10.1016/j.nut.2004.11.013

Jones CI (2016) Platelet function and ageing. Mamm Genome 27:358–366. https://doi.org/10.1007/s00335-016-9629-8

Kato K, Odamaki T, Mitsuyama E, Sugahara H, Xiao JZ, Osawa R (2017) Age-related changes in the composition of gut Bifidobacterium species. Curr Microbiol 74:987–995. https://doi.org/10.1007/s00284-017-1272-4

Kazemi A, Noorbala AA, Azam K, Eskandari MH, Djafarian K (2019) Effect of probiotic and prebiotic vs placebo on psychological outcomes in patients with major depressive disorder: a randomized clinical trial. Clin Nutr 38:522–528. https://doi.org/10.1016/j.clnu.2018.04.010

Kim J, Parish AL (2017) Polypharmacy and medication management in older adults. Nurs Clin North Am 52:457–468. https://doi.org/10.1016/j.cnur.2017.04.007

Lahaye C, Tatar Z, Dubost JJ, Soubrier M (2015) Overview of biologic treatments in the elderly. Joint Bone Spine 82:154–160. https://doi.org/10.1016/j.jbspin.2014.10.012

Lahaye C, Tatar Z, Dubost JJ, Tournadre A, Soubrier M (2019) Management of inflammatory rheumatic conditions in the elderly. Rheumatology (Oxford) 58:748–764. https://doi.org/10.1093/rheumatology/key165

La-Ongkham O, Nakphaichit M, Nakayama J, Keawsompong S, Nitisinprasert S (2020) Age-related changes in the gut microbiota and the core gut microbiome of healthy Thai humans. 3 Biotech 10(6):276. https://doi.org/10.1007/s13205-020-02265-7

Ladanza E, Fabbri R, Bašić-ČiČak D, Amedei A, Telalovic JH (2020) Gut microbiota and artificial intelligence approaches: a scoping review. Heal Technol 10:1343–1358. https://doi.org/10.1007/s12553-020-00486-7

Lee J, Venna VR, Durgan DJ, Shi H, Hudobenko J, Putluri N et al (2020a) Young versus aged microbiota transplants to germ-free mice: increased short-chain fatty acids and improved cognitive performance. Gut Microbes 8:1–14. https://doi.org/10.1080/19490976.2020.1814107

Lee SH, Yoon SH, Jung Y, Kim N, Min U, Chun J, Choi I (2020b) Emotional Well-being and gut microbiome profiles by enterotype. Sci Rep 10:20736. https://doi.org/10.1038/s41598-020-77673-z

Liang SY (2016) Sepsis and other infectious disease emergencies in the elderly. Emerg Med Clin North Am 34:501–522. https://doi.org/10.1016/j.emc.2016.04.005

Lin D, Peters BA, Friedlander C, Freiman HJ, Goedert JJ, Sinha R et al (2018) Association of dietary fibre intake and gut microbiota in adults. Br J Nutr 120:1014–1022. https://doi.org/10.1017/S0007114518002465

Little MO (2018) Updates in nutrition and polypharmacy. Curr Opin Clin Nutr Metab Care 21:4–9. https://doi.org/10.1097/MCO.0000000000000425

Liubakka A, Vaughn BP (2016) *Clostridium difficile* infection and fecal microbiota transplant. AACN Adv Crit Care 27:324–337. https://doi.org/10.4037/aacnacc2016703

Madhumita P, Gupta N (2011) Clinical and bacteriological spectrum of community-acquired acute bacterial meningitis in adults at a tertiary care hospital in northern India. Int J Nutr Pharmacol Neurol Dis 1:194–200. https://doi.org/10.4103/2231-0738.84214

Maher RL, Hanlon J, Hajjar ER (2014) Clinical consequences of polypharmacy in elderly. Expert Opin Drug Saf 13:57–65. https://doi.org/10.1517/14740338.2013.827660

Martínez-Martínez MI, Calabuig-Tolsá R, Cauli O (2017) The effect of probiotics as a treatment for constipation in elderly people: a systematic review. Arch Gerontol Geriatr 71:142–149. https://doi.org/10.1016/j.archger.2017.04.004

Maynard C, Weinkove D (2018) The gut microbiota and ageing. Subcell Biochem 90:351–371. https://doi.org/10.1007/978-981-13-2835-0_12

McArthur DB (2019) Emerging infectious diseases. Nurs Clin North Am 54:297–311. https://doi.org/10.1016/j.cnur.2019.02.006

Miller LE, Lehtoranta L, Lehtinen MJ (2019) Short-term probiotic supplementation enhances cellular immune function in healthy elderly: systematic review and meta-analysis of controlled studies. Nutr Res 64:1–8. https://doi.org/10.1016/j.nutres.2018.12.011

Moayyedi P, Surette MG, Kim PT, Libertucci J, Wolfe M, Onischi C et al (2015) Fecal microbiota transplantation induces remission in patients with active ulcerative colitis in a randomized controlled trial. Gastroenterology 149:102–109.e6. https://doi.org/10.1053/j.gastro.2015.04.001

Moser A, Panczak R, Zwahlen M, Clough-Gorr KM, Spoerri A, Stuck AE, Egger M (2014) Swiss National Cohort. What does your neighbourhood say about you? A study of life expectancy in 1.3 million Swiss neighbourhoods. J Epidemiol Community Health 68:1125–1132. https://doi.org/10.1136/jech-2014-204352

Naseribafrouei A, Hestad K, Avershina E, Sekelja M, Linløkken A, Wilson R, Rudi K (2014) Correlation between the human fecal microbiota and depression. Neurogastroenteral Motil 26:1155–1162. https://doi.org/10.1111/nmo.12378

Navarro-López V, Martínez-Andrés A, Ramírez-Boscá A, Ruzafa-Costas B, Núñez-Delegido E, Carrión-Gutiérrez MA et al (2019) Efficacy and safety of Oral Administration of a Mixture of probiotic strains in patients with psoriasis: a randomized controlled clinical trial. Acta Derm Venereol 99:1078–1084. https://doi.org/10.2340/00015555-3305

Nikolakakis N, Dragioti E, Paritsis N, Tsamakis K, Christodoulou NG, Rizos EN (2019) Association between happiness and psychopathology in an elderly regional rural population in Crete. Psychiatriki 30:299–310. https://doi.org/10.22365/jpsych.2019.304.299

Olyaei AJ, Bennett WM (2009) Drug dosing in the elderly patients with chronic kidney disease. Clin Geriatr Med 25:459–527. https://doi.org/10.1016/j.cger.2009.04.004

O'Toole PW, Jeffery IB (2018) Microbiome-health interactions in older people. Cell Mol Life Sci 75:119–128. https://doi.org/10.1007/s00018-017-2673-z

Paalanne N, Husso A, Salo J, Pieviläinen O, Tejesvi MV, Koivusaari P et al (2018) Intestinal microbiome as a risk factor for urinary tract infections in children. Eur J Clin Microbiol Infect Dis 37:1881–1891. https://doi.org/10.1007/s10096-018-3322-7

Przewłócka K, Folwarski M, Kaźmierczak-Siedlecka K, Skonieczna-Żydecka K, Kaczor JJ (2020) Gut-muscle AxisExists and may affect skeletal muscle adaptation to training. Nutrients 125:1451. https://doi.org/10.3390/nu12051451

Ramírez-Carrillo E, Gaona O, Nieto J, Sánchez-Quinto A, Cerqueda-García D, Falcón LI et al (2020) Disturbance in human gut microbiota networks by parasites and its implications in the incidence of depression. Sci Rep 10:3680. https://doi.org/10.1038/s41598-020-60562-w

Rossi MI, Young A, Maher R, Rodriguez KL, Appelt CJ, Perera S, Hajjar ER, Hanlon JT (2007) Polypharmacy and health beliefs in older outpatients. Am J Geriatr Pharmacother 5:317–323. https://doi.org/10.1016/j.amjopharm.2007.12.001

Sachdeva A (2017) Pneumococcal conjugate vaccine introduction in India's universal immunization program. Indian Pediatr 54:445–446. https://doi.org/10.1007/s13312-017-1044-z

Salive ME (2013) Multimorbidity in older adults. Epidemiol Rev 35:75–83. https://doi.org/10.1093/epirev/mxs009
Scher JU, Nayak RR, Ubeda C, Turnbaugh PJ, Abramson SB (2020) Pharmacomicrobiomics in inflammatory arthritis: gut microbiome as modulator of therapeutic response. Nat Rev Rheumatol 16:282–292. https://doi.org/10.1038/s41584-020-0395-3
Sebastiani P, Thyagarajan B, Sun F, Schupf N, Newman AB, Montano M, Perls TT (2017) Biomarker signatures of ageing. Ageing. Cell 16:329–338. https://doi.org/10.1111/acel.12557
Severance EG, Gressitt KL, Stallings CR, Katsafanas E, Schweinfurth LA, Savage CLG et al (2017) Probiotic normalization of Candida albicans in schizophrenia: a randomized, placebo-controlled, longitudinal pilot study. Brain Behav Immun 62:41–45. https://doi.org/10.1016/j.bbi.2016.11.019
Song WJ, Won HK, An J, Kang SY, Jo EJ, Chang YS et al (2019) Chronic cough in the elderly. Pulm Pharmacol Ther 56:63–68. https://doi.org/10.1016/j.pupt.2019.03.010
Sosnowski K, Akarapipad P, Yoon JY (2020) The future of microbiome analysis: biosensor methods for big data collection and clinical diagnostics. Med Devices Sens 3:e1008. https://doi.org/10.1002/mds3.10085
Stewart EJ (2012) Growing unculturable bacteria. J Bacteriol 194(16):4151–4160. https://doi.org/10.1128/JB.00345-12
Strandwitz P, Kim KH, Terekhova D, Liu JK, Sharma A, Levering J et al (2019) GABA-modulating bacteria of the human gut microbiota. Nat Microbiol 4:396–403. https://doi.org/10.1038/s41564-018-0307-3
Takahasi MK, Tan X, Dy AJ, Braff D, Akana RT, Furuta Y et al (2018) A low-cost paper-based synthetic biology platform for analyzing gut microbiota and host biomarkers. Nat Commun 3347:1–12
Tamtaji OR, Heidari-Soureshjani R, Mirhosseini N, Kouchaki E, Bahmani F, Aghadavod E et al (2019a) Probiotic and selenium co-supplementation, and the effects on clinical, metabolic and genetic status in Alzheimer's disease: a randomized, double-blind, controlled trial. Clin Nutr 38:2569–2575. https://doi.org/10.1016/j.clnu.2018.11.034
Tamtaji OR, Taghizadeh M, DaneshvarKakhaki R, Kouchaki E, Bahmani F, Borzabadi S et al (2019b) Clinical and metabolic response to probiotic administration in people with Parkinson's disease: a randomized, double-blind, placebo-controlled trial. Clin Nutr 38:1031–1035. https://doi.org/10.1016/j.clnu.2018.05.018
Templier V, Roux A, Roupioz Y, Livache T (2016) Ligands for label-free detection of whole bacteria on biosensors: a review. TrAC 79:71–79. https://doi.org/10.1016/j.trac.2015.10.015
Thigpen MC, Whitney CG, Messonnier NE, Zell ER, Lynfield R, Hadler JL et al (2011) Emerging infections programs network. Bacterial meningitis in the United States, 1998-2007. N Engl J Med 364:2016–2025. https://doi.org/10.1056/NEJMoa1005384
Ticinesi A, Milani C, Lauretani F, Nouvenne A, Mancabelli L, Lugli GA et al (2017) Gut microbiota composition is associated with polypharmacy in elderly hospitalized patients. Sci Rep 7(1):11102. https://doi.org/10.1038/s41598-017-10734-y
Tiihonen K, Ouwehand AC, Rautonen N (2009) Human intestinal microbiota and healthy ageing. Ageing Res Rev 9:107–116. https://doi.org/10.1016/j.arr.2009.10.004
Tillisch K, Mayer EA, Gupta A, Gill Z, Brazeilles R, Le Nevé B et al (2017) Brain structure and response to emotional stimuli as related to gut microbial profiles in healthy women. Psychosom Med 79:905–913. https://doi.org/10.1097/PSY.0000000000000493
Valdes-Rodriguez R, Stull C, Yosipovitch G (2015) Chronic pruritus in the elderly: pathophysiology, diagnosis and management. Drugs Ageing 32:201–215. https://doi.org/10.1007/s40266-015-0246-0
Velasquez MT, Ramezani A, Manal A, Raj DS (2016) Trimethylamine N-oxide: the good, the bad and the unknown. Toxins (Basel) 8:326. https://doi.org/10.3390/toxins8110326
Violi F, Loffredo L, Carnevale R, Pignatelli P, Pastori D (2017) Atherothrombosis and oxidative stress: mechanisms and management in elderly. Antioxid Redox Signal 27:1083–1124. https://doi.org/10.1089/ars.2016.6963

Wilson ID, Nicholson JK (2017) Gut microbiome interactions with drug metabolism, efficacy, and toxicity. Transl Res 179:204–222. https://doi.org/10.1016/j.trsl.2016.08.002

Wu GD, Chen J, Hoffmann C, Bittinger K, Chen YY, Keilbaugh SA et al (2011) Linking long-term dietary patterns with gut microbial enterotypes. Science 334:105–108. https://doi.org/10.1126/science.1208344

Zeppa SD, Agostini D, Piccoli G, StocchiV SP (2020) Gut microbiota status in COVID-19: an unrecognized player? Front Cell Infect Microbiol 10:742. https://doi.org/10.3389/fcimb.2020.576551

Zhuang ZQ, Shen LL, Li WW, Fu X, Zeng F, Gui L et al (2018) Gut microbiota is altered in patients with Alzheimer's disease. J Alzheimers Dis 63:1337–1346. https://doi.org/10.3233/JAD-180176

Chapter 10
Virome: Sentinels or Marauders in the Microbiome

Joydeep Nag, Nisha Asok Kumar, Reshma Koolaparambil Mukesh, Umerali Kunnakkadan, and John Bernet Johnson

1 Introduction

The human virome is a part of one of the most complex ecosystems in the world, namely the microbiome. It represents a repertoire of all the viruses that inhabit the human body which includes the eukaryotic viruses, the virus-derived elements that are inserted into host chromosomes (endogenous retroviruses) and the bacteriophages that are capable of infecting the inhabitant bacteria and archaea (prokaryotic viruses) (Virgin 2014). A large number of viruses have been reported to inhabit humans, while only half of them are pathogenic (Parker 2016). Non-pathogenic viruses are referred to as "commensal" viruses as they survive by either integrating into the host chromosome or by infecting bacteria without causing any clinical outcome. However, pathogenic ones have been shown to affect human health by causing acute, persistent, or latent infections which in many instances are detrimental to the host. The bulk of the healthy human virome comprises of the bacteriophages, which are capable of infecting bacteria present in the intestine and other parts of the human body. A study on the diversity of gut virome in 1-year-old infants has found a strong correlation between diversity and the manner of birth (McCann et al. 2018). Several pathogenic viruses are also transmitted to the newborn during vaginal delivery as well as during breastfeeding. The gut of the healthy neonates is devoid of any viruses at the time of birth. The direct transmission of the virus strains from the mother to the infant was validated by sequencing the fecal sample of the baby and the breast milk of the mother. Human milk viruses are important in shaping the gut virome of the infants and are also important in the overall immune development (Mohandas and Pannaraj 2020).

J. Nag · N. A. Kumar · R. K. Mukesh · U. Kunnakkadan · J. B. Johnson (✉)
Viral Disease Biology, Rajiv Gandhi Centre for Biotechnology, Thiruvananthapuram, Kerala, India
e-mail: johnbjohnson@rgcb.res.in

S. Thomas (ed.), *Human Microbiome*,
https://doi.org/10.1007/978-981-16-7672-7_10

Over the past decade, several studies have shed light on the link between the microbiome and the various states of human health. Nevertheless, a better understanding of the virome is still lacking when compared to the other players of the microbiome. This lag in knowledge is attributed to the absence of a universal viral sequence similar to the 16S rRNA present in the bacteria. Most often, the virus-enriched preparations do not align with any reference sequences and hence are represented as the viral "dark matter" (Roux et al. 2015). The dark matter hence might comprise several novel and highly divergent viruses, which collectively constitute the "viral assemblage".

2 Factors Influencing the Distribution and Diversity of Virome

The microbiota of the human body is enriched with distinct microbial flora, which accommodates the growth of a variety of viruses. Several factors are known to affect the distribution and diversity of the human virome (Fig. 10.1). A major constituent of the human virome is bacteriophage, and the type and distribution of the microbial community has a significant impact on the distribution and diversity of the virome. Marked differences exist within the microbiome based on the anatomical site of distribution, and parallels can be drawn likewise with the virome (Abeles et al. 2014; Reyes et al. 2010). The establishment of human virome inside the body is closely associated with the initial colonization. Various viruses, including Zika virus, HIV, rubella virus, herpes simplex virus, and human papillomavirus, are

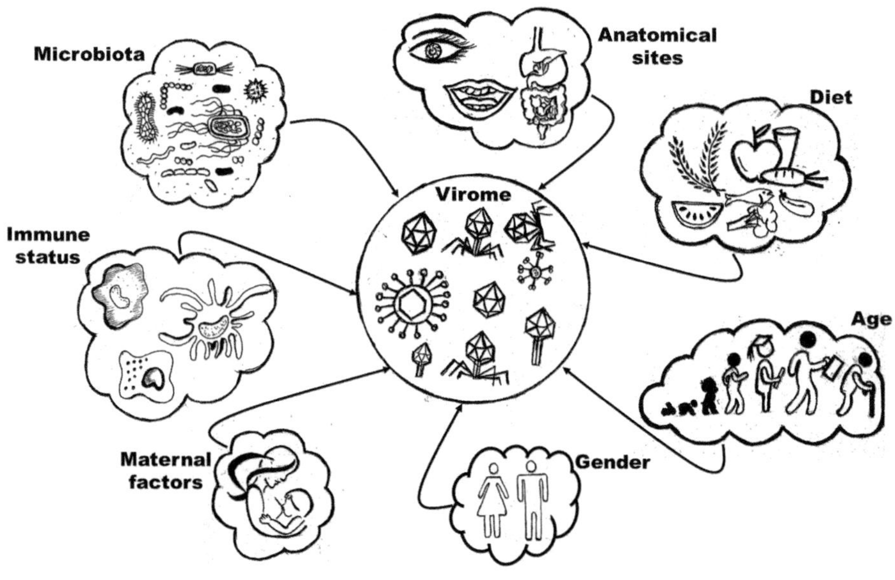

Fig. 10.1 Factors influencing virome diversity and distribution in humans

known to adopt a vertical mode of transfer during pregnancy (Leeper and Lutzkanin 3rd 2018; Arora et al. 2017). According to Breitbart et al., neonates lack virome at birth but are gradually colonized after a week of age (Breitbart et al. 2008). This is supported by yet another study which identified a high diversity of gut virome in the neonates after just few weeks of birth (Lim et al. 2015). Breastfeeding is an important factor in shaping the gut viral community in the neonates. An abundance in viral population was observed in infants who were exclusively fed formula than those dependent entirely or partially on breast milk (Liang et al. 2020b). Considering that breast milk is a rich source of maternal antibodies, immune cells, lactoferrin, mucin, and milk oligosaccharides which can restrict the invasion of a wide variety of viruses including influenza virus, rotavirus, enterovirus, norovirus, and SARS-CoV, breast milk while protecting neonates from deadly viruses can aid in the adaptation and colonization of beneficial viruses (Lang et al. 2011; Turin and Ochoa 2014; Simister 2003; Pou et al. 2019; Albrecht and Arck 2020; Berlutti et al. 2011; Wicinski et al. 2020).

The diet, age, and sex of a person can also modulate the diversity of the virome (Fig. 10.1). It has been reported that oral viromes were similar in people with the same diet or oral bacterial population and between people from the same household or family (Robles-Sikisaka et al. 2013). This is compounded by a study that indicated an inter-individual variation of the gut microbiome in response to diet changes (Minot et al. 2011). Within the oral cavity, significant variation in phage communities has been observed in the saliva, dental plaque, and subgingival and supragingival biofilms (Wang et al. 2016a, b). Sex-specific variation inside the oral viral community has been reported, wherein it was identified that the genotype of the oral virome in an individual is highly personalized and gender-specific (Abeles et al. 2014). The age and immune status of an individual are also key factors which can affect the richness of viral communities. Gregory et al. showed that the enrichment of eukaryotic viruses most importantly human Annelloviruses is high during infancy and then decreases with childhood and remains constant and low through the rest of life which corresponds to the fact that the patterns in viral diversity is age dependent (Gregory et al. 2020).

There is also a direct correlation between the abundance of the viral population and an individual's immune status. Studies have shown that boosting immunity is an effective strategy in enhancing anti-viral immunity in the gut. Metagenomic analysis of viral population in the gut of an X-linked severe combined immunodeficiency patient revealed a viral population rich in adenovirus and bocaviruses and upon immune reconstitution, the gut microbiota was normalized and the viral infections were cleared (Clarke et al. 2018a). A positive effect of bacteria or bacterial components in restricting viral infection has also been demonstrated wherein it was shown that flagellin exposure activated the immune response and restricted rotavirus infection in mice (Zhang et al. 2014). Alternatively, the administration of recombinant IFNλ could effectively clear persistent norovirus infection (Nice et al. 2015). These observations highlight that effective cross-talk exists between microbiome components and the host which is pivotal in clearing pathogenic viruses which are also part of the virus assemblage.

Host genetics also contributes to the virome composition and diversity. Studies have demonstrated that monozygotic twins have a similar microbiome compared to dizygotic twins (Goodrich et al. 2016; Goodrich et al. 2014). However, few studies have identified that the environment of an individual has a significant role in shaping the microbiome rather than host genetics (Rothschild et al. 2018). Several studies have also claimed a significant variation in human virome associated with geographic location (Holtz et al. 2014). However, it is not always true, for instance, Polyomavirus species collected from individuals of different geographic regions showed very low genetic diversity (Foulongne et al. 2012; Rascovan et al. 2016). Children with diarrhea from two locations within Australia have shown a significant variation of eukaryotic viromes with a differential prevalence of Adenoviridae and Picornaviridae (Holtz et al. 2014). Thus it is clear that viral abundance and diversity is dependent on multiple factors. As more and more systematic studies are carried out, it would emerge that many other factors besides those described above also contribute to the assemblage of viruses in the human virome.

3 System-Wise Distribution of the Human Virome

3.1 Ocular Virome

Ocular surface (OS) microbiome constitutes the microbiota that resides on the surface of the conjunctiva and the cornea (micro-organisms that colonize eyelids are considered as a part of skin microbiota) (Lu and Liu 2016). Investigations into the ocular microbiome are a relatively new and emerging area and most of the studies are designed to investigate the prokaryotic residents as opposed to the larger, more inclusive microbial community including virome and mycobiome (Fig. 10.2). Culture-independent metagenomic studies on OS have revealed that, unlike skin or other mucosal tissues, the healthy OS microbiome is sparsely colonized (~100 times less than that of the facial skin or the buccal mucosa) (Doan et al. 2016). Analysis of metagenomic data of OS microbiome from 90 adult healthy individuals showed that approximately 98% of the microbial reads were of bacterial origin while viral and fungal sequences accounted for <1% each (Wen et al. 2017). Doan T. et al. employed biome representational in silico karyotyping (BRiSK), a deep sequencing technique that achieves unbiased representation of all DNA-based metagenomic constituents and uncovered the presence of viruses such as torque teno virus (TTV), multiple sclerosis-associated retrovirus (MRSV), and human endogenous retrovirus K (HERV-K) in the conjunctiva of healthy volunteers. Although less frequent, sequences pertaining to human papillomavirus (HPV), Merkel Cell Polyomavirus (MCV), and Abelson murine leukemia virus were also retrieved by BRiSK. A noteworthy observation however was the high PCR positivity rate for TTV which was as high as 65% in all the conjunctiva samples tested, suggesting that TTV might be a homeostatic resident on ocular surfaces of healthy humans (Doan et al. 2016).

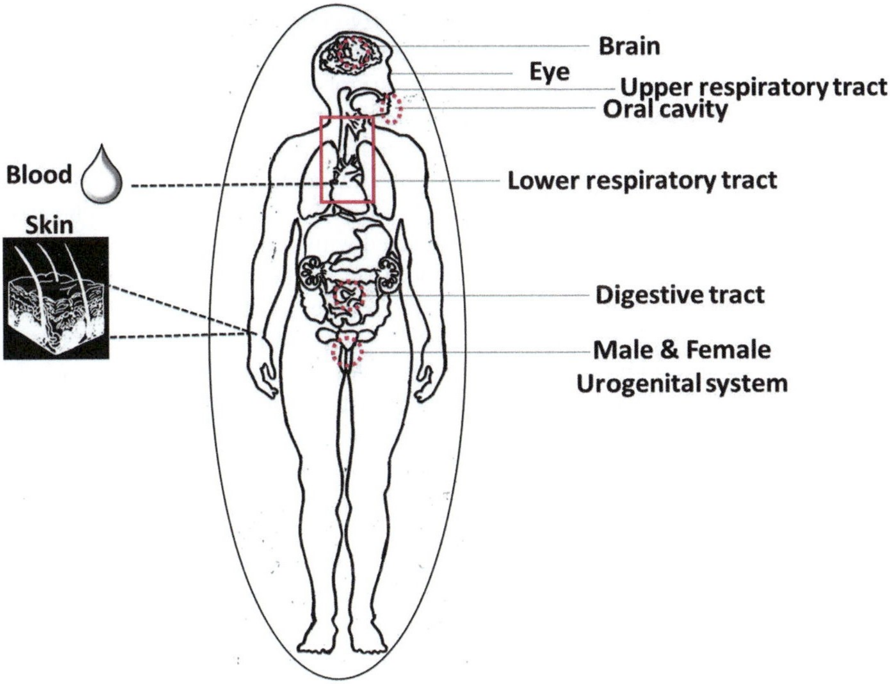

Fig. 10.2 System-wise presence of virus niches in the human body

However, in an unrelated study, deep sequencing of vitreous biopsies from patients with presumed culture-negative infectious endophthalmitis could identify TTV in all culture-negative samples, although a direct association or causation was not claimed (Lee et al. 2015). Though metagenomic deep sequencing allows a comprehensive analysis of microbial or host genetic materials in the sample, the choice of the genetic material (RNA vs DNA) can limit the power of the microbiome studies in the context of "virome." Both the metagenomic studies highlighted above used DNA as their starting material which subsequently restrains or impacts the "virome" analysis as the approach is technically "blind" to the group of viruses with RNA as their genetic material.

3.2 *Oral Virome*

Microbial communities in the oral cavity are diverse comprising fungi, bacteria, and viruses. Traditionally, oral microbiome sampling was limited to saliva; but recent studies have identified distinct virus communities from other oral microenvironments like dental plaques also. Saliva collected from healthy volunteers, when subjected to SYBR-gold nucleic acid staining post sequential filtration, showed ~10^8

virus-like particles (VLPs) per milliliter, majorly constituting lysogenic bacteriophages. Among the genomes analyzed, more than 90% accounted for prophages, which largely outnumber bacteriophages. Bacteriophages of the order caudovirales, belonging to the families *Siphoviridae* and *Myoviridae,* and *Podoviridae* are the most abundant. *Herpesviridae*, *Papillomaviridae*, *Anelloviridae*, and *Redondoviridae* families of eukaryotic viruses are also frequently encountered. Unlike many eukaryotic viruses that are asymptomatic in healthy individuals (Table 10.1) (Wylie et al. 2014), redondoviruses are found to be associated with periodontitis in the oral cavity, and their abundance in the oro-respiratory tract has been implicated with worse prognosis in patients admitted in critical care facilities (Abbas et al. 2019a; Perez-Brocal and Moya 2018). The dental plaque has ~10^8 VLPs per milligram sample and shares a significant proportion of viral homologs with the salivary virome suggesting the presence of some of the viruses in both the niches (Naidu et al. 2014). The oral virome is influenced by the living environment and its composition is temporally regulated (Pride et al. 2012; Robles-Sikisaka et al. 2013).

The oral virome is dynamic in nature and it has been suggested that these viruses might serve as a reservoir of pathogenic gene functions, imparting virulence to the resident bacteria of the buccal cavity. The non-exclusive coexistence of both bacteriophages and their hosts in the same ecological niche points toward the existence of both positive and negative interactions between them (Pride et al. 2012). The lysogenic lifestyle of the siphoviruses in the oral cavity, in a dynamic equilibrium with their prokaryotic hosts, makes them excellent vehicles for horizontal gene transfer, potentially imparting antibiotic resistance to the host. On the other hand, myoviruses and podoviruses being predominantly lytic are responsible for the elimination of 20–80% of the oral bacteria. This arms race between phages and bacteria prevent the successful establishment of novel species of bacteria or phages in the oral cavity (Baker et al. 2017).

Table 10.1 Viral diversity across different anatomical sites of the human body

Location/ microenvironment	Predominant virus families
Eye	*Anelloviridae, Retroviridae, Papillomaviridae, Polyomaviridae*
Oral cavity	*Siphoviridae, Myoviridae, Podoviridae, Herpesviridae, Papillomaviridae, Anelloviridae, Redondoviridae*
Gut	*Siphoviridae, Microviridae, Myoviridae, crass like phages, Anelloviridae, Herpesviridae, Adenoviridae, Papillomaviridae, Polyomaviridae*
Respiratory tract	*Inoviridae, Microviridae, Anelloviridae, Redondoviridae, Adenoviridae, Papillomaviridae, Herpesviridae*
Central nervous system	*Siphoviridae, Myoviridae, Podoviridae, Herpesviridae*
Blood	*Phycodnaviridae, Picornaviridae, Mimiviridae, Marseilleviridae*
Skin	*Microviridae, Siphoviridae, Papilloma, Polyomaviridae, Circoviridae, poxviridae*
Urogenital tract	*Papillomaviridae, Anelloviridae, Herpesviridae*

3.3 Gut Virome

The collective population of both the eukaryotic and prokaryotic viruses colonizing the human gut comprises the human gut virome (Fig. 10.2). Although the total viral loads in the human gut vary from subject to subject, they can range anywhere between 2.2×10^8 and 8.4×10^{10} genome copies per gram of feces (Shkoporov et al. 2019). The human Gut Virome database, a compilation of numerous microbial metagenomic studies across the continents, uncovered 33,242 potentially unique viral population found in the human gut (Gregory et al. 2020). The viruses that infect bacteria (bacteriophages) predominate (>97%) the gut virome and evidences point to a temperate lifestyle exhibited by majority of the phages within the gut ecosystem (Ogilvie and Jones 2015). Most common bacteriophages in the gut virome belonged to the family *Siphoviridae*, *Microviridae*, and *Myoviridae* (Table 10.1). Another family of bacteriophage known as crAssphage and its expansive group of crAss-like phages are the most abundant human-associated virus found in ~50% of the human gut samples, which often comprises 90% of the annotated sequence reads in gut virome-specific metagenome (Shkoporov et al. 2018; Yutin et al. 2018). Though minimal in proportion, several DNA and RNA eukaryotic viruses have also been detected in the feces sample. Most commonly associated RNA viruses included enterovirus, parechovirus, tombamovirus, sapovirus, calciviruses, astroviruses, and picornaviruses (Lim et al. 2015). Members from the family *Anelloviridae*, *Herpesviridae*, *Adenoviridae*, *Papillomaviridae*, and *Polyomaviridae* are the main eukaryotic DNA viruses associated with human intestinal virome (Rampelli et al. 2017). Although rare, contigs matching the sequences of megavirome such as Mimivirus and Marseillevirus (Colson et al. 2013), an archaeal virus family (*Lipothrixviridae*) (Lim et al. 2015), and several plant pathogenic viruses have also been identified in the fecal metagenome (Table 10.1) (Zhang et al. 2006). Similar to their bacterial counterparts in the microbiome, the composition of the gut virome is also dynamic and is mostly shaped during the early years of the development (Lim et al. 2015). In a longitudinal study of the intestinal virome in infants by Lim et al., it was found that the bacteriophage diversity was maximum at the earliest time point tested (month 0) and it eventually decreased with age. They also reported a noticeable shift in the composition of phage community trending toward a relative increase in the abundance of *Microviridae* family of bacteriophages by the age of 24 months. Interestingly, a parallel analysis of gut bacterial diversity revealed an inverse correlation with the richness of the associated virome in an age-dependent manner indicating dynamic interplay during the early years of life (Lim et al. 2015). Studies with monozygotic twins indicated that, although co-twins and their respective mothers shared similar virome profiles, nevertheless, each subject harbored a distinct and unique individual virome irrespective of the genetic relatedness (Lim et al. 2015; Reyes et al. 2010). Despite high interpersonal variations in fecal virome among the subjects, intrapersonal diversity within individual subjects across time was very low. In a longitudinal study by Reyes et al., it was demonstrated that >95% of the virome was retained over a period of 1 year (Minot et al. 2013). Yet another study suggested that nearly 80% of the virotypes persisted in the stool samples of a subject throughout the study period of 2.5 years (Minot et al. 2011) indicating remarkable long-term genetic stability of the member species. This unusual genomic stability exhibited by the virome despite a hallmark error-prone viral replication system is due to the fact that majority of the phages in the virome exhibit a temperate lifestyle with low mutation rates mainly because the viral genome maintenance involves replication by high fidelity bacterial DNA polymerase. Nevertheless, few members of the viral community (lytic bacteriophages such as members from *Microviridae* family) had very high substitution rates that propelled the evolution of some long-term virome members over time contributing to interpersonal virome diversity (Minot et al. 2013). It is well established that diet is a key modifier influencing the composition of gut bacterial community. Congruently, Minot et al. showed that a controlled diet can significantly alter gut virome composition where individuals on the same diet converged and showed a tendency toward more similar virome, however not identical (Minot et al. 2011).

3.4 Skin Virome

The human skin is the largest and the most exposed organ in the body which facilitates the inception of a complex ecosystem of cutaneous flora containing bacteria, fungi, and viruses. Investigations into the bacterial and fungal flora of the skin microbiota and their role in health and disease were extensively carried out in the last two decades, however, the studies related to their viral counterpart, the "virobiota," is still in its infancy. The human skin virobiota is highly diverse. Much like the composition of oral and the gut virome, bacteriophages from the order Caudovirales (mainly *Microviridae* and *Siphoviridae* family) largely predominate the niche compared to the viruses that are potent human pathogens (Foulongne et al. 2012). Among the other abundant bacteriophages of skin included *Staphylococcus* and *Propionibacterium* phages. Although the relative proportion of phages varied across different anatomical sites and skin microenvironment, >85% of the phages were predicted to exhibit temperate lifestyle (Hannigan et al. 2015). The eukaryotic DNA viruses in the skin include members of the family *Papillomaviridae, Polyomaviridae, Circoviridae*, and *Poxviridae* including sequences related to beta and gamma-papillomaviruses (Table 10.1), human polyomavirus 6, 7, and 9, and Merkel cell polyomavirus (MCPyV) (Foulongne et al. 2012; Hannigan et al. 2015). Majority of the reads pertaining to RNA virome from the skin and the nasal swabs from a cohort of patients with primary immunodeficiency could be mapped to DNA viruses of the family *Papillomaviridae* and *Polyomaviridae*, suggesting actively replicating DNA viruses in the skin (Tirosh et al. 2018). While most studies focused on the compositional analysis of the whole viral metagenome, a study by Hannigan et al. focused on the variability of the skin virome in the evolutionary context. They identified 106 and 465 hypervariable loci in human papillomavirus (HPV) and *Staphylococcus* phages respectively which mapped to genes involved in host tropism, immune evasion, virulent gene expression, and utilization of host resources. However, these hypervariable loci expressed low non-synonymous to synonymous ratio thereby suggesting purifying selection with a propensity to maintain the consensus protein sequences (Hannigan et al. 2017). Several factors affect the diversity of the viral communities in the skin, this includes the moisture content and the occlusion status of different anatomic sites. Significant differences in both alpha (within-sample) and beta (between-samples) diversities in the virome and the whole metagenome structure have been reported which strongly depended on the skin microenvironment. The virome structure, but not the whole metagenome, significantly differed at each anatomical site paired over a period of 1 month for each sample, suggesting greater longitudinal stability of whole microbiome when compared to viral communities (Hannigan et al. 2015). Cutaneous microflora is known to influence host immunity in a myriad of ways (reviewed in (Belkaid and Tamoutounour 2016)). However, the immune status of the host also plays a significant role in shaping the ecological constituents of the skin. A remarkable increase in the relative abundance of eukaryotic viruses in the skin of DOC8-deficient patients with primary immunodeficiency compared to healthy controls has been observed (Tirosh et al. 2018). Unbiased high-throughput sequencing has also revealed the association of specific virus with certain diseases, for instance, approximately 25% of the patients with Merkel cell carcinoma (MCC) were identified with a strain of polyomavirus almost identical to human polyomavirus 9 as compared to 0.9% of the healthy controls without MCC (Sauvage et al. 2011). Yet another study reportedly confirmed relatively high levels of MCPyV DNA in the skin samples of patients with MCC as compared to the age-matched healthy controls and other cutaneous cancer patients (Hashida et al. 2016). Metagenomic analysis of skin virome from a single subject with widespread warts could map 30% of all the reads to HPV genomes, particularly HPV2 (Landini et al. 2015). Thus the skin is enriched with viruses that are integral to the microbiome and these viruses can be either symbiotic or pathogenic.

3.5 Respiratory Tract Virome

The human respiratory tract is a dynamic site of interaction of diverse air-borne viruses that cocirculate in the space, forming an ecological niche, facilitating several interspecific interactions (Fig. 10.2). Interactions within this niche are often maintained in a balanced state, failure of which affects the equilibrium and may turn out to be harmful to the host (Young et al. 2015; Willner et al. 2009; Wylie et al. 2012; Clarke et al. 2018b; Abbas et al. 2017, 2019b). Although there are extensive evidences for bacteria-bacteria and bacteria-virus interactions, knowledge on the occurrence of virus-virus interactions is limited (Bosch et al. 2013; Weinberger et al. 2015; McCullers and Rehg 2002). Many viruses have demonstrated selectivity in their attachment and replication in specific regions of the respiratory tract, which largely depends on the type and availability of specific receptors in those regions. Pathogenic viruses that target the respiratory tract have devastated mankind; the most notable ones include the 1918 pandemic flu caused by H1N1 influenza virus and more recent COVID-19 caused by SARS-CoV-2.

The human respiratory virome consists of all the eukaryotic viruses and bacteriophages that are found in the upper respiratory tract and within the lungs. Sampling is the foremost challenge in defining the respiratory tract virome. For example, sampling from the lower respiratory tract requires invasive procedures. Bronchoalveloar lavages are only possible in symptomatic individuals for which we do not have a normal healthy control (Wylie et al. 2012). Studies in patients with cystic fibrosis even showed that there was distinct virus population in different regions of the lung (Willner et al. 2012). Comprehensive studies so far have shown that like the gut virome, many species of phages including those belonging to the order caudovirales and the *Inoviridae*, and the *Microviridae* families are present in the lungs. The most pliable source of these phages is either the oral cavity or is derived from the upper respiratory tract. The most notable and widely prevalent families of viruses include those of *Anelloviridae*, *Redondoviridae*, *Adenoviridae*, *Papillomaviridae*, *Herpesviridae*, *Picornaviridae*, and *Paramyxoviridae* (Table 10.1) (Willner et al. 2009; Young et al. 2015; Wylie et al. 2012; Clarke et al. 2018b; Abbas et al. 2017, 2019b).

4 Other Sites

4.1 The Central and Peripheral Nervous System Virome

The difficulty involved in studying the resident populations of viruses in the nervous system can at least in part be attributed to the difficulty in obtaining samples from normal and healthy individuals. Available literature on the virome of the nervous system is based on the analysis of cerebrospinal fluid. Viruses at times employ multiple mechanisms to breach the blood-brain barrier and can cause acute infection or can integrate into the genome of neuronal cells and be reactivated later. Apart from the pathogenic opportunistic viruses, regions of the nervous system are home to viral communities which included those belonging to *Herpesviridae* family but mostly predominated by phages. The phages in the CNS were found to be of *Siphoviridae*, *Podoviridae*, and *Myoviridae* families (Ghose et al. 2019).

4.2 Blood Virome

The human blood maybe least expected to harbor viral communities because of the presence of innumerable immune components including an array of immune cells, antibodies, and the complement system. It is known that the human blood is exploited by many pathogenic viruses and viremia, although can serve many times as an indicator of the presence of viruses in the blood stream, is a tool used by these viruses to disseminate to distal sites. Interestingly, certain classes of viruses have been shown to co-habit the blood stream and the well-documented viruses belong to *Phycodnaviridae, Picornaviridae, Mimiviridae, Marseilleviridae,* and *Herpesviridae* families (Table 10.1) (Moustafa et al. 2017; Liu et al. 2018). The most compelling evidence was obtained in the case of viruses of *Anelloviridae* family, which was confirmed by electron microscopy besides the use of other genomic tools (Breitbart and Rohwer 2005). What is the precise role of these viruses in the blood and the nature of their very existence and source are intriguing questions that require systematic investigation.

4.3 Urogenital Tract Virome

The urogenital system of both the male and the female are also found to host a myriad of viruses. The richness of virome in this system is highlighted by the observation that urine samples of healthy individuals contain approximately 1×10^7 virus-like particles per milliliter with phages and papilloma viruses contributing to the overall bulk of viruses (Santiago-Rodriguez et al. 2015; Garretto et al. 2019). Both the male and the female urogenital tracts are home to resident viruses. Analysis of the seminal fluid of healthy men showed that the predominant viral families include *Anelloviridae*, *Herpesviridae*, and *Papillomaviridae* while vaginal swabs from healthy women showed the presence of an abundance of double-stranded DNA phages (Table 10.1) (Jakobsen et al. 2020; Li et al. 2020).

5 Dynamics of Virome–Host Interactions

Viruses are largely believed to be obligate parasites. With technological advancements in the last two decades, we have come to a point where we now know that they are not just parasites that are detrimental to the host, but many of them play dynamic roles in maintaining tissue homeostasis. A “healthy virome” is heterogenous in nature and consists of three components (a) the viruses that systemically enter the human body mainly through food but however do not replicate, (b) viruses that infect prokaryotes and probably the unicellular eukaryotes that comprise the healthy human microbiome, and (c) the viruses that can essentially replicate and persist in humans. Several of them are essential parts of the ecosystem and coexist either temporarily or forever as symbionts. The mode of interaction is governed by multiple factors such as environment, diet, lifestyle, host and virus population structure, and the general health and immunity of the individuals. Due to the influence of the aforesaid factors and due to co-evolution of viruses with the host, the nature of the virus-host relationship does not follow a single pattern, but ranges from aggressive antagonism to mutualism. Therefore, the borderline that dictates the time and the conditions due to which the coexisting viruses may turn to be harmful cannot be well defined. Reports also suggest that the symbiotic association between persistent viruses and the host especially the retroviruses has contributed significantly to the constructive evolution of the host.

Growing evidence suggests that the human virome can act as an immune-modulator and thereby facilitating either the protection from or initiation of diseases. Viruses present inside the human body can interact with both the microbiome and host cells, ultimately resulting in the immune variation inside the host (Shi and Gewirtz 2018). For example, during vertical transmission of mouse mammary tumor virus through maternal milk, the presence of MMTV-bound LPS complex activated TLR-4/MyD88 pathway resulting in the production of the immunosuppressive cytokine IL-10 in the pups, which facilitated the establishment of infection (Kane et al. 2011). Enteric bacteria promote murine norovirus infection of B cells by reducing the efficacy of IFN-λ mediated viral clearance, and depletion of intestinal microbiota by antibiotic treatment reduced mouse norovirus replication in vivo (Jones et al. 2014; Baldridge et al. 2015). Analogous to this, ablation of microbiota negatively affected the initial infectivity of rotavirus and enhanced specific humoral immunity (Uchiyama et al. 2014). Human endogenous retroviruses (HERVs) have been found to be integrated into the human genome and play critical roles in modulating host immunity even in the absence of functional viral proteins (Grandi and Tramontano 2018). The association between human and HERVs represents a typical symbiosis, which can have both beneficial and detrimental effects on the host. HERV-K is a bonafide member of the healthy virome and is found to suppress the spread of invasive melanoma (Singh et al. 2020). Parvovirus B19 (B19) infection in adults is asymptomatic in nature but with a prevalence rate of 25% in human skin biopsies (Bonvicini et al. 2010). However, they are capable of arresting hematopoiesis resulting in anemia and less frequently neutropenia, which may even be life-threatening in immune-compromised patients (Shehi et al. 2020).

Potential beneficial role of viruses in a holobiont is well noted in instances where viruses can even impede further infection or pathogenesis. Classic examples are where the Hepatitis G virus can slow down the progression of HIV infection to acquired immune deficiency syndrome (AIDS) (Tillmann et al. 2001) and that of latent herpes virus having a protective role against *Listeria monocytogenes* and *Yersinia pestis* infections (Barton et al. 2007). On the contrary, cytomegalovirus (CMV) exhibits the potential to promote *Pneumocystis jiroveci* infections (an opportunistic pathogen causing severe pulmonary infections) in immunocompromised patients (Lee et al. 2020). Several trans-kingdom interactions have also been reported. For example., influenza virus by virtue of its neuraminidase activity exposes several bacterial receptors on the cell surface which in turn augments the super infection with *Streptococcus pneumoniae* or *Staphylococcus aureus* (Bosch et al. 2013).

Human anelloviruses (AV) represent a group of highly diverse and omnipresent commensal viruses. Their presence in blood, tears, saliva, semen, breast milk, nasal secretion, bile, etc. suggest that these viruses exhibit a broad range of tropism (Kaczorowska and van der Hoek 2020). Children are mostly infected with AVs months after birth, but these viruses have also been detected in children and adults of all ages (Vasilyev et al. 2009; Brassard et al. 2015). Pegiviruses cluster tightly with hepaciviruses which include the hepatitic C virus (HCV), which is a major human pathogen. The incidence rate of pegivirus infection in humans is 5%. Although suspected to be a causative agent of diarrhea, the prototypic pegivirus, formerly known as the hepatitis G virus (HGV), apparently is not linked to any pathology like HCV (Hartlage et al. 2016; Stapleton et al. 2011). They can readily infect and grow in human cell lines and are implicated in fighting the burden of AIDS (Greenhalgh et al. 2019). Another example is that of Picobirnaviruses. Isolation and culturing of these viruses in the lab have proved difficult, and hence the true identity of the host still remains unknown. They have been isolated from the stool samples of individuals with diarrhea of unknown cause (Ganesh et al. 2012; Ganesh et al. 2014) and are suspected to infect bacteria populating the mammalian enteric tract (Krishnamurthy and Wang 2018). Also, infection with human cytomegalovirus can suppress superinfection with HIV-1 (King et al. 2006); and infection with hepatitis A can also suppress hepatitis C virus infection (Deterding et al. 2006). Similarly, hepatitis C can suppress hepatitis B virus replication (Murai et al. 2020). Human papillomaviruses (HPVs) often cause warts which are cleared by the immune system. But a small fraction of the individuals infected with HPVs (HPV-16 and HPV-18) develop cervical cancer. High-risk HPVs, despite being carcinogenic, are also considered to be a symbiont, as the immunity produced against these commensal papillomaviruses offers protection against skin cancers (Strickley et al. 2019).

The human intestine offers a conducive environment for both bacteria and viruses to thrive and this site of residence provides room for cross-talk between bacteria and viruses. The intestine nurtures both symbionts and commensals; however, their biodiversity diversifies during the various stages of development and aging. It also varies along the length of the gut and is influenced by the diet and several other environmental factors. Further, any immunodeficiencies and other host genetic factors affecting the immune regulation in the intestine can tip the balance towards pathological conditions in the intestine (Cadwell et al. 2010; Handley et al. 2016). The bulk of the virome in the intestine consists of bacteriophages followed by DNA viruses such as anelloviruses and herpesvirus and other endogenous retroviruses. However, the benefits and detrimental effects of the resident enteric viruses in healthy individuals and in the diseased need further investigation.

A plethora of diseases has been associated with the dysbiosis in the human gut microbiome (Carding et al. 2015); however, studies that specifically link the gut virome with human diseases have only emerged recently. Inflammatory bowel disease (IBD) is a group of conditions that are characterized by chronic inflammation of the gut. As inflammation is directly related to a heightened immune response of the host, its effect should reflect on the composition of the gut virome in chronic cases of IBD. As expected, alteration in the gut viromes was observed in multiple cohorts of IBD. Patients with IBD had more rich and diverse taxa of bacteriophages, particularly of the order *Caudovirales* as compared to the controls which were usually accompanied by a significant reduction in the overall bacterial diversity in the IBD fecal microbiome (Norman et al. 2015). Whether the reduction in the diversity of bacterial species in the IBD patients is a causal effect of the increase or abundance of the bacteriophages is unknown. A significant increase in the relative levels of bacteriophages was also observed in patients with type 2 diabetes. Members of the family *Podoviridae, Myoviridae, Siphoviridae,* and yet unclassified families from the order *Caudovirales* were significantly abundant in type 2 diabetes when compared to the control group (Ma et al. 2018). Interestingly, a reverse trend was observed in the intestinal virome of type 1 diabetes patients, wherein the overall virome and bacteriophage diversity was less in the patients when compared to the matched control groups. The change in the virome was also associated with the development of autoimmunity which is typically observed in the case with type 1 diabetes (Zhao et al. 2017). The stratification of the gut virome has also been used as a biomarker in the case of hypertension. Han et al. identified a group of 11 and 8 viruses that can act as biomarkers to discriminate the cases with hypertension from the normal control groups respectively. Stratification of hypertension and prehypertension group from the control groups based on the gut virome composition was validated to be superior and more accurate than the gut bacteriome (Han et al. 2018). Dysbiosis of the enteric virome has been reported in colorectal cancer (CRC), wherein the metagenomic samples from CRC patients were associated with significant increase in the diversity of gut bacteriophages primarily exhibiting temperate lifestyle. Overall 22 viral taxa were identified which could be used as biomarkers to discriminate CRC cases from the controls. In this study, it was also suggested that a subgroup of 4 taxonomic markers was found to be strongly associated with poor prognosis and survival outcomes in colorectal cancer (Hannigan et al. 2018; Nakatsu et al. 2018). Although none of the studies linked or confirmed the cause of the condition/disease to the alteration of the gut virome, these incidental studies can be used to predict the status or progression of the disease for better disease management in the future.

6 Translational Prospects of Virome Research

Metagenomic analysis and next-generation sequencing studies have increased the knowledge about human-associated viruses and many studies on the characterization of these viruses revealed their association with several disease phenotypes. For example, a metagenomic analysis in a patient with a respiratory tract infection revealed the presence of gamma papillomavirus; however, it was demonstrated to be unrelated to the disease (Canuti et al. 2014). Large data may or may not provide real-time evidence of the virus populations in a region, and the beneficial or pathogenic nature of viruses thus identified especially in terms of disease may be correlative.

Identification and characterization of novel viruses in niches of microbiota will help us to determine their role in the host and this information can further be used to develop new strategies for disease prevention and therapies. As mentioned earlier the virome as a whole or viral diversity between individuals can be unique. Thus personalized treatment strategies that takes into consideration the virome of an individual can help target the disease effectively and thereby aid in improving the health status of an individual.

It is well documented that the bacteriome of an individual has a direct effect on the abundance of the phage community, and vice versa. Introduction of bacteria derived from the human gut into germ-free mice followed by inoculation of virus-like particles enriched from human feces showed a concomitant decrease of the bacterial population in the host and the levels of bacteria could stabilize only when the phage abundance decreased (Reyes et al. 2013). Phage therapy is widely used to eliminate multi-drug-resistant bacterial population as an alternative to antibiotics (Morozova et al. 2018). One more exciting aspect of phage therapy is that genetically modified phages can control nutrient biosynthesis and degradation especially in obese and dysmetabolic patients (Scarpellini et al. 2015).

The gut microbiome in its entirety is thought to function as an "organ" which coordinates with other bodily functions and has a potentially beneficial effect on human health. Dysbiosis of the gut microbiome composition and function are linked to many pathophysiological conditions and restoration of the same through fecal microbiota transplantation (FMT) has surfaced as a highly effective alternative treatment for patients with recurrent *Clostridioides difficile* infections (rCDI) and pediatric ulcerative colitis (Fujimoto et al. 2021; Broecker et al. 2016; Nusbaum et al. 2018). The success of the FMT is positively correlated with restructuring of the recipient's gut bacterial community to levels that are either identical or resemble closely that of the healthy donor (Nusbaum et al. 2018; Fujimoto et al. 2021). Emerging evidence suggests the possible role of the inherent viral community in the outcome of the FMT treatment. An alternative approach to FMT is the treatment with sterile fecal filtrate (residual after removing the bacterial counterparts by a series of centrifugation and filtration steps), which was sufficient to restore normal bowel movements and eliminated all symptoms in patients with rCDI. Interestingly, a subject in the study who was initially treated with FMT with limited to no success, upon treatment with sterile fecal filtrate obtained from the same donor showed better recovery and resolution of the symptoms (Ott et al. 2017). Since serious adverse events including deaths have been reported in FMT (reviewed in Wang et al. 2016a, b), fecal filtrate transfer and thereby the viral assemblage could be a viable alternative to FMT for the treatment of patients with rCDI, particularly in the case of immune-compromised individuals.

The effects that viruses have on human health are highly dependent on their anatomical location inside the host and the interaction with other cells and microbes. All these factors have a direct influence on whether the virus has an advantageous, deleterious, or neutral impact on the host. Studies have shown that human pegivirus (HPgV) infection resulted in the survival of individuals infected with the human immunodeficiency virus (HIV). Greenhalgh et al. showed that HPgV vaccination resulted in reducing morbidity and mortality associated with HIV/AIDS (Greenhalgh et al. 2019). All these studies highlight that there is a pressing need for understanding the interactions of the virome with the host cells and other microbes. This will open up new avenues in targeting disease and improving health.

7 The Human Virome: The Extremes

Viral pathogens are innumerable and diverse in nature and are known to cause potential infections in humans which involve one or many sites. Local transmission results in virus replication and dissemination to distal sites by hematogenous route and results in viremia. Some viruses tap peripheral nerve endings to traverse across and gain access to the CNS. Integration of viral genes in the host genome may be a property of the virus but is also one of the features adopted by the viruses to evade the host immune system and to maintain a steady pool of viral genes which can later be used during virus reactivation. Enrichment of a particular group of virus in the virome has been implicated in disease progression or severity. Such a phenomenon is well documented in Crohn's disease and ulcerative colitis. Enhanced levels of *Caudovirales*, *Hepadnaviridae,* and *Hepeviridae* were found to be associated with Crohn's disease or ulcerative colitis or both (Norman et al. 2015; Fernandes et al. 2019; Ungaro et al. 2019; Zuo et al. 2019). A shift in the ratio of *Microviridae* to *Caudovirales* has been implicated in the early-onset of inflammatory bowel disease (Liang et al. 2020a). Similarly, a shift in the population dynamics of virulent and temperate phages was observed in Crohn's disease patients (Clooney et al. 2019). Thus from the available data it is clear that unprecedented increase in specific families of viruses can contribute to disease induction or exacerbation. Besides Crohn's disease, an increase in enterovirus has been attributed to coeliac disease autoimmunity, while a prevalence of Picobarnoviruse*s* and Tobamoviruses has been reported in pregnant women with type 1 diabetes (Lindfors et al. 2020; Wook Kim et al. 2019). Similarly, the presence of certain phages (Erwinia phage ϕEaH2, Lactococcus phage 1706) in the gut contributed to the development of hypertension (Monaco et al. 2016; Han et al. 2018).

Although outside the context of this chapter, it should be pointed out that viruses are also known to exploit the human microbiome to productively infect the host and also in many instances to evade the host immune system. Bacteria or bacterial components are exploited by many members of the *Picornaviridae* family including polio virus, coxsackievirus A21, coxsackievirus B5, and echovirus 30 to gain stability in their hostile environment to infect the host (Kuss et al. 2011; Waldman et al. 2017). Enhancement in viral infectivity utilizing the host bacteria has also been reported in the case of rotavirus and reovirus (Uchiyama et al. 2014; Berger et al. 2017). Poliovirus has been shown to exploit the lipopolysaccharide (LPS) and peptidoglycan of the resident microbiome to replicate efficiently in the host (Robinson et al. 2014; Erickson et al. 2018). Such bacterial-virus interaction has been found to enhance poliovirus's receptor interaction predominantly mediated by LPS and promote poliovirus co-infection (Robinson et al. 2014; Erickson et al. 2018; Kuss et al. 2011). It is therefore quite evident that viruses can utilize the microbiome for their advantage but in some instances they are also known to disrupt the bacterial flora. A comparative analysis of microbiome from children with astrovirus, norovirus, rotavirus, and adenovirus infection showed that a substantial decline in Bifidobacterium and the overall microbiome diversity in the astrovirus-infected cases (Ma et al. 2011).

Reports on the beneficial role of the virome to the host are relatively limited in number, but accumulating evidences from both clinical and animal studies are quite compelling. At least two reports highlight the importance of virome diversity in limiting disease. An overall reduction in viral diversity was identified as one of the contributing factors in type 1 diabetes and acute malnutrition (Zhao et al. 2017; Terho et al. 1983). Especially in the gut which is known to host phages, there exists a constant tussle between the phages and resident bacteria. As described earlier, a shift in the phage property from virulent to temperate supported Crohn's disease (Clooney et al. 2019). Such a disparity in the virus-bacteria interaction also resulted in exacerbation of ulcerative colitis and growth stunting in children (Khan Mirzaei et al. 2020; Desai et al. 2020). Thus it is clear that the resident phages are important for restricting pathogenic bacteria and supporting the host. Experimental studies on gnotobiotic mice demonstrated that the effects of the lack in bacteria are compensated by murine norovirus that had chronically infected these mice (Kernbauer et al. 2014). Type III interferon induction in the intestine by murine astrovirus could effectively keep the enteric norovirus at bay (Ingle et al. 2019). Thus it can be confidently asserted that the resident human virome can have a positive effect on the overall health status of the host.

8 Conclusions and Future Perspectives

Microbes predominate the host cells in number and were often considered to be associated with human diseases. Technological advancements in the field of genomics have helped us foray into unchartered territories like the human microbiome in which significant progress has been made. It now emerges that the microbiome is an integral part of the host, albeit in a personalized manner. Studies suggest that the role of the human microbiome is multifaceted, playing a major role in the overall physiology. With the emergence of the virus assemblage as a part of the microbiome family, many questions have come to the forefront. Some of the viruses are known to spill over into humans from other species or are transmitted between humans and are either cleared by the host or can have devastating effects on the host. Yet others are known to integrate into the host genome and remain dormant until ambient conditions arise. So the big question is what constitutes the "virus assemblage"? It is now better understood that many viruses reside in the host at specific sites and their relative abundance is dependent on multiple factors. This is quite evident in the case of the neonates where the mother plays an important role in the initial virome enrichment. The biggest conundrum in the field of virome is the source of the viruses in the assemblage as to whether they are resident or are contaminants. The inability to recover viruses, to culture and characterize them, the non-existence of reliable molecular markers and sample contamination, all compound to the problem, adding to the concept of "viral dark matter". However, of much promise is the evidence, that like the bacterial entities of the microbiome, the virome also plays a pivotal role in contributing to normal host functions and in some cases also adversely affecting the host, resulting in disease conditions. All this points to the fact that the field is in its infancy and serious investigations including setting up of standardization parameters are required to repeal the controversies associated with this field. The cross-talk between organ systems which is highlighted as axes, including the gut-liver axis, gut-brain axis, etc., are noted check points in maintaining normal physiology and homeostasis which is well appreciated. The role of the microbiome in limiting pathogenic bacteria and thereby alleviating major disorders like liver cirrhosis is well appreciated. To what extent the liver virome is involved in these axes remains under-studied and requires in-depth investigation.

The relevance of the resident microbiome in the overall health of an individual is gaining wider recognition. Supplementation of pro-biotics during antibiotic administration and managing of certain metabolic and immune disorders using FMT are emerging as established clinical practices. Clinically, the contribution of virome to the well-being of an individual remains under-appreciated partly due to the limited studies in this area. Alongside the diet and other factors, the existence of cross-talk between the virome and bacteriome is gaining more and more traction. Evidences point to a role of certain viral populations in aggravating disease while others abrogate the symptoms as in the case of Crohn's disease. Viral pathogens like HIV and H1N1, post-infection generate conditions conducive for secondary infections. One school of thought is the potent immune response initiated by such pathogens can mediate depletion of beneficial bacteria and viruses which indirectly facilitates the survival and spread of the pathogens. More recently, gut bacteriome dysbiosis due to SARS-CoV-2 infection has been attributed as a contributing factor for disease severity and poor recovery. A similar scenario is possible with anti-viral administration, which can deplete the healthy virome besides targeting the pathogen; however, detailed and systematic studies are required to gain further knowledge in this area. Thus the field of virome is riddled with exciting mysteries that require careful unraveling which may have significant implications in understanding the virus niches in the human body and their interaction with the host.

References

Abbas AA, Diamond JM, Chehoud C, Chang B, Kotzin JJ, Young JC et al (2017) The perioperative lung transplant Virome: torque Teno viruses are elevated in donor lungs and show divergent dynamics in primary graft dysfunction. Am J Transplant 17(5):1313–1324. https://doi.org/10.1111/ajt.14076

Abbas AA, Taylor LJ, Dothard MI, Leiby JS, Fitzgerald AS, Khatib LA et al (2019a) Redondoviridae, a family of small, circular DNA viruses of the human Oro-respiratory tract associated with periodontitis and critical illness. Cell Host Microbe 25(5):719–729. e714. https://doi.org/10.1016/j.chom.2019.04.001

Abbas AA, Young JC, Clarke EL, Diamond JM, Imai I, Haas AR et al (2019b) Bidirectional transfer of Anelloviridae lineages between graft and host during lung transplantation. Am J Transplant 19(4):1086–1097. https://doi.org/10.1111/ajt.15116

Abeles SR, Robles-Sikisaka R, Ly M, Lum AG, Salzman J, Boehm TK et al (2014) Human oral viruses are personal, persistent and gender-consistent. ISME J 8(9):1753–1767. https://doi.org/10.1038/ismej.2014.31

Albrecht M, Arck PC (2020) Vertically transferred immunity in neonates: mothers, mechanisms and mediators. Front Immunol 11:555. https://doi.org/10.3389/fimmu.2020.00555

Arora N, Sadovsky Y, Dermody TS, Coyne CB (2017) Microbial vertical transmission during human pregnancy. Cell Host Microbe 21(5):561–567. https://doi.org/10.1016/j.chom.2017.04.007

Baker JL, Bor B, Agnello M, Shi W, He X (2017) Ecology of the Oral microbiome: beyond bacteria. Trends Microbiol 25(5):362–374. https://doi.org/10.1016/j.tim.2016.12.012

Baldridge MT, Nice TJ, McCune BT, Yokoyama CC, Kambal A, Wheadon M et al (2015) Commensal microbes and interferon-λ determine persistence of enteric murine norovirus infection. Science 347(6219):266–269. https://doi.org/10.1126/science.1258025

Barton ES, White DW, Cathelyn JS, Brett-McClellan KA, Engle M, Diamond MS et al (2007) Herpesvirus latency confers symbiotic protection from bacterial infection. Nature 447(7142):326–329. https://doi.org/10.1038/nature05762

Belkaid Y, Tamoutounour S (2016) The influence of skin microorganisms on cutaneous immunity. Nat Rev Immunol 16(6):353–366. https://doi.org/10.1038/nri.2016.48

Berger AK, Yi H, Kearns DB, Mainou BA (2017) Bacteria and bacterial envelope components enhance mammalian reovirus thermostability. PLoS Pathog 13(12):e1006768. https://doi.org/10.1371/journal.ppat.1006768

Berlutti F, Pantanella F, Natalizi T, Frioni A, Paesano R, Polimeni A et al (2011) Antiviral properties of lactoferrin--a natural immunity molecule. Molecules 16(8):6992–7018. https://doi.org/10.3390/molecules16086992

Bonvicini F, La Placa M, Manaresi E, Gallinella G, Gentilomi GA, Zerbini M et al (2010) Parvovirus b19 DNA is commonly harboured in human skin. Dermatology 220(2):138–142. https://doi.org/10.1159/000277431

Bosch AA, Biesbroek G, Trzcinski K, Sanders EA, Bogaert D (2013) Viral and bacterial interactions in the upper respiratory tract. PLoS Pathog 9(1):e1003057. https://doi.org/10.1371/journal.ppat.1003057

Brassard J, Gagne MJ, Leblanc D, Poitras E, Houde A, Boras VF et al (2015) Association of age and gender with torque Teno virus detection in stools from diarrheic and non-diarrheic people. J Clin Virol 72:55–59. https://doi.org/10.1016/j.jcv.2015.08.020

Breitbart M, Haynes M, Kelley S, Angly F, Edwards RA, Felts B et al (2008) Viral diversity and dynamics in an infant gut. Res Microbiol 159(5):367–373. https://doi.org/10.1016/j.resmic.2008.04.006

Breitbart M, Rohwer F (2005) Method for discovering novel DNA viruses in blood using viral particle selection and shotgun sequencing. BioTechniques 39(5):729–736. https://doi.org/10.2144/000112019

Broecker F, Klumpp J, Schuppler M, Russo G, Biedermann L, Hombach M et al (2016) Long-term changes of bacterial and viral compositions in the intestine of a recovered Clostridium difficile patient after fecal microbiota transplantation. Cold Spring Harb Mol Case Stud 2(1):a000448. https://doi.org/10.1101/mcs.a000448

Cadwell K, Patel KK, Maloney NS, Liu TC, Ng AC, Storer CE et al (2010) Virus-plus-susceptibility gene interaction determines Crohn's disease gene Atg16L1 phenotypes in intestine. Cell 141(7):1135–1145. https://doi.org/10.1016/j.cell.2010.05.009

Canuti M, Deijs M, Jazaeri Farsani SM, Holwerda M, Jebbink MF, de Vries M et al (2014) Metagenomic analysis of a sample from a patient with respiratory tract infection reveals the presence of a gamma-papillomavirus. Front Microbiol 5:347. https://doi.org/10.3389/fmicb.2014.00347

Carding S, Verbeke K, Vipond DT, Corfe BM, Owen LJ (2015) Dysbiosis of the gut microbiota in disease. Microb Ecol Health Dis 26:26191. https://doi.org/10.3402/mehd.v26.26191

Clarke EL, Connell AJ, Six E, Kadry NA, Abbas AA, Hwang Y et al (2018a) T cell dynamics and response of the microbiota after gene therapy to treat X-linked severe combined immunodeficiency. Genome Med 10(1):70. https://doi.org/10.1186/s13073-018-0580-z

Clarke EL, Lauder AP, Hofstaedter CE, Hwang Y, Fitzgerald AS, Imai I et al (2018b) Microbial lineages in sarcoidosis. A metagenomic analysis tailored for low-microbial content samples. Am J Respir Crit Care Med 197(2):225–234. https://doi.org/10.1164/rccm.201705-0891OC

Clooney AG, Sutton TDS, Shkoporov AN, Holohan RK, Daly KM, O'Regan O et al (2019) Whole-Virome analysis sheds light on viral dark matter in inflammatory bowel disease. Cell Host Microbe 26(6):764–778 e765. https://doi.org/10.1016/j.chom.2019.10.009

Colson P, Fancello L, Gimenez G, Armougom F, Desnues C, Fournous G et al (2013) Evidence of the megavirome in humans. J Clin Virol 57(3):191–200. https://doi.org/10.1016/j.jcv.2013.03.018

Desai C, Handley SA, Rodgers R, Rodriguez C, Ordiz MI, Manary MJ et al (2020) Growth velocity in children with environmental enteric dysfunction is associated with specific bacterial and viral taxa of the gastrointestinal tract in Malawian children. PLoS Negl Trop Dis 14(6):e0008387. https://doi.org/10.1371/journal.pntd.0008387

Deterding K, Tegtmeyer B, Cornberg M, Hadem J, Potthoff A, Boker KH et al (2006) Hepatitis a virus infection suppresses hepatitis C virus replication and may lead to clearance of HCV. J Hepatol 45(6):770–778. https://doi.org/10.1016/j.jhep.2006.07.023

Doan T, Akileswaran L, Andersen D, Johnson B, Ko N, Shrestha A et al (2016) Paucibacterial microbiome and resident DNA Virome of the healthy conjunctiva. Invest Ophthalmol Vis Sci 57(13):5116–5126. https://doi.org/10.1167/iovs.16-19803

Erickson AK, Jesudhasan PR, Mayer MJ, Narbad A, Winter SE, Pfeiffer JK (2018) Bacteria facilitate enteric virus co-infection of mammalian cells and promote genetic recombination. Cell Host Microbe 23(1):77–88 e75. https://doi.org/10.1016/j.chom.2017.11.007

Fernandes MA, Verstraete SG, Phan TG, Deng X, Stekol E, LaMere B et al (2019) Enteric Virome and bacterial microbiota in children with ulcerative colitis and Crohn disease. J Pediatr Gastroenterol Nutr 68(1):30–36. https://doi.org/10.1097/MPG.0000000000002140

Foulongne V, Sauvage V, Hebert C, Dereure O, Cheval J, Gouilh MA et al (2012) Human skin microbiota: high diversity of DNA viruses identified on the human skin by high throughput sequencing. PLoS One 7(6):e38499. https://doi.org/10.1371/journal.pone.0038499

Fujimoto K, Kimura Y, Allegretti JR, Yamamoto M, Zhang YZ, Katayama K et al (2021) Functional restoration of Bacteriomes and Viromes by fecal microbiota transplantation. Gastroenterology 160(6):2089–2102 e2012. https://doi.org/10.1053/j.gastro.2021.02.013

Ganesh B, Banyai K, Martella V, Jakab F, Masachessi G, Kobayashi N (2012) Picobirnavirus infections: viral persistence and zoonotic potential. Rev Med Virol 22(4):245–256. https://doi.org/10.1002/rmv.1707

Ganesh B, Masachessi G, Mladenova Z (2014) Animal picobirnavirus. Virusdisease 25(2):223–238. https://doi.org/10.1007/s13337-014-0207-y

Garretto A, Miller-Ensminger T, Wolfe AJ, Putonti C (2019) Bacteriophages of the lower urinary tract. Nat Rev Urol 16(7):422–432. https://doi.org/10.1038/s41585-019-0192-4

Ghose C, Ly M, Schwanemann LK, Shin JH, Atab K, Barr JJ et al (2019) The Virome of cerebrospinal fluid: viruses where we once thought there were none. Front Microbiol 10:2061. https://doi.org/10.3389/fmicb.2019.02061

Goodrich JK, Davenport ER, Beaumont M, Jackson MA, Knight R, Ober C et al (2016) Genetic determinants of the gut microbiome in UK twins. Cell Host Microbe 19(5):731–743. https://doi.org/10.1016/j.chom.2016.04.017

Goodrich JK, Waters JL, Poole AC, Sutter JL, Koren O, Blekhman R et al (2014) Human genetics shape the gut microbiome. Cell 159(4):789–799. https://doi.org/10.1016/j.cell.2014.09.053

Grandi N, Tramontano E (2018) Human endogenous retroviruses are ancient acquired elements still shaping innate immune responses. Front Immunol 9:2039. https://doi.org/10.3389/fimmu.2018.02039

Greenhalgh S, Schmidt R, Day T (2019) Fighting the public health burden of AIDS with the human pegivirus. Am J Epidemiol 188(9):1586–1594. https://doi.org/10.1093/aje/kwz139

Gregory AC, Zablocki O, Zayed AA, Howell A, Bolduc B, Sullivan MB (2020) The gut Virome database reveals age-dependent patterns of Virome diversity in the human gut. Cell Host Microbe 28(5):724–740 e728. https://doi.org/10.1016/j.chom.2020.08.003

Han M, Yang P, Zhong C, Ning K (2018) The human gut Virome in hypertension. Front Microbiol 9:3150. https://doi.org/10.3389/fmicb.2018.03150

Handley SA, Desai C, Zhao G, Droit L, Monaco CL, Schroeder AC et al (2016) SIV infection-mediated changes in gastrointestinal bacterial microbiome and Virome are associated with immunodeficiency and prevented by vaccination. Cell Host Microbe 19(3):323–335. https://doi.org/10.1016/j.chom.2016.02.010

Hannigan GD, Duhaime MB, MTT R, Koumpouras CC, Schloss PD (2018) Diagnostic potential and interactive dynamics of the colorectal cancer Virome. mBio 9(6). https://doi.org/10.1128/mBio.02248-18

Hannigan GD, Meisel JS, Tyldsley AS, Zheng Q, Hodkinson BP, SanMiguel AJ et al (2015) The human skin double-stranded DNA virome: topographical and temporal diversity, genetic enrichment, and dynamic associations with the host microbiome. mBio 6(5):e01578–e01515. https://doi.org/10.1128/mBio.01578-15

Hannigan GD, Zheng Q, Meisel JS, Minot SS, Bushman FD, Grice EA (2017) Evolutionary and functional implications of hypervariable loci within the skin virome. PeerJ 5:e2959. https://doi.org/10.7717/peerj.2959

Hartlage AS, Cullen JM, Kapoor A (2016) The strange, expanding world of animal Hepaciviruses. Annu Rev Virol 3(1):53–75. https://doi.org/10.1146/annurev-virology-100114-055104

Hashida Y, Nakajima K, Nakajima H, Shiga T, Tanaka M, Murakami M et al (2016) High load of Merkel cell polyomavirus DNA detected in the normal skin of Japanese patients with Merkel cell carcinoma. J Clin Virol 82:101–107. https://doi.org/10.1016/j.jcv.2016.07.011

Holtz LR, Cao S, Zhao G, Bauer IK, Denno DM, Klein EJ et al (2014) Geographic variation in the eukaryotic virome of human diarrhea. Virology 468-470:556–564. https://doi.org/10.1016/j.virol.2014.09.012

Ingle H, Lee S, Ai T, Orvedahl A, Rodgers R, Zhao G et al (2019) Viral complementation of immunodeficiency confers protection against enteric pathogens via interferon-lambda. Nat Microbiol 4(7):1120–1128. https://doi.org/10.1038/s41564-019-0416-7

Jakobsen RR, Haahr T, Humaidan P, Jensen JS, Kot WP, Castro-Mejia JL et al (2020) Characterization of the vaginal DNA Virome in health and Dysbiosis. Viruses 12(10). https://doi.org/10.3390/v12101143

Jones MK, Watanabe M, Zhu S, Graves CL, Keyes LR, Grau KR et al (2014) Enteric bacteria promote human and mouse norovirus infection of B cells. Science 346(6210):755–759. https://doi.org/10.1126/science.1257147

Kaczorowska J, van der Hoek L (2020) Human anelloviruses: diverse, omnipresent and commensal members of the virome. FEMS Microbiol Rev 44(3):305–313. https://doi.org/10.1093/femsre/fuaa007

Kane M, Case LK, Kopaskie K, Kozlova A, MacDearmid C, Chervonsky AV et al (2011) Successful transmission of a retrovirus depends on the commensal microbiota. Science 334(6053):245–249. https://doi.org/10.1126/science.1210718

Kernbauer E, Ding Y, Cadwell K (2014) An enteric virus can replace the beneficial function of commensal bacteria. Nature 516(7529):94–98. https://doi.org/10.1038/nature13960

Khan Mirzaei M, Khan MAA, Ghosh P, Taranu ZE, Taguer M, Ru J et al (2020) Bacteriophages isolated from stunted children can regulate gut bacterial communities in an age-specific manner. Cell Host Microbe 27(2):199–212 e195. https://doi.org/10.1016/j.chom.2020.01.004

King CA, Baillie J, Sinclair JH (2006) Human cytomegalovirus modulation of CCR5 expression on myeloid cells affects susceptibility to human immunodeficiency virus type 1 infection. J Gen Virol 87(Pt 8):2171–2180. https://doi.org/10.1099/vir.0.81452-0

Krishnamurthy SR, Wang D (2018) Extensive conservation of prokaryotic ribosomal binding sites in known and novel picobirnaviruses. Virology 516:108–114. https://doi.org/10.1016/j.virol.2018.01.006

Kuss SK, Best GT, Etheredge CA, Pruijssers AJ, Frierson JM, Hooper LV et al (2011) Intestinal microbiota promote enteric virus replication and systemic pathogenesis. Science 334(6053):249–252. https://doi.org/10.1126/science.1211057

Landini MM, Borgogna C, Peretti A, Doorbar J, Griffin H, Mignone F et al (2015) Identification of the skin virome in a boy with widespread human papillomavirus-2-positive warts that completely regressed after administration of tetravalent human papillomavirus vaccine. Br J Dermatol 173(2):597–600. https://doi.org/10.1111/bjd.13707

Lang J, Yang N, Deng J, Liu K, Yang P, Zhang G et al (2011) Inhibition of SARS pseudovirus cell entry by lactoferrin binding to heparan sulfate proteoglycans. PLoS One 6(8):e23710. https://doi.org/10.1371/journal.pone.0023710

Lee AY, Akileswaran L, Tibbetts MD, Garg SJ, Van Gelder RN (2015) Identification of torque Teno virus in culture-negative endophthalmitis by representational deep DNA sequencing. Ophthalmology 122(3):524–530. https://doi.org/10.1016/j.ophtha.2014.09.001

Lee S, Park Y, Kim SG, Ko EJ, Chung BH, Yang CW (2020) The impact of cytomegalovirus infection on clinical severity and outcomes in kidney transplant recipients with pneumocystis jirovecii pneumonia. Microbiol Immunol 64(5):356–365. https://doi.org/10.1111/1348-0421.12778

Leeper C, Lutzkanin A 3rd (2018) Infections during pregnancy. Prim Care 45(3):567–586. https://doi.org/10.1016/j.pop.2018.05.013

Li Y, Altan E, Pilcher C, Hartogensis W, Hecht FM, Deng X et al (2020) Semen virome of men with HIV on or off antiretroviral treatment. AIDS 34(6):827–832. https://doi.org/10.1097/QAD.0000000000002497

Liang G, Conrad MA, Kelsen JR, Kessler LR, Breton J, Albenberg LG et al (2020a) Dynamics of the stool Virome in very early-onset inflammatory bowel disease. J Crohns Colitis 14(11):1600–1610. https://doi.org/10.1093/ecco-jcc/jjaa094

Liang G, Zhao C, Zhang H, Mattei L, Sherrill-Mix S, Bittinger K et al (2020b) The stepwise assembly of the neonatal virome is modulated by breastfeeding. Nature 581(7809):470–474. https://doi.org/10.1038/s41586-020-2192-1

Lim ES, Zhou Y, Zhao G, Bauer IK, Droit L, Ndao IM et al (2015) Early life dynamics of the human gut virome and bacterial microbiome in infants. Nat Med 21(10):1228–1234. https://doi.org/10.1038/nm.3950

Lindfors K, Lin J, Lee HS, Hyoty H, Nykter M, Kurppa K et al (2020) Metagenomics of the faecal virome indicate a cumulative effect of enterovirus and gluten amount on the risk of coeliac disease autoimmunity in genetically at risk children: the TEDDY study. Gut 69(8):1416–1422. https://doi.org/10.1136/gutjnl-2019-319809

Liu S, Huang S, Chen F, Zhao L, Yuan Y, Francis SS et al (2018) Genomic analyses from non-invasive prenatal testing reveal genetic associations, patterns of viral infections, and Chinese population history. Cell 175(2):347–359 e314. https://doi.org/10.1016/j.cell.2018.08.016

Lu LJ, Liu J (2016) Human microbiota and ophthalmic disease. Yale J Biol Med 89(3):325–330

Ma C, Wu X, Nawaz M, Li J, Yu P, Moore JE et al (2011) Molecular characterization of fecal microbiota in patients with viral diarrhea. Curr Microbiol 63(3):259–266. https://doi.org/10.1007/s00284-011-9972-7

Ma Y, You X, Mai G, Tokuyasu T, Liu C (2018) A human gut phage catalog correlates the gut phageome with type 2 diabetes. Microbiome 6(1):24. https://doi.org/10.1186/s40168-018-0410-y

McCann A, Ryan FJ, Stockdale SR, Dalmasso M, Blake T, Ryan CA et al (2018) Viromes of one year old infants reveal the impact of birth mode on microbiome diversity. PeerJ 6:e4694. https://doi.org/10.7717/peerj.4694

McCullers JA, Rehg JE (2002) Lethal synergism between influenza virus and Streptococcus pneumoniae: characterization of a mouse model and the role of platelet-activating factor receptor. J Infect Dis 186(3):341–350. https://doi.org/10.1086/341462

Minot S, Bryson A, Chehoud C, Wu GD, Lewis JD, Bushman FD (2013) Rapid evolution of the human gut virome. Proc Natl Acad Sci U S A 110(30):12450–12455. https://doi.org/10.1073/pnas.1300833110

Minot S, Sinha R, Chen J, Li H, Keilbaugh SA, Wu GD et al (2011) The human gut virome: inter-individual variation and dynamic response to diet. Genome Res 21(10):1616–1625. https://doi.org/10.1101/gr.122705.111

Mohandas S, Pannaraj PS (2020) Beyond the bacterial microbiome: virome of human Milk and effects on the developing infant. Nestle Nutr Inst Workshop Ser 94:86–93. https://doi.org/10.1159/000504997

Monaco CL, Gootenberg DB, Zhao G, Handley SA, Ghebremichael MS, Lim ES et al (2016) Altered Virome and bacterial microbiome in human immunodeficiency virus-associated acquired immunodeficiency syndrome. Cell Host Microbe 19(3):311–322. https://doi.org/10.1016/j.chom.2016.02.011

Morozova VV, Vlassov VV, Tikunova NV (2018) Applications of bacteriophages in the treatment of localized infections in humans. Front Microbiol 9:1696. https://doi.org/10.3389/fmicb.2018.01696

Moustafa A, Xie C, Kirkness E, Biggs W, Wong E, Turpaz Y et al (2017) The blood DNA virome in 8,000 humans. PLoS Pathog 13(3):e1006292. https://doi.org/10.1371/journal.ppat.1006292

Murai K, Hikita H, Kai Y, Kondo Y, Fukuoka M, Fukutomi K et al (2020) Hepatitis C virus infection suppresses hepatitis B virus replication via the RIG-I-like helicase pathway. Sci Rep 10(1):941. https://doi.org/10.1038/s41598-020-57603-9

Naidu M, Robles-Sikisaka R, Abeles SR, Boehm TK, Pride DT (2014) Characterization of bacteriophage communities and CRISPR profiles from dental plaque. BMC Microbiol 14:175. https://doi.org/10.1186/1471-2180-14-175

Nakatsu G, Zhou H, Wu WKK, Wong SH, Coker OO, Dai Z et al (2018) Alterations in enteric Virome are associated with colorectal cancer and survival outcomes. Gastroenterology 155(2):529–541 e525. https://doi.org/10.1053/j.gastro.2018.04.018

Nice TJ, Baldridge MT, McCune BT, Norman JM, Lazear HM, Artyomov M et al (2015) Interferon-lambda cures persistent murine norovirus infection in the absence of adaptive immunity. Science 347(6219):269–273. https://doi.org/10.1126/science.1258100

Norman JM, Handley SA, Baldridge MT, Droit L, Liu CY, Keller BC et al (2015) Disease-specific alterations in the enteric virome in inflammatory bowel disease. Cell 160(3):447–460. https://doi.org/10.1016/j.cell.2015.01.002

Nusbaum DJ, Sun F, Ren J, Zhu Z, Ramsy N, Pervolarakis N et al (2018) Gut microbial and metabolomic profiles after fecal microbiota transplantation in pediatric ulcerative colitis patients. FEMS Microbiol Ecol 94(9). https://doi.org/10.1093/femsec/fiy133

Ogilvie LA, Jones BV (2015) The human gut virome: a multifaceted majority. Front Microbiol 6:918. https://doi.org/10.3389/fmicb.2015.00918

Ott SJ, Waetzig GH, Rehman A, Moltzau-Anderson J, Bharti R, Grasis JA et al (2017) Efficacy of sterile fecal filtrate transfer for treating patients with Clostridium difficile infection. Gastroenterology 152(4):799–811 e797. https://doi.org/10.1053/j.gastro.2016.11.010

Parker MT (2016) An ecological framework of the human Virome provides classification of current knowledge and identifies areas of forthcoming discovery. Yale J Biol Med 89(3):339–351

Perez-Brocal V, Moya A (2018) The analysis of the oral DNA virome reveals which viruses are widespread and rare among healthy young adults in Valencia (Spain). PLoS One 13(2):e0191867. https://doi.org/10.1371/journal.pone.0191867

Pou C, Nkulikiyimfura D, Henckel E, Olin A, Lakshmikanth T, Mikes J et al (2019) The repertoire of maternal anti-viral antibodies in human newborns. Nat Med 25(4):591–596. https://doi.org/10.1038/s41591-019-0392-8

Pride DT, Salzman J, Haynes M, Rohwer F, Davis-Long C, White RA 3rd et al (2012) Evidence of a robust resident bacteriophage population revealed through analysis of the human salivary virome. ISME J 6(5):915–926. https://doi.org/10.1038/ismej.2011.169

Rampelli S, Turroni S, Schnorr SL, Soverini M, Quercia S, Barone M et al (2017) Characterization of the human DNA gut virome across populations with different subsistence strategies and geographical origin. Environ Microbiol 19(11):4728–4735. https://doi.org/10.1111/1462-2920.13938

Rascovan N, Monteil Bouchard S, Grob JJ, Collet-Villette AM, Gaudy-Marqueste C, Penicaud M et al (2016) Human Polyomavirus-6 infecting lymph nodes of a patient with an Angiolymphoid hyperplasia with eosinophilia or Kimura disease. Clin Infect Dis 62(11):1419–1421. https://doi.org/10.1093/cid/ciw135

Reyes A, Haynes M, Hanson N, Angly FE, Heath AC, Rohwer F et al (2010) Viruses in the faecal microbiota of monozygotic twins and their mothers. Nature 466(7304):334–338. https://doi.org/10.1038/nature09199

Reyes A, Wu M, McNulty NP, Rohwer FL, Gordon JI (2013) Gnotobiotic mouse model of phage-bacterial host dynamics in the human gut. Proc Natl Acad Sci U S A 110(50):20236–20241. https://doi.org/10.1073/pnas.1319470110

Robinson CM, Jesudhasan PR, Pfeiffer JK (2014) Bacterial lipopolysaccharide binding enhances virion stability and promotes environmental fitness of an enteric virus. Cell Host Microbe 15(1):36–46. https://doi.org/10.1016/j.chom.2013.12.004

Robles-Sikisaka R, Ly M, Boehm T, Naidu M, Salzman J, Pride DT (2013) Association between living environment and human oral viral ecology. ISME J 7(9):1710–1724. https://doi.org/10.1038/ismej.2013.63

Rothschild D, Weissbrod O, Barkan E, Kurilshikov A, Korem T, Zeevi D et al (2018) Environment dominates over host genetics in shaping human gut microbiota. Nature 555(7695):210–215. https://doi.org/10.1038/nature25973

Roux S, Hallam SJ, Woyke T, Sullivan MB (2015) Viral dark matter and virus-host interactions resolved from publicly available microbial genomes. elife 4. https://doi.org/10.7554/eLife.08490

Santiago-Rodriguez TM, Ly M, Bonilla N, Pride DT (2015) The human urine virome in association with urinary tract infections. Front Microbiol 6:14. https://doi.org/10.3389/fmicb.2015.00014

Sauvage V, Foulongne V, Cheval J, Ar Gouilh M, Pariente K, Dereure O et al (2011) Human polyomavirus related to African green monkey lymphotropic polyomavirus. Emerg Infect Dis 17(8):1364–1370. https://doi.org/10.3201/eid1708.110278

Scarpellini E, Ianiro G, Attili F, Bassanelli C, De Santis A, Gasbarrini A (2015) The human gut microbiota and virome: potential therapeutic implications. Dig Liver Dis 47(12):1007–1012. https://doi.org/10.1016/j.dld.2015.07.008

Shehi E, Ghazanfar H, Fortuzi K, Gonzalez E, Zeana C (2020) A rare Case of parvovirus B19 infection manifesting as chronic aplastic anemia and neutropenia in a human immunodeficiency virus-infected patient. Cureus 12(12):e12174. https://doi.org/10.7759/cureus.12174

Shi Z, Gewirtz AT (2018) Together forever: bacterial-viral interactions in infection and immunity. Viruses 10(3). https://doi.org/10.3390/v10030122

Shkoporov AN, Clooney AG, Sutton TDS, Ryan FJ, Daly KM, Nolan JA et al (2019) The human gut Virome is highly diverse, stable, and individual specific. Cell Host Microbe 26(4):527–541 e525. https://doi.org/10.1016/j.chom.2019.09.009

Shkoporov AN, Khokhlova EV, Fitzgerald CB, Stockdale SR, Draper LA, Ross RP et al (2018) PhiCrAss001 represents the most abundant bacteriophage family in the human gut and infects Bacteroides intestinalis. Nat Commun 9(1):4781. https://doi.org/10.1038/s41467-018-07225-7

Simister NE (2003) Placental transport of immunoglobulin G. Vaccine 21(24):3365–3369. https://doi.org/10.1016/s0264-410x(03)00334-7

Singh M, Cai H, Bunse M, Feschotte C, Izsvak Z (2020) Human endogenous retrovirus K rec forms a regulatory loop with MITF that opposes the progression of melanoma to an invasive stage. Viruses 12(11). https://doi.org/10.3390/v12111303

Stapleton JT, Foung S, Muerhoff AS, Bukh J, Simmonds P (2011) The GB viruses: a review and proposed classification of GBV-A, GBV-C (HGV), and GBV-D in genus Pegivirus within the family Flaviviridae. J Gen Virol 92(Pt 2):233–246. https://doi.org/10.1099/vir.0.027490-0

Strickley JD, Messerschmidt JL, Awad ME, Li T, Hasegawa T, Ha DT et al (2019) Immunity to commensal papillomaviruses protects against skin cancer. Nature 575(7783):519–522. https://doi.org/10.1038/s41586-019-1719-9

Terho EO, Lindstrom P, Mantyjarvi R, Tukiainen H, Wager O (1983) Circulating immune complexes and rheumatoid factors in patients with farmer's lung. Allergy 38(5):347–352. https://doi.org/10.1111/j.1398-9995.1983.tb04129.x

Tillmann HL, Heiken H, Knapik-Botor A, Heringlake S, Ockenga J, Wilber JC et al (2001) Infection with GB virus C and reduced mortality among HIV-infected patients. N Engl J Med 345(10):715–724. https://doi.org/10.1056/NEJMoa010398

Tirosh O, Conlan S, Deming C, Lee-Lin SQ, Huang X, Program NCS et al (2018) Expanded skin virome in DOCK8-deficient patients. Nat Med 24(12):1815–1821. https://doi.org/10.1038/s41591-018-0211-7

Turin CG, Ochoa TJ (2014) The role of maternal breast Milk in preventing infantile diarrhea in the developing world. Curr Trop Med Rep 1(2):97–105. https://doi.org/10.1007/s40475-014-0015-x

Uchiyama R, Chassaing B, Zhang B, Gewirtz AT (2014) Antibiotic treatment suppresses rotavirus infection and enhances specific humoral immunity. J Infect Dis 210(2):171–182. https://doi.org/10.1093/infdis/jiu037

Ungaro F, Massimino L, Furfaro F, Rimoldi V, Peyrin-Biroulet L, D'Alessio S et al (2019) Metagenomic analysis of intestinal mucosa revealed a specific eukaryotic gut virome signature in early-diagnosed inflammatory bowel disease. Gut Microbes 10(2):149–158. https://doi.org/10.1080/19490976.2018.1511664

Vasilyev EV, Trofimov DY, Tonevitsky AG, Ilinsky VV, Korostin DO, Rebrikov DV (2009) Torque Teno virus (TTV) distribution in healthy Russian population. Virol J 6:134. https://doi.org/10.1186/1743-422X-6-134

Virgin HW (2014) The virome in mammalian physiology and disease. Cell 157(1):142–150. https://doi.org/10.1016/j.cell.2014.02.032

Waldman P, Meseguer A, Lucas F, Moulin L, Wurtzer S (2017) Interaction of human enteric viruses with microbial compounds: implication for virus persistence and disinfection treatments. Environ Sci Technol 51(23):13633–13640. https://doi.org/10.1021/acs.est.7b03875

Wang J, Gao Y, Zhao F (2016a) Phage-bacteria interaction network in human oral microbiome. Environ Microbiol 18(7):2143–2158. https://doi.org/10.1111/1462-2920.12923

Wang S, Xu M, Wang W, Cao X, Piao M, Khan S et al (2016b) Systematic review: adverse events of fecal microbiota transplantation. PLoS One 11(8):e0161174. https://doi.org/10.1371/journal.pone.0161174

Weinberger DM, Klugman KP, Steiner CA, Simonsen L, Viboud C (2015) Association between respiratory syncytial virus activity and pneumococcal disease in infants: a time series analysis of US hospitalization data. PLoS Med 12(1):e1001776. https://doi.org/10.1371/journal.pmed.1001776

Wen X, Miao L, Deng Y, Bible PW, Hu X, Zou Y et al (2017) The influence of age and sex on ocular surface microbiota in healthy adults. Invest Ophthalmol Vis Sci 58(14):6030–6037. https://doi.org/10.1167/iovs.17-22957

Wicinski M, Sawicka E, Gebalski J, Kubiak K, Malinowski B (2020) Human Milk oligosaccharides: health benefits, potential applications in infant formulas, and pharmacology. Nutrients 12(1). https://doi.org/10.3390/nu12010266

Willner D, Furlan M, Haynes M, Schmieder R, Angly FE, Silva J et al (2009) Metagenomic analysis of respiratory tract DNA viral communities in cystic fibrosis and non-cystic fibrosis individuals. PLoS One 4(10):e7370. https://doi.org/10.1371/journal.pone.0007370

Willner D, Haynes MR, Furlan M, Hanson N, Kirby B, Lim YW et al (2012) Case studies of the spatial heterogeneity of DNA viruses in the cystic fibrosis lung. Am J Respir Cell Mol Biol 46(2):127–131. https://doi.org/10.1165/rcmb.2011-0253OC

Wook Kim K, Allen DW, Briese T, Couper JJ, Barry SC, Colman PG et al (2019) Distinct gut Virome profile of pregnant women with type 1 diabetes in the ENDIA study. Open forum. Infect Dis 6(2):ofz025. https://doi.org/10.1093/ofid/ofz025

Wylie KM, Mihindukulasuriya KA, Sodergren E, Weinstock GM, Storch GA (2012) Sequence analysis of the human virome in febrile and afebrile children. PLoS One 7(6):e27735. https://doi.org/10.1371/journal.pone.0027735

Wylie KM, Mihindukulasuriya KA, Zhou Y, Sodergren E, Storch GA, Weinstock GM (2014) Metagenomic analysis of double-stranded DNA viruses in healthy adults. BMC Biol 12:71. https://doi.org/10.1186/s12915-014-0071-7

Young JC, Chehoud C, Bittinger K, Bailey A, Diamond JM, Cantu E et al (2015) Viral metagenomics reveal blooms of anelloviruses in the respiratory tract of lung transplant recipients. Am J Transplant 15(1):200–209. https://doi.org/10.1111/ajt.13031

Yutin N, Makarova KS, Gussow AB, Krupovic M, Segall A, Edwards RA et al (2018) Discovery of an expansive bacteriophage family that includes the most abundant viruses from the human gut. Nat Microbiol 3(1):38–46. https://doi.org/10.1038/s41564-017-0053-y

Zhang B, Chassaing B, Shi Z, Uchiyama R, Zhang Z, Denning TL et al (2014) Viral infection. Prevention and cure of rotavirus infection via TLR5/NLRC4-mediated production of IL-22 and IL-18. Science 346(6211):861–865. https://doi.org/10.1126/science.1256999

Zhang T, Breitbart M, Lee WH, Run JQ, Wei CL, Soh SW et al (2006) RNA viral community in human feces: prevalence of plant pathogenic viruses. PLoS Biol 4(1):e3. https://doi.org/10.1371/journal.pbio.0040003

Zhao G, Vatanen T, Droit L, Park A, Kostic AD, Poon TW et al (2017) Intestinal virome changes precede autoimmunity in type I diabetes-susceptible children. Proc Natl Acad Sci U S A 114(30):E6166–E6175. https://doi.org/10.1073/pnas.1706359114

Zuo T, Lu XJ, Zhang Y, Cheung CP, Lam S, Zhang F et al (2019) Gut mucosal virome alterations in ulcerative colitis. Gut 68(7):1169–1179. https://doi.org/10.1136/gutjnl-2018-318131

Chapter 11
Unlocking the Mysteries of the Human Microbiome to Combat COVID-19

Pushpanathan Muthuirulan, Meenakshi Bandyopadhyay, Sireesha Mamillapalli, and Pooja Sharma

1 Introduction

The recent global dissemination of the novel coronavirus SARS-CoV-2 and the subsequent COVID-19 pandemic have galvanized the scientific community around the central goal of developing therapeutics for immediate and long-term treatment (Malinis et al. 2020; Shi et al. 2020; Pascarella et al. 2020; Shah et al. 2020). While global research efforts are being directed toward development of effective treatment strategies against COVID-19, the possible connection between the human microbiome and COVID-19, which may influence the outcome of the clinical manifestation, should be considered and investigated. Human-microbe associations and their roles in influencing host physiology and immunity have been well known since the early nineteenth century (Hooper et al. 2012; Belkaid and Hand 2014; Quigley 2013; Young 2017). Microbial evolution and colonization within the human host has led to the establishment of an important biological interface between human health and diseases (O'Hara and Shanahan 2006; Fan and Pedersen 2021). Humans

Pushpanathan Muthuirulan, Meenakshi Bandyopadhyay, Sireesha Mamillapalli and Pooja Sharma contributed equally with all other contributors.

P. Muthuirulan (✉)
Department of Human Evolutionary Biology, Harvard University, Cambridge, MA, USA
e-mail: muthuirulanp@fas.harvard.edu

M. Bandyopadhyay
Mithibai College, Mumbai, India

S. Mamillapalli
Geisinger Commonwealth School of Medicine, Scranton, Pennsylvania, USA

P. Sharma
Department of Biology, The Catholic University of American, Washington, DC, USA

S. Thomas (ed.), *Human Microbiome*,
https://doi.org/10.1007/978-981-16-7672-7_11

host a highly diverse group of microbes, which consists mainly of ecological communities of commensals, symbionts, and opportunistic pathogens that reside within different parts of the body, including the gastrointestinal tract (GI), to perform life-sustaining functions. Commensals and symbionts constitute a major portion of the diverse microbial group, while opportunistic pathogens are relatively few and less abundant. Commensals are beneficial to humans and they provide colonization resistance to pathogens (Thursby and Juge 2017; Dekaboruah et al. 2020).

A balance within the innate microbiota with reduced populations or complete elimination of pathogenic microbes is expected in a healthy individual (Belkaid and Hand 2014). The overall balance in the structure and composition of microbiota is important to ensure a healthy well-being and quality of life. Dysbiosis of the microbiota induced by certain risk factors such as infectious diseases, dietary changes, hypertension, cholesterol, diabetes, stress, aging, lack of physical activity, and use of antibiotics exerts a profound impact on human health (DeGruttola et al. 2016; Riccio and Rossano 2018; Frohlich et al. 2016). An overview of the relationships between microbial dysbiosis, risk factors, and COVID-19 disease is shown in Fig. 11.1. Several studies have demonstrated the remarkable association between human diseases and dysbiosis of the microbiota, and have shown that subtle alterations in the human microbiota can cause severe health complications such as diabetes, eczema, allergies, acne, diarrhea, autism, cancer, gastric ulcer, cardiovascular diseases, obesity, rheumatoid arthritis, muscular dystrophy, multiple sclerosis, and other disorders, suggesting that the microbiome may serve as a key regulator of human health and disease development (Kesh et al. 2020; Lee et al. 2018, 2019; Pascal et al. 2018; Saffouri et al. 2019; Pulikkan et al. 2018; Sheflin et al. 2014; Bruno et al. 2018; Lau et al. 2017; Amabebe et al. 2020; Correa et al. 2019; Picca et al. 2018; Kirby and Ochoa-Reparaz 2018). With an aim to circumvent an aggressive immunological response to pathogenic infections like COVID-19, a microbiome may be pivotal in maintaining a host physiology and immunity to prevent an array of excessive physiological reactions that eventually become detrimental to vital organs (e.g., lungs, heart among others) in the human body.

Certain additional factors, such as excessive use of antibiotics and dietary changes, have been proven to cause disruption of the human microbiome which serves as a major risk factor for the development of several diseases (Francino 2015; Vangay et al. 2015; Dudek-Wicher et al. 2018). An excessive use of antibiotics considerably disrupts the ecology of the human microbiome. Unlike the innate microbiome, dysbiotic microbiota possesses a relatively less potential to afford protection against pathogens that may result in serious health issues associated with metabolic, immunological, and developmental disorders. The excessive use of antibiotics may also accelerate the evolution of drug resistance (Francino 2015; Vangay et al. 2015; Dudek-Wicher et al. 2018; Neuman et al. 2018; Magana et al. 2020). Despite the fact that antibiotics do not treat or prevent viral infections like COVID-19, antibiotic usage during COVID-19 has dramatically increased, which may exacerbate the current global status of antimicrobial resistance. Diet is one of the key factors influencing the composition and diversity of the human microbiota (Brown et al. 2012; Hills Jr. et al. 2019; Chan et al. 2013). Further studies are necessary to examine the

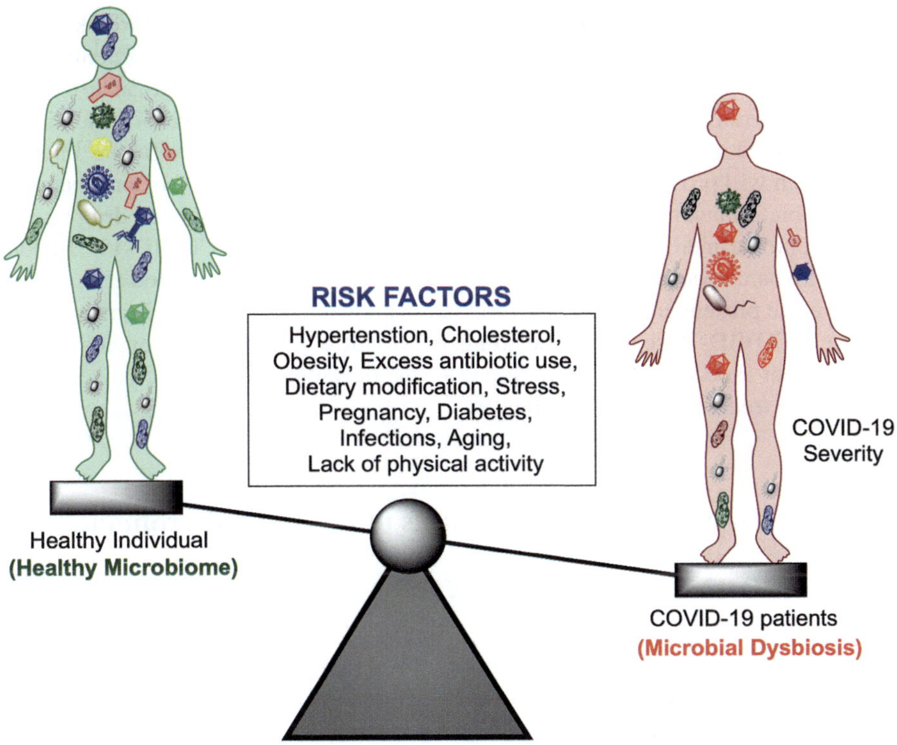

Fig. 11.1 An overview of the relationships between microbial dysbiosis, risk factors, and COVID-19. Risk factors such as hypertension, cholesterol, obesity, diabetes, excessive use of antibiotics, stress, infections, aging, pregnancy, and lack of physical activity could induce microbial dysbiosis in humans which might contribute to the progression of COVID-19 disease

mechanisms by which dietary changes and lifestyle modifications during COVID-19 influence the composition of the human microbiome, which may indicate the potential of therapeutic dietary strategies used for modulation of the microbial composition, diversity, and stability in terms of preventing COVID-19. Pregnancy-induced microbial dysbiosis is often associated with cesarean delivery and is caused by complications, such as preterm birth, extremes of maternal body mass index (BMI), infection, extremes of infant size, and gestational diabetes (Neu and Rushing 2011). The inflammatory and immune changes mediated by pregnancy alter the maternal microbiome and contribute to long-term negative consequence for both the mother and child. Much remains to be discovered on this aspect; however, most studies are focused only on the healthy desired microbial changes during pregnancy. Future research is warranted to elucidate precise roles and mechanisms of the microbiota associated with pregnancy-related complications (Nuriel-Ohayon et al. 2016; Edwards et al. 2017).

A better understanding of the host-microbiome interaction is also important for the development of diagnostic approaches and for the treatment of diseases caused by dysbiosis of the microbiota (Varghese et al. 2020; Casadevall and Pirofski 2000; Lebeer and Spacova 2019). Recent advances in high-throughput sequencing technologies offer deeper understanding of host-microbe interactions that can reveal the core characteristics of the microbiome interactions, including microbial identification, classification, profile prediction, and mechanisms of host-pathogen interaction, which will provide new avenues to gain deeper insights into the consequences of microbial imbalance with the potential to identify novel therapeutic drug targets or microbiome-mediated interventions for the treatment of COVID-19 (Baddal 2019; Hovhannisyan and Gabaldon 2019; Malla et al. 2018; Greenwood et al. 2016).

Here, we present an account of the existing knowledge linking the human microbiome to COVID-19 severity. The aim is to provide a foundation for exploration of the different aspects of the microbiome for the development of personalized interventions to treat or prevent COVID-19.

2 COVID-19-Associated Dysbiosis of the Host Microbiome

2.1 Gut Dysbiosis and COVID-19

The human gut harbors a large repertoire of microorganisms and exerts a marked influence on host homeostasis and disease pathophysiology. Most microbial members of the gut predominantly belong to the phyla Bacteroidetes, Firmicutes, Actinobacteria, Proteobacteria, and Verrucomicrobia (Konstantinidis et al. 2020; Ferreira et al. 2020; Kim et al. 2017). Gut dysbiosis induced by several risk factors has worsened human health, leading to the development of common respiratory diseases including asthma, chronic obstructive pulmonary disease (COPD), cystic fibrosis (CF), lung cancer, and other respiratory infections (Chunxi et al. 2020). COVID-19 patients also exhibit extrapulmonary distress, such as gastrointestinal tract infections and bleeding, vomiting, nausea, loss of appetite, abdominal pain/discomfort, diarrhea, and ulcerative colitis (Olaimat et al. 2020; From the American Association of Neurological Surgeons et al. 2018). Notably, the patients presenting with respiratory disorders are at increased risk, wherein a reduction in the population of *Lactobacillus* and *Bifidobacterium* has been observed, along with an increase in the number of opportunistic pathogens, thereby highlighting the negative effects exerted by microbial dysbiosis on pulmonary functions (Din et al. 2020; Ferreira et al. 2020).

Impaired gastrointestinal function and detection of the SARS-CoV-2 in stools of the affected individuals may hint at a fecal–oral route of transmission. Reports from the US and China highlight the SARS-CoV-2 multiplication ability in both respiratory and digestive tracts. Additionally, fecal samples obtained from infected patients showed the presence of the SARS-CoV-2 RNA even when respiratory samples showed the absence of the viral RNA. Thus, COVID-19 infection negatively affects the gastrointestinal (GI) tract and gut microbiota diversity. Studies also indicate that the growth of opportunistic pathogens and reduction in the population of beneficial bacteria in the gut are positively correlated with the severity of COVID-19 infections (Olaimat et al. 2020). Based on meta-analysis reports of patients from Wuhan, 20% of the COVID-19 patients showed GI symptoms, including diarrhea. The SARS-CoV-2 virus has been detected in anal swabs and stool samples of almost 50% of the affected patients. The virus utilizes the angiotensin-converting enzyme 2 (ACE2) receptors for cellular entry and these receptors are reportedly expressed in respiratory and GI tracts, enterocytes, renal tubules, gallbladder, cardiomyocytes, male reproductive cells, placental trophoblasts, ductal cells, eye, and vasculature (Hikmet et al. 2020; Zuo et al. 2020).

The presence of *Collinsella aerofaciens*, *C. tanakaei, Streptococcus infantis,* and *Morganella morganii* has been reported in fecal samples of patients with a high burden of SARS-CoV-2 infection. On the contrary, fecal samples enriched with *Parabacteroides merdae, Bacteroides stercoris, Alistipes onderdonkii,* and Lachnospiraceae members demonstrated negligible or absence of the SARS-CoV-2 viral load. Notably, the elderly population is the most vulnerable group to COVID-19-associated mortality, and this may be attributed to the gut microbiota dysbiosis and impaired immune system usually observed in the elderly. Such a dysbiosis is also responsible for depression, increase in inflammatory markers, and development of cognitive deficits in the elderly individuals. Additionally, decrease in the Firmicutes to Bacteroidetes ratio and alterations in the abundance of Bacteroides, *Clostridium,* and *Lactobacillus* have been reported in the elderly. Thus, it can be implied that reduction in the gut microbial diversity may exacerbate the already impaired immune system which is observed in the elderly people, and this may increase mortality rates of such individuals (Villapol 2020). It can also be inferred that advancing age is a major factor responsible for gut microbiota dysbiosis, and measures should be undertaken to replenish the gut microbiota using microbiome-directed strategies.

COVID-19 has also threatened the mental health of the public, causing problems such as stress, panic, depression, anxiety, sleep disorders, lower mental well-being, and even suicide (Roy et al. 2020; Rajkumar 2020; Shinu et al. 2020). Mask wearing is another key precautionary measure that can protect us from contracting COVID-19 disease, but it can also provoke significant psychological responses that might cause life-long health consequences. One important aspect to be considered while discussing COVID-19-associated dysbiosis is the impact of psychological stress during the pandemic. It has been proven that the human gut microbiome plays an important role in human health and well-being, including mental health. Especially, the gut microbiota can cooperate with the hosts to regulate the development and function of the immune system, metabolic and nervous systems through dynamic bidirectional communication along the gut–brain axis. Disruption of microbial communities influencing central nervous system components (gut–brain axis) has been implicated in several neurological disorders (Morais et al. 2020). In addition, the COVID-19 pandemic has also caused decline in the physical health of individuals due to lack of exercise, ingestion of improper food, the effect of quarantine in the deterioration the mental health, which all can severely affect the human gut microbial composition. Exercise is one of the best ways to optimize human physical and mental health, and lack of exercise during COVID-19 pandemic can put individuals at higher risk of infection. Prolonged exercise has beneficial effects and has been reported to increase intestinal permeability, compromising gut-barrier function and resulting in bacterial translocation from the colon (Peters et al. 2001; Gisolfi 2000). Probiotics have been reported to restore proper life balance and act as "psychobiotics," thereby serving as an alternate therapeutic option for COVID-19 (Rishi et al. 2020). The utilization of psychobiotics to manage serious problems related to psychological responses during this pandemic is almost unavoidable. Many microorganisms have been proposed as potential psychotropic agents to relieve anxiety and stress including *S. thermophiles*, *B. animalis*, *B. bifidum*, *B. longum*, *Lactobacillus bulgaricus*, *L. lactis*, *L. acidophilus*, *L. plantarum*, *L. reuteri*, *L. paracasei*, *L. helveticus*, *L. rhamnosus*, *Bacillus coagulans*, *Clostridium butyricum*, and others (de Araujo and Farias 2020). Recent evidence also hints at the mechanism by which high levels of stress increase gut permeability via increase in the corticotropin-releasing hormone levels, thereby altering gut microbial composition and leading to dysbiosis and possible susceptibility to SARS-CoV-2 infections (Anderson and Reiter 2020). Thus, focusing on the interaction of COVID-19 with gut–brain axis would allow us to evaluate the basic mechanisms involved in clinical manifestation of COVID-19 and would help endorse in the advancement of prophylactic and treatment strategies.

Social distancing is another key component of the expert-recommended guidelines to prevent the spread of SARS-CoV-2 infections. According to the World Health Organization, the transmission of SARS-CoV-2 virus primarily occurs through saliva or airborne respiratory droplets. Protective precautions to reduce the chances of being infected or spreading COVID-19 include wearing masks, hand sanitation, and social distancing from other people. Recent study has demonstrated the potential connection between social isolation and reduced bacterial diversity. Severe disruption of bacterial diversity caused by social distancing and other stress-related tension can lead to gut microbiota dysbiosis, which is associated with reduced numbers of protective bacteria. Such reduced numbers of protective bacteria can lead to higher risk of opportunistic infections and it has also been shown to increase the risk of influenza infections in the lung. Recent study has also suggested that a human microbiota can influence response to COVID-19, and that COVID-19 patients do possess increased risk of dysbiosis than healthy individual (Domingues et al. 2020). Further, the strict isolation and lockdown protocols implemented by different countries also play an important role in dysbiosis. While lockdown protocols were necessary for containment of the virus, this approach was observed to be a double-edged sword; as complete lack of human contact potentially reduces the dissemination of pathogens and helps to curb the pandemic, it also affects the microbial profile of an individual and reduces the microbial diversity, thereby increasing susceptibility to the SARS-CoV-2 owing to microbial dysbiosis (Domingues et al. 2020).

The involvement of gastrointestinal milieu in COVID-19 makes the gut microbiota a potential target in COVID-19 management and transmission (Chan et al.

2004). Moreover, COVID-19 infection is more severe among individuals with high blood pressure, diabetes, and obesity, conditions that are known to be associated with changes in the composition of the gut microbiota (Sattar et al. 2020; Rodgers and Gibbons 2020; Lim et al. 2021). Understanding the possible connections between the gut microbiota and COVID-19 severity would help to develop a novel and targeted approach to modulate harmful gut microbiota, that may represent a new therapeutic strategy against COVID-19 and its morbidities. Further, understanding of the host-microbial perturbations that underlie SARS-CoV-2 infections would also enable us to utilize the gut microbiota as an indicator for diagnosis of COVID-19 severity. Additionally, improving the composition of the gut microbiota and the proportion of metabolites produced therein through probiotics and personalized nutrition may enhance immunity and minimize the impact of COVID-19 severity in the elderly and immunocompromised patients (Olaimat et al. 2020).

In a study examining the role played by the gut microbiota in COVID-19 severity, a blood proteomic risk score (PRS) was used. Normal, non-infected susceptible individuals and patients with COVID-19 were screened using proteome data and via analysis of inflammatory biomarkers present in blood, to verify the PRS association with the risk of developing COVID-19 in healthy individuals. Studies on the core gut microbiota characteristics, such as gut microbiota metabolites produced and biosynthesis pathways involved, and fecal metabolomics were conducted. Demographics, lifestyle, and socioeconomic background of the patients and healthy individuals were also considered. The study demonstrated the involvement of the biosynthesis pathways for aminoacyl-tRNA, arginine, valine, leucine, and isoleucine, and highlighted the fact that alterations in the pathways could be used to differentiate between healthy and infected individuals, thereby indicating the utility of proteome data and inflammatory parameters to assess the severity of COVID-19 (Gou et al. 2020). Thus, tapping into the potential of the gut microbiome would help to identify potentially safe and affordable approaches for the prevention and treatment of COVID-19 and other viral respiratory diseases (Sadiq 2021; Donati Zeppa et al. 2020). However, more clinical and evidence-based trials are warranted to determine the appropriate strategy to fight against SARS-CoV-2 infections.

2.2 *Lung Dysbiosis and Susceptibility to Viral Infections*

Lung microbiota is defined as the pulmonary microbial community that harbors a diverse group of microbes and is considered to be in close contact with the exogeneous microbes on a daily basis. This feature indicates that the lungs are one of the vital systems whose structure and functionality should be maintained for health and survival. The upper respiratory tract (URT) and lower respiratory tract (LRT) reportedly shelter similar microbial populations, although denser communities have been observed in the former versus the latter. The URT interconnected system predominantly consists of Actinobacteria (*Corynebacterium* and *Propionibacterium* species), Firmicutes (*Staphylococcus* species), Proteobacteria, and Bacteroidetes, including *Streptococcus, Neisseria, Haemophilus*, and *Lachnospira* species. A commensal population including *Streptococcus pneumoniae, Neisseria meningitides*, and *Haemophilus influenzae* is native to the URT (Frank et al. 2010; Lemon et al. 2010; Bassis et al. 2014; Charlson et al. 2011; Yi et al. 2014; Ling et al. 2013; Allen et al. 2014). Microbial populations are relatively less diverse in the LRT (Dickson et al. 2017; Abreu et al. 2012; Bassis et al. 2015; Venkataraman et al. 2015), although phyla including Bacteroidetes and Firmicutes, which mainly include *Prevotella, Veillonella*, and *Streptococcus* species are found in lungs (Morris et al. 2013; Segal et al. 2013; Dickson et al. 2015). Relative abundance of certain species in LRT are often attributed to chronic airway diseases such as COPD and cystic fibrosis (Morris et al. 2013). Inadequate respiratory tract clearance due to increased mucus production and reduced ciliary beat frequency also leads to altered viral and bacterial clearance.

Influenza A virus (IAV) is known to cause flu infections posing a serious public health challenge, resulting in reduced annual workforce and an economic burden. Frequent antigenic substitution often referred to as an antigenic drift contributes to challenges in vaccine design. Alterations in healthy respiratory microbial populations are found to be associated with IAV infection. *Streptococcus* colonization, as evidenced in a mouse model, resulted in decreased susceptibility to IAV infection. Elevated H1 immunoglobulin (IgA) titers in an inoculation study of young adults by attenuated influenza vaccine were positively associated with *Streptococcus infantis* (Short et al. 2012; Diavatopoulos et al. 2010; McCullers and Rehg 2002). In contrast, *Prevotella* species abundance is associated with increased susceptibility to Influenza B viral infection, tuberculosis, and COPD. Children are more susceptible to IAV than young adults and varied reasons, including frequent exposure and lack robust immune development at young age, are attributed to the observed effect (Langevin et al. 2017; Hui et al. 2013; Cheung et al. 2013). Earlier study has demonstrated that pretreatment of mice with antibiotics disrupts the innate and adaptive immune systems (Ichinohe et al. 2011). It has also been reported that an altered microbiome results in the loss of lipopolysaccharides and pattern recognition receptors for activation of toll-like receptors and it thus reduces immune action by type I and II interferons (Ichinohe et al. 2011; Abt et al. 2012). Immunity is at the forefront in discerning the severity of the disease. Though several studies indicate a relationship between microbial populations and viral infections, comprehensive interventions involving animal and human subjects remain to be conducted to address the true effect of the respiratory microbiome and its susceptibility to viral infections and to exclude an altered immune response (Khatiwada and Subedi 2020).

Lung microbiome has received greater attention in recent times due to its association with immunity and respiratory diseases, including COVID-19. Lung microbiome plays an important role in activating an innate and adaptive immune response, which can potentially reduce the risk and consequences of COVID-19 (Khatiwada and Subedi 2020). Only a few studies have examined the relationship between COVID-19 and the lung microbiome. Shen et al. investigated the bronchoalveolar lavage fluid and found significant difference in microbial composition between COVID-19 patients and healthy control. COVID-19 patients showed enrichment of pathogenic bacteria indicating, the degree of microbial imbalance in diseased states (Shen et al. 2020). In another study, Fan et al. have investigated the lung microbiome from the lung post-mortem biopsies from deceased COVID-19 patients. This study has reported the presence of most common bacterial (*Acinetobacter, Chryseobacterium, Burkholderia, Brevundimonas, Sphingobium,* and *Enterobacteriaceae*) and fungal genera (*Cutaneotrichosporon, Issatchenkia, Wallemia, Cladosporium, Alternaria, Dipodascus, Mortierella, Aspergillus, Naganishia, Diutina,* and *Candida*), indicating that bacterial and fungal infections are prevalent in COVID-19 patients (Fan et al. 2020). Overall, there is less substantial information available to explains the relationship between lung microbiome and COVID-19. Further studies are required to understand the role of lung microbiome in COVID-19 severity.

The gut and the lungs are the dominant locations for hosting the microbiota; however, the gut microbiota diversity and microbial population are remarkably higher than those observed in the lungs. Evidence indicates the presence of the gut–lung axis and a bidirectional crosstalk between the gut and the lungs. It has been hypothesized that inflammation of the gut also leads to lung inflammation through this axis. According to previous reports, it has been observed that the gut microbiome dysbiosis is linked with several respiratory disorders; further, in several respiratory diseases, the lung microbiota composition shifts toward the gut microbiota. Several factors have been proposed for this phenomenon. One of the factors hints at migration of the gut microbiota toward the lungs owing to increased permeability of the GI tract (Olaimat et al. 2020). To date, there is no direct evidence that describes the role of the lung microbiome in influencing COVID-19; however, related human and animal studies have shown that the human microbiome can play critical role in immune response development against viral infections. Future studies are necessary to investigate the relationship between the lung microbiome and COVID-19.

2.3 Pregnancy, Human Microbiota, and COVID-19

The inflammatory and immune changes mediated by pregnancy alter maternal gut function and microbial composition. The maternal gut microbiome composition significantly contributes to obstetric outcomes with long-term health consequences for both the mother and the child. The hormones such as estrogen and progesterone contribute to a shift in the human microbiota and impact gut function, especially during the prenatal period (Edwards et al. 2017). Several studies have shown that the microbiome can be vertically transmitted from parents to offspring, and it is plausible that the maternal–infant microbiome transfer may influence the early stages of infant health (Yang et al. 2016; Dunn et al. 2017). The overall risk of developing complications associated with COVID-19 in pregnant women is low (Maleki Dana et al. 2020). However, recent data highlight the increased risk for severe COVID-19 during pregnancy. According to the Centers for Disease Control and Prevention (CDC), pregnant women are 5.4 times more likely to be hospitalized, 1.5 times more likely to be subjected to intensive care, and 1.7 times more likely to require mechanical ventilation than non-pregnant women (Zambrano et al. 2020). Certain studies suggest that premature birth is more likely to be observed in pregnant women with COVID-19 and their babies are more likely to be admitted to a neonatal unit (Maleki Dana et al. 2020; Yang et al. 2020). One study has suggested that newborns rarely acquire COVID-19 from SARS-CoV-2 positive or suspected SARS-CoV-2-infected mothers. Over 800 newborns reported, the incidence of vertical transmission has proven to be low, indicating that adverse clinical outcomes in newborn seem to be due to maternal disease status in the small subset of newborns with critically ill mothers, rather than illness due to SARS-CoV-2 infection (Kyle et al. 2020). One another study has confirmed that COVID-19 infection in pregnant women resembles the SARS-CoV-2 infection in non-pregnant adult population, with possibly less chance for adverse maternal or perinatal outcome (Elshafeey et al. 2020). All these studies suggest that there is no vertical transmission of COVID-19 from the mother to the fetus; however, certain studies indicate such a pattern of transmission, but additional convincing evidence regarding the same remains to be reported (Dashraath et al. 2020; Chen et al. 2020). Further studies are necessary to understand the COVID-19-mediated microbiome alteration and maternal microbial transmission during pregnancy which may help explain the mechanisms of microbiome alterations associated with COVID-19 that impact fetal growth and development.

3 Antimicrobial Resistance in the Era of COVID-19

3.1 Host Gut Microbiome Dysbiosis Exacerbated by Use of Antibiotics

Excessive and long-term use of antibiotics can trigger microbiome dysbiosis. Studies on vancomycin have reported long-lasting shifts in the gut microbiome, with expansion of less abundant bacterial populations (Kim et al. 2017). It has been reported that excessive antibiotic usage can lead to altered GI tract anatomy and physiology; this may play a role in the migration of gut microbes toward the lungs and lead to altered microbial diversity (Olaimat et al. 2020). Considerable evidence has demonstrated an association between antibiotic usage during the first year of life and development of asthma by the 6th–7th year of life (Becattini et al. 2016). A recent study has shown that antibiotic-naive patients with COVID-19 demonstrated presence of bacteremia-causing opportunistic pathogens, such as *Clostridium hathewayi*, *Actinomyces viscosus*, and *Bacteroides nordii* compared to healthy individuals. Antibiotic-treated COVID-19 patients showed depletion of beneficial microbes including *Faecalibacterium prausnitzii*, *Lachnospiraceae bacterium 5_1_63FAA*, *Eubacterium rectale*, *Ruminococcus obeum*, and *Dorea formicigenerans* compared with antibiotic-naive patients with COVID-19. Bacteroides species, including *Bacteroides dorei*, *Bacteroides thetaiotaomicron*, *Bacteroides massiliensis*, and *Bacteroides ovatus*, showed inverse correlation with the fecal SARS-CoV-2 load; notably, these species were associated with decreased ACE2 expression in the murine colon, indicating that *Bacteroides* species might play a protective role in combating SARS-CoV-2 through ACE2 expression. The highest SARS-CoV-2 mortality and morbidity rates have been reported in older patients and in those with underlying comorbidities. Notably, a less abundant population of the *Bacteroides* species was observed in such patients, indicating that an individual's gut microbiome might affect the immunological response to SARS-CoV-2 infection (Zuo et al. 2020).

3.2 Antibiotic Prescription, Over-sanitation, and Antimicrobial Resistance During the COVID-19 Pandemic

With regard to symptomatic cases, individuals infected with the SARS-CoV-2 usually present with fever, respiratory distress, and pneumonia; in extreme cases, they present with multiple gastrointestinal, renal, neurological, and cardiac issues, wherein hospitalization is deemed necessary (Ferreira et al. 2020). Generally, individuals exhibiting upper or lower respiratory tract diseases are prescribed with antibiotics. However, as per findings of a recent study, 72% of the patients received antibiotics, among which only 8% were diagnosed with bacterial or fungal co-infections. As per WHO reports, treatments using azithromycin and hydroxychloroquine have been rampantly prescribed irrespective of any conclusive evidence from COVID-19 clinical trials. Considering the indiscriminate and injudicious use of antibiotics during COVID-19, which may lead to subsequent development of antimicrobial resistance, the WHO has outlined specific antibiotic usage guidelines along with antibiotic stewardship principles. In the absence of any underlying bacterial infection, the guidelines explicitly deter individuals from opting for an antibiotic therapy or antibiotic-mediated prophylaxis for moderate COVID-19 symptoms. The guidelines also recommend consideration of epidemiology, host factors, and routine clinical assessments prior to antibiotic prescription. Only older patients residing in long-term care facilities and children below 5 years of age exhibiting moderate COVID-19 symptoms can be treated with antibiotics prescribed for bacterial pneumonia (Getahun et al. 2020).

Increased mortality rates of patients with COVID-19 seem to be associated with excessive antibiotic usage and gut microbial dysbiosis (Din et al. 2020). A majority of the respiratory tract infection (RTI) cases are erroneously treated with antibiotics, regardless of the presence of a bacterial etiology. Considering this, the Choosing Wisely campaigns have been initiated to disseminate appropriate information on antibiotic usage. The campaigns propagate avoidance of antibiotic usage in cases of viral origin (influenza-like illness), upper respiratory infections, and self-limiting sinusitis. A recent study has further proposed that COVID-19/influenza-like symptoms and common cold cases should not be treated with antibiotics or symptomatic management is sufficient; further examination, in-person visits, bacterial culture tests, vital sign abnormalities, and increase/decrease in symptoms should be considered before prescribing antibiotics for acute otitis media, pharyngitis, sinusitis, COPD, and suspected pneumonia cases (Leis et al. 2020).

A recent study conducted on the antibiotic usage in the initial period of the COVID-19 pandemic reported a biphasic pattern. Antibiotic prescription and consumption increased through March and April 2020. In March 2020, during the first peak, amoxicillin/clavulanate was recommended for patients with COVID-19 and administration or prescription of antibiotics increased gradually. In April 2020, during the second peak, broad-spectrum antibiotics (cefepime, piperacillin/tazobactam, meropenem, imipenem, and ertapenem) were prescribed with reduced prescription of amoxicillin/clavulanate. The first peak and antibiotic prescription pattern coincided with increased hospitalization rates. The second peak coincided with increase in severity of cases and probable development of nosocomial infections, thereby demonstrating increased prescription of broad-spectrum antibiotics (Abelenda-Alonso et al. 2020).

To provide a more accurate explanation of antimicrobial resistance, the term "resistome" is frequently used. The resistome comprises antimicrobial resistance genes (ARGs) of the pathogenic and non-pathogenic gut bacteria. The dissemination of ARGs via horizontal gene transfer and mobile genetic elements increases the risk of antimicrobial resistance within the intestinal microbiome (Konstantinidis et al. 2020). The risk is increased further with consumption of antibiotics. In a study involving pigs fed with a diet supplemented with antibiotics, findings showed that ARG abundance increased in the porcine microbiota, which led to the development of tolerance against drugs to which they were not exposed. In another study, it was observed that approximately 40% of the bacterial members within hosts harbored quinolone-resistance genes, even in those who had never been exposed to the drugs. In a study involving Finnish children, early use of macrolides demonstrated a microbial profile in which depletion of Actinobacteriaceae and an increased population of Bacteroidetes and Proteobacteria was observed along with ARG induction (Becattini et al. 2016).

Since use of disinfectant and over-sanitation have the capacity to alter the microbial diversity, increased exposure to hand sanitizer, disinfectants, and household cleaning products during this pandemic could be associated with disturbance of human microbiota. Moreover, the emerging links between over-sanitization and occurrence of non-communicable diseases and antimicrobial resistance have involved the human microbiome. The disruption of gut microbiota induced by disinfectants and over-sanitation have life-long health consequences. Regarding the evidence-based reduction in exposure to non-pathogenic commensal bacteria and gut dysbiosis, further study is warranted to investigate the effects of massive use of disinfectants or sanitizers during the COVID-19 pandemic. In this context, recommendations to consume probiotics, pychobiotics, and fermented foods might reverse the consequences by alleviating dysbiosis (Ejtahed et al. 2020). Altogether, the above-mentioned findings highlight the importance of judicious use of antibiotics, hand sanitizer, and household cleaning products to curb antimicrobial resistance and microbial dysbiosis, which many seem to consider as a collateral damage of the COVID-19 pandemic.

4 Dietary Changes and Human Microbiota

Diet is one of the most important regulators of the human microbiome; however, the precise mechanisms by which diet induces microbiome variations remain elusive. Health benefits attained by following an optimal diet are evident as per previous findings and also provide a concrete foundation for leading a healthy lifestyle in the future. The COVID-19 pandemic has affected the global population, thereby emphasizing the need for awareness among communities to adopt safe practices in terms of food hygiene and consumption. Several governmental and non-governmental organizations have recognized the necessity of specific guidelines for the prognosis of COVID-19. A recent study has reported that implementation of the lockdown during COVID-19 has resulted in the practice of consumption of home-cooked, healthy meals that enrich beneficial microflora in the gut, which may have resulted in better prognosis of COVID-19 patients in India compared to those in western countries (Rishi et al. 2020). Nutritional modulation is vital for individuals of different ages, with chronic health conditions, and for therapy and management of several health issues. Nutritional excess or deficiency has been associated with immunodeficiency, and therefore adequate nutrition is critically important for homeostasis and for optimal functioning of the immune system to fight against SARS-CoV-2 infection, as well as for the development of an efficient immune system to combat other pathogenic viruses and microorganisms (Chaari et al. 2020).

Considerable cultural and geographical differences also play a role in varied global food consumption patterns, thereby making nutritional optimization a challenging yet a necessary task. Several dietary recommendations were made during the initial phases of the pandemic and have been implemented as a part of the treatment and prevention strategy against COVID-19. Fresh fruits and vegetables rich in nutrients and water were recommended by most studies to boost the intake of micronutrients. Vitamins and minerals contribute toward healthy maintenance of physical barrier organs including the skin, mucus membrane, respiratory tract, and gastrointestinal tract to prevent viral infections. Vitamins A, C, D, E, B6, and B12 help to maintain cell division, proliferation, and functional aspects of immune cells. They provide support in inflammatory response and antibody production of T and B cells (https://www.eufic.org/en/food-safety). A special emphasis has been laid on vitamins C and D, supporting the significance of the former in individuals who are at risk of developing respiratory tract infections. Antioxidant properties of both vitamins C and D have been well established in lowering the pulmonary-associated infections. Vitamin D status is also associated with the severity of COVID-19 (McCartney and Byrne 2020; Mansur 2020). Minerals such as zinc and selenium are known to exhibit antioxidant properties as evidenced by suppression of oxidative stress and augmentation of host immune responses (Beck et al. 2003; Read et al.

2019; Lee 2018). Mice with selenium deficiency subjected to influenza viral challenge showed an enhanced pathology in the lungs (Beck et al. 2003). Adequate hydration is necessary for maintaining body homeostasis, kidney function, appropriate cognitive senses, and cardiovascular function (El-Sharkawy et al. 2015). Hypohydration leads to exhibition of adverse health effects over varying age groups. Gut commensal populations like *Bifidobacterium* and *Lactobacillus* and pathogenic bacteria like *Bacteroides fragilis* and *Clostridium perfringens* were shown to be increased and decreased respectively via consumption of whey and pea protein extracts (Swiatecka et al. 2011). Consumption of whole-grain food rich in nondigestible carbohydrates reduced proinflammatory cytokines IL-6 and insulin resistance (Keim and Martin 2014). Increased levels of IL-10 (an anti-inflammatory cytokine) were observed with the intake of butyrated maize starch (West et al. 2017). Fermented foods rich in live microorganisms *Lactobacillus* and *Bifidobacterium* that include many different strains such as *L. fermentum, L. reuteri, L. paracasei, L. rhamnosus, L. acidophilus, L. plantarum, B. longum, B. breve, B. bifidum*, and *B. animalis* were shown to reduce enteropathogens *E. coli* and *Helicobacter pylori* (Yang and Sheu 2012). Treg cells, which are downregulators of allergic response, were shown to be induced by consumption of probiotics (Feleszko et al. 2007). It has been demonstrated that diet-microbiome interactions are personalized, suggesting that diet-microbiome studies should either include longitudinal sampling within individuals to identify personalized responses to dietary changes or should consider adequate number of participants spanning a wide range of microbiome types to study more generalized responses (Johnson et al. 2020). Although the dietary guidelines for the COVID-19 pandemic represent generic information based on healthy personnel, it would be beneficial to formulate dietary recommendations based on patients' requirements. A range of tolerable intake levels of nutrients with respect to varied chronic conditions are desirable to provide specific information rather than a "one-size-fits-all" approach. However, extensive research should be performed to understand the role of dietary changes on human microbiome alterations to develop better diagnosis and therapeutic dietary strategies for COVID-19 patients.

5 Microbiome-Based Interventions

Host-microbe interactions play a key role in determining the health and disease status in humans. Microbial imbalance is related to a plethora of diseases, including COVID-19. A better understanding of the host-microbe interaction is important to develop efficient diagnosis and treatment strategies for these ailments (Varghese et al. 2020; Casadevall and Pirofski 2000; Lebeer and Spacova 2019). By precisely modulating the host microbiome, either by removing the pathogenic taxa or by reintroducing missing beneficial taxa, development of new therapeutic approaches for treatment of diseases associated with the dysbiosis of microbiota can be realized. Culturing of large microbial communities in the laboratory is impossible using traditional microbiology approaches. Consequently, it is difficult to comprehensively profile individual microbes comprising a specific microbiome, and to understand their complex, multipartite interactions (Forbes et al. 2017). Microbiome-based interventions should be considered to formulate strategies in the source tracking and monitoring of microbial communities. Tools such as FEAST, PHASTER, PHASTEST, and Source Tracker are utilized to conduct source tracking to determine the origins of microbial agents, especially those implicated in diseases. In a previous study based on analysis of sequencing datasets of fecal samples obtained from patients with COVID-19, highlighting alterations of the gut microbiota, FEAST was used to conduct source tracking of the patients, and the results showed extremely high accuracy of source tracking. Thus, using such approaches, the microbiome can be used to determine sources and patterns of dissemination, divergence, and variations of pathogens (Han et al. 2020).

The structure of the microbial community observed in patients with COVID-19 and those with community-acquired pneumonia is reportedly similar. The oral microbiota and its dysbiosis have been implicated in multiple diseases, including COVID-19, type 2 diabetes, hypertension, and cardiovascular disease; notably, the comorbidities mentioned herein increase the risk of COVID-19-associated mortality. Modulation of the human gut microbiota diversity has been reported to ameliorate conditions like enteritis and ventilator-associated pneumonia. These findings indicate the crucial role played by the microbiome in various diseases and the potential of the microbiome to be altered to mitigate disease conditions. Apart from the microbiome, probiotics have garnered considerable attention to combat COVID-19. Several studies have highlighted the role of probiotics in reduction of serum lipid levels and augmentation of immunity; thus, probiotic-based approaches may be used to modulate the host microbiome and to elicit a remarkable immune response against SARS-CoV-2. Maintaining a moderate exercise regimen may also be beneficial to maintain the homeostasis of the gut microbiome (Han et al. 2020).

Recent advances in the next-generation sequencing (NGS) technology and availability of state-of-the-art bioinformatics tools have enabled investigation of the microbiome, defying the need for cultivation (Hiergeist et al. 2015). NGS is now becoming a mainstream option for most researchers in the fight against the viral pandemic and they provide key insights into comprehensive structure of the microbiome and microbiome-host metabolic signal disruption in humans that would help us gain advanced knowledge on the impact of microbial imbalance and the role of microbial communities in human health and diseases. Several methods such as 16S rRNA sequencing and metagenome shotgun sequencing are available to explore the structural and functional composition of human microbiome. OMICs technologies (transcriptomics, proteomics, and metabolomics) offer newfound analytical opportunities to understand the mechanisms by which these microbial communities function and relate to their environment (Jiang et al. 2019; Hiergeist et al. 2015). Utilization of these technologies in COVID-19 research will improve our ability to rapidly and reproducibly characterize the microbial changes associated with COVID-19 severity, and it also offers an opportunity to develop fundamentally new diagnostic biomarkers (microbiome signatures) and therapeutics for COVID-19 (Fig. 11.2).

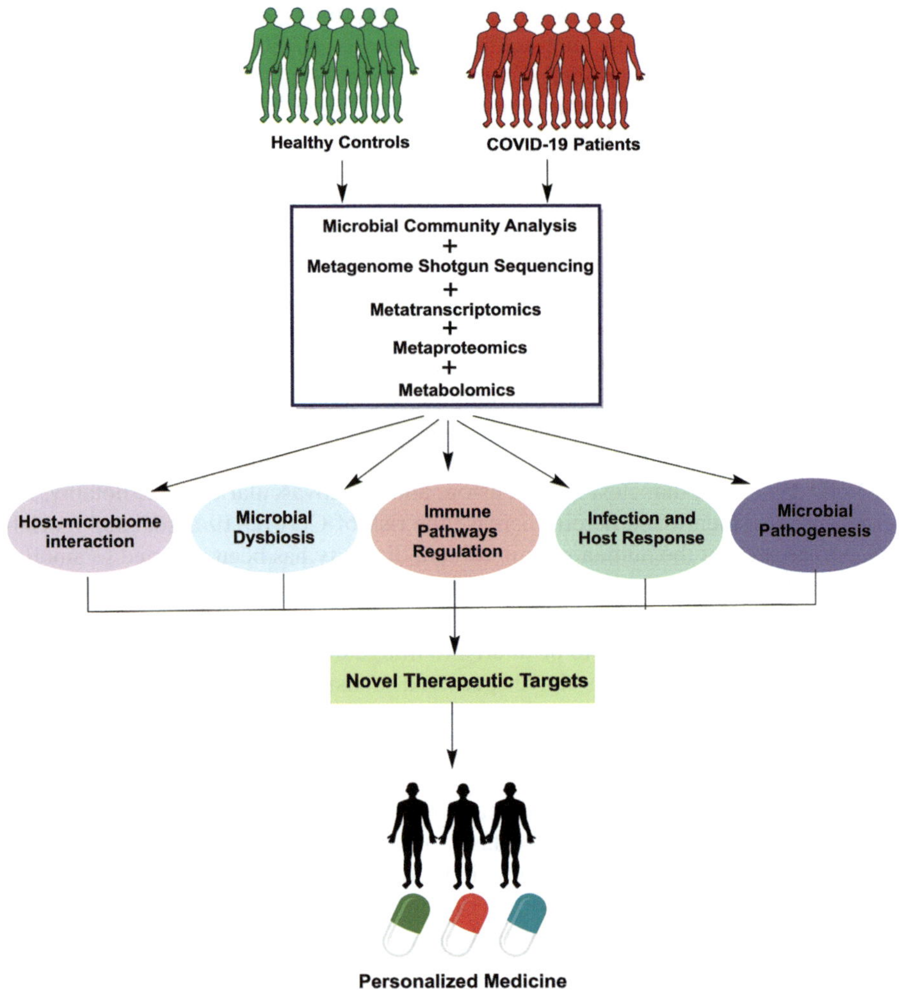

Fig. 11.2 Strategies for development of personalized therapy to treat COVID-19. Utilization of OMICS technologies in COVID-19 research will improve our ability to investigate microbial changes associated with COVID-19 severity and offers an opportunity to develop personalized therapeutics for the treatment of COVID-19

6 Conclusion

Based on the above-mentioned information, it can be inferred that the human microbiome can have a profound impact on the susceptibility to SARS-CoV-2 infection and COVID-19 severity. The involvement of gastrointestinal symptoms and respiratory illness in COVID-19 makes the gut and lung microbiota a potential target in COVID-19 management. In addition, public health actions such as social distancing, mask wearing, quarantine, and lack of physical activities have threatened the mental health of the public which impose major negative impact on the human microbiome. The hope is that the new class of medicines "psychobiotics" will eventually provide powerful treatment for depression and other mental illness that arise during this pandemic. Pregnant women are at increased risk of developing complications due to COVID-19; however, risk of neonatal infection via perinatal/postnatal transmission is low, suggesting that vertical transmission of microbiome from infected mother to newborn may not affect the fetal growth and development. This is one area where further study is warranted. Besides these, dietary changes, lifestyle modification, over-sanitation, and excessive use of antibiotics during this pandemic can cause severe microbiota dysbiosis. In this context, recommendations to consume probiotics, pychobiotics, fermented foods, and judicious use of antibiotics/disinfectant might reverse the consequences by alleviating dysbiosis. Thus, the microbiome is a key regulator of human health and diseases and it is essentially important for us to protect our microbiome from harmful risk factors to promote disease-free life. Future studies should investigate the human microbiome and correlate findings with the severity of COVID-19. Identification of the beneficial and harmful microbial components and their roles in early development of disease may help in the design of novel strategies for alteration of the microbiota to reduce disease severity. We are therefore confident that future microbiome studies will provide useful clinical knowledge, as well as offer a broader understanding of COVID-19 progression which will aid in the development of necessary tools and approaches to better diagnose, treat, and prevent this disease.

References

Abelenda-Alonso G, Padulles A, Rombauts A, Gudiol C, Pujol M, Alvarez-Pouso C, Jodar R, Carratala J (2020) Antibiotic prescription during the COVID-19 pandemic: a biphasic pattern. Infect Control Hosp Epidemiol 41(11):1371–1372. https://doi.org/10.1017/ice.2020.381

Abreu NA, Nagalingam NA, Song Y, Roediger FC, Pletcher SD, Goldberg AN, Lynch SV (2012) Sinus microbiome diversity depletion and Corynebacterium tuberculostearicum enrichment mediates rhinosinusitis. Sci Transl Med 4(151):151ra124. https://doi.org/10.1126/scitranslmed.3003783

Abt MC, Osborne LC, Monticelli LA, Doering TA, Alenghat T, Sonnenberg GF, Paley MA, Antenus M, Williams KL, Erikson J, Wherry EJ, Artis D (2012) Commensal bacteria calibrate the activation threshold of innate antiviral immunity. Immunity 37(1):158–170. https://doi.org/10.1016/j.immuni.2012.04.011

Allen EK, Koeppel AF, Hendley JO, Turner SD, Winther B, Sale MM (2014) Characterization of the nasopharyngeal microbiota in health and during rhinovirus challenge. Microbiome 2:22. https://doi.org/10.1186/2049-2618-2-22

Amabebe E, Robert FO, Agbalalah T, Orubu ESF (2020) Microbial dysbiosis-induced obesity: role of gut microbiota in homoeostasis of energy metabolism. Br J Nutr 123(10):1127–1137. https://doi.org/10.1017/S0007114520000380

Anderson G, Reiter RJ (2020) COVID-19 pathophysiology: interactions of gut microbiome, melatonin, vitamin D, stress, kynurenine and the alpha 7 nicotinic receptor: treatment implications. Melatonin Res 3(3):322–345. https://doi.org/10.32794/mr11250066

Baddal B (2019) Next-generation technologies for studying host-pathogen interactions: a focus on dual transcriptomics, CRISPR/Cas9 screening and organs-on-chips. Pathog Dis 77(6). https://doi.org/10.1093/femspd/ftz060

Bassis CM, Erb-Downward JR, Dickson RP, Freeman CM, Schmidt TM, Young VB, Beck JM, Curtis JL, Huffnagle GB (2015) Analysis of the upper respiratory tract microbiotas as the source of the lung and gastric microbiotas in healthy individuals. mBio 6(2):e00037. https://doi.org/10.1128/mBio.00037-15

Bassis CM, Tang AL, Young VB, Pynnonen MA (2014) The nasal cavity microbiota of healthy adults. Microbiome 2:27. https://doi.org/10.1186/2049-2618-2-27

Becattini S, Taur Y, Pamer EG (2016) Antibiotic-induced changes in the intestinal microbiota and disease. Trends Mol Med 22(6):458–478. https://doi.org/10.1016/j.molmed.2016.04.003

Beck MA, Levander OA, Handy J (2003) Selenium deficiency and viral infection. J Nutr 133(5 Suppl 1):1463S–1467S. https://doi.org/10.1093/jn/133.5.1463S

Belkaid Y, Hand TW (2014) Role of the microbiota in immunity and inflammation. Cell 157(1):121–141. https://doi.org/10.1016/j.cell.2014.03.011

Brown K, DeCoffe D, Molcan E, Gibson DL (2012) Diet-induced dysbiosis of the intestinal microbiota and the effects on immunity and disease. Nutrients 4(8):1095–1119. https://doi.org/10.3390/nu4081095

Bruno G, Rocco G, Zaccari P, Porowska B, Mascellino MT, Severi C (2018) Helicobacter pylori infection and gastric Dysbiosis: can probiotics administration be useful to treat this condition? Can J Infect Dis Med Microbiol 2018:6237239. https://doi.org/10.1155/2018/6237239

Casadevall A, Pirofski LA (2000) Host-pathogen interactions: basic concepts of microbial commensalism, colonization, infection, and disease. Infect Immun 68(12):6511–6518. https://doi.org/10.1128/iai.68.12.6511-6518.2000

Chaari A, Bendriss G, Zakaria D, McVeigh C (2020) Importance of dietary changes during the coronavirus pandemic: how to upgrade your immune response. Front Public Health 8:476. https://doi.org/10.3389/fpubh.2020.00476

Chan KH, Poon LL, Cheng VC, Guan Y, Hung IF, Kong J, Yam LY, Seto WH, Yuen KY, Peiris JS (2004) Detection of SARS coronavirus in patients with suspected SARS. Emerg Infect Dis 10(2):294–299. https://doi.org/10.3201/eid1002.030610

Chan YK, Estaki M, Gibson DL (2013) Clinical consequences of diet-induced dysbiosis. Ann Nutr Metab 63(Suppl 2):28–40. https://doi.org/10.1159/000354902

Charlson ES, Bittinger K, Haas AR, Fitzgerald AS, Frank I, Yadav A, Bushman FD, Collman RG (2011) Topographical continuity of bacterial populations in the healthy human respiratory tract. Am J Respir Crit Care Med 184(8):957–963. https://doi.org/10.1164/rccm.201104-0655OC

Chen H, Guo J, Wang C, Luo F, Yu X, Zhang W, Li J, Zhao D, Xu D, Gong Q, Liao J, Yang H, Hou W, Zhang Y (2020) Clinical characteristics and intrauterine vertical transmission potential of COVID-19 infection in nine pregnant women: a retrospective review of medical records. Lancet 395(10226):809–815. https://doi.org/10.1016/S0140-6736(20)30360-3

Cheung MK, Lam WY, Fung WY, Law PT, Au CH, Nong W, Kam KM, Kwan HS, Tsui SK (2013) Sputum microbiota in tuberculosis as revealed by 16S rRNA pyrosequencing. PLoS One 8(1):e54574. https://doi.org/10.1371/journal.pone.0054574

Chunxi L, Haiyue L, Yanxia L, Jianbing P, Jin S (2020) The gut microbiota and respiratory diseases: new evidence. J Immunol Res 2020:2340670. https://doi.org/10.1155/2020/2340670

Correa JD, Fernandes GR, Calderaro DC, Mendonca SMS, Silva JM, Albiero ML, Cunha FQ, Xiao E, Ferreira GA, Teixeira AL, Mukherjee C, Leys EJ, Silva TA, Graves DT (2019) Oral microbial dysbiosis linked to worsened periodontal condition in rheumatoid arthritis patients. Sci Rep 9(1):8379. https://doi.org/10.1038/s41598-019-44674-6

Dashraath P, Wong JLJ, Lim MXK, Lim LM, Li S, Biswas A, Choolani M, Mattar C, Su LL (2020) Coronavirus disease 2019 (COVID-19) pandemic and pregnancy. Am J Obstet Gynecol 222(6):521–531. https://doi.org/10.1016/j.ajog.2020.03.021

de Araujo FF, Farias DP (2020) Psychobiotics: an emerging alternative to ensure mental health amid the COVID-19 outbreak? Trends Food Sci Technol 103:386–387. https://doi.org/10.1016/j.tifs.2020.07.006

DeGruttola AK, Low D, Mizoguchi A, Mizoguchi E (2016) Current understanding of Dysbiosis in disease in human and animal models. Inflamm Bowel Dis 22(5):1137–1150. https://doi.org/10.1097/MIB.0000000000000750

Dekaboruah E, Suryavanshi MV, Chettri D, Verma AK (2020) Human microbiome: an academic update on human body site specific surveillance and its possible role. Arch Microbiol 202(8):2147–2167. https://doi.org/10.1007/s00203-020-01931-x

Diavatopoulos DA, Short KR, Price JT, Wilksch JJ, Brown LE, Briles DE, Strugnell RA, Wijburg OL (2010) Influenza a virus facilitates Streptococcus pneumoniae transmission and disease. FASEB J 24(6):1789–1798. https://doi.org/10.1096/fj.09-146779

Dickson RP, Erb-Downward JR, Freeman CM, McCloskey L, Beck JM, Huffnagle GB, Curtis JL (2015) Spatial variation in the healthy human Lung microbiome and the adapted island model of Lung biogeography. Ann Am Thorac Soc 12(6):821–830. https://doi.org/10.1513/AnnalsATS.201501-029OC

Dickson RP, Erb-Downward JR, Freeman CM, McCloskey L, Falkowski NR, Huffnagle GB, Curtis JL (2017) Bacterial topography of the healthy human lower respiratory tract. mBio 8(1). https://doi.org/10.1128/mBio.02287-16

Din AU, Mazhar M, Waseem M, Ahmad W, Bibi A, Hassan A, Ali N, Gang W, Qian G, Ullah R, Shah T, Ullah M, Khan I, Nisar MF, Wu J (2020) SARS-CoV-2 microbiome dysbiosis linked disorders and possible probiotics role. Biomed Pharmacother 133:110947. https://doi.org/10.1016/j.biopha.2020.110947

Domingues CPF, Rebelo JS, Dionisio F, Botelho A, Nogueira T (2020) The social distancing imposed to contain COVID-19 can affect our microbiome: a double-edged sword in human health. mSphere 5(5). https://doi.org/10.1128/mSphere.00716-20

Donati Zeppa S, Agostini D, Piccoli G, Stocchi V, Sestili P (2020) Gut microbiota status in COVID-19: an unrecognized player? Front Cell Infect Microbiol 10:576551. https://doi.org/10.3389/fcimb.2020.576551

Dudek-Wicher RK, Junka A, Bartoszewicz M (2018) The influence of antibiotics and dietary components on gut microbiota. Prz Gastroenterol 13(2):85–92. https://doi.org/10.5114/pg.2018.76005

Dunn AB, Jordan S, Baker BJ, Carlson NS (2017) The maternal infant microbiome: considerations for labor and birth. MCN Am J Matern Child Nurs 42(6):318–325. https://doi.org/10.1097/NMC.0000000000000373

Edwards SM, Cunningham SA, Dunlop AL, Corwin EJ (2017) The maternal gut microbiome during pregnancy. MCN Am J Matern Child Nurs 42(6):310–317. https://doi.org/10.1097/NMC.0000000000000372

Ejtahed HS, Hasani-Ranjbar S, Siadat SD, Larijani B (2020) The most important challenges ahead of microbiome pattern in the post era of the COVID-19 pandemic. J Diabetes Metab Disord:1–3. https://doi.org/10.1007/s40200-020-00579-0

El-Sharkawy AM, Sahota O, Lobo DN (2015) Acute and chronic effects of hydration status on health. Nutr Rev 73(Suppl 2):97–109. https://doi.org/10.1093/nutrit/nuv038

Elshafeey F, Magdi R, Hindi N, Elshebiny M, Farrag N, Mahdy S, Sabbour M, Gebril S, Nasser M, Kamel M, Amir A, Maher Emara M, Nabhan A (2020) A systematic scoping review of COVID-19 during pregnancy and childbirth. Int J Gynaecol Obstet 150(1):47–52. https://doi.org/10.1002/ijgo.13182

Fan J, Li X, Gao Y, Zhou J, Wang S, Huang B, Wu J, Cao Q, Chen Y, Wang Z, Luo D, Zhou T, Li R, Shang Y, Nie X (2020) The lung tissue microbiota features of 20 deceased patients with COVID-19. J Infect 81(3):e64–e67. https://doi.org/10.1016/j.jinf.2020.06.047

Fan Y, Pedersen O (2021) Gut microbiota in human metabolic health and disease. Nat Rev Microbiol 19(1):55–71. https://doi.org/10.1038/s41579-020-0433-9

Feleszko W, Jaworska J, Rha RD, Steinhausen S, Avagyan A, Jaudszus A, Ahrens B, Groneberg DA, Wahn U, Hamelmann E (2007) Probiotic-induced suppression of allergic sensitization and airway inflammation is associated with an increase of T regulatory-dependent mechanisms in a murine model of asthma. Clin Exp Allergy 37(4):498–505. https://doi.org/10.1111/j.1365-2222.2006.02629.x

Ferreira C, Viana SD, Reis F (2020) Gut microbiota Dysbiosis-immune Hyperresponse-inflammation triad in coronavirus disease 2019 (COVID-19): impact of pharmacological and nutraceutical approaches. Microorganisms 8(10). https://doi.org/10.3390/microorganisms8101514

Forbes JD, Knox NC, Ronholm J, Pagotto F, Reimer A (2017) Metagenomics: the next culture-independent game changer. Front Microbiol 8:1069. https://doi.org/10.3389/fmicb.2017.01069

Francino MP (2015) Antibiotics and the human gut microbiome: Dysbioses and accumulation of resistances. Front Microbiol 6:1543. https://doi.org/10.3389/fmicb.2015.01543

Frank DN, Feazel LM, Bessesen MT, Price CS, Janoff EN, Pace NR (2010) The human nasal microbiota and Staphylococcus aureus carriage. PLoS One 5(5):e10598. https://doi.org/10.1371/journal.pone.0010598

Frohlich EE, Farzi A, Mayerhofer R, Reichmann F, Jacan A, Wagner B, Zinser E, Bordag N, Magnes C, Frohlich E, Kashofer K, Gorkiewicz G, Holzer P (2016) Cognitive impairment by antibiotic-induced gut dysbiosis: analysis of gut microbiota-brain communication. Brain Behav Immun 56:140–155. https://doi.org/10.1016/j.bbi.2016.02.020

From the American Association of Neurological Surgeons ASoNC, Interventional Radiology Society of Europe CIRACoNSESoMINTESoNESOSfCA, Interventions SoIRSoNS, World Stroke O, Sacks D, Baxter B, BCV C, Carpenter JS, Cognard C, Dippel D, Eesa M, Fischer U, Hausegger K, Hirsch JA, Shazam Hussain M, Jansen O, Jayaraman MV, Khalessi AA, Kluck BW, Lavine S, Meyers PM, Ramee S, Rufenacht DA, Schirmer CM, Vorwerk D (2018) Multisociety consensus quality improvement revised consensus statement for endovascular therapy of acute ischemic stroke. Int J Stroke 13(6):612–632. https://doi.org/10.1177/1747493018778713

Getahun H, Smith I, Trivedi K, Paulin S, Balkhy HH (2020) Tackling antimicrobial resistance in the COVID-19 pandemic. Bull World Health Organ 98(7):442–442A. https://doi.org/10.2471/BLT.20.268573

Gisolfi CV (2000) Is the GI system built for exercise? News Physiol Sci 15:114–119. https://doi.org/10.1152/physiologyonline.2000.15.3.114

Gou W, Fu Y, Yue L, Chen GD, Cai X, Shuai M, Xu F, Yi X, Chen H, Zhu YJ, Xiao ML (2020) Gut microbiota may underlie the predisposition of healthy individuals to COVID-19. MedRxiv. https://doi.org/10.1101/2020.04.22.20076091

Greenwood JM, Ezquerra AL, Behrens S, Branca A, Mallet L (2016) Current analysis of host-parasite interactions with a focus on next generation sequencing data. Zoology (Jena) 119(4):298–306. https://doi.org/10.1016/j.zool.2016.06.010

Han M, Zha Y, Chong H, Zhong C, Ning K (2020) Utilizing microbiome approaches to assist source tracking, treatment and prevention of COVID-19: review and assessment. Comput Struct Biotechnol J 18:3615–3622. https://doi.org/10.1016/j.csbj.2020.11.027

Hiergeist A, Glasner J, Reischl U, Gessner A (2015) Analyses of intestinal microbiota: culture versus sequencing. ILAR J 56(2):228–240. https://doi.org/10.1093/ilar/ilv017

Hikmet F, Mear L, Edvinsson A, Micke P, Uhlen M, Lindskog C (2020) The protein expression profile of ACE2 in human tissues. Mol Syst Biol 16(7):e9610. https://doi.org/10.15252/msb.20209610

Hills RD Jr, Pontefract BA, Mishcon HR, Black CA, Sutton SC, Theberge CR (2019) Gut microbiome: profound implications for diet and disease. Nutrients 11(7). https://doi.org/10.3390/nu11071613

Hooper LV, Littman DR, Macpherson AJ (2012) Interactions between the microbiota and the immune system. Science 336(6086):1268–1273. https://doi.org/10.1126/science.1223490

Hovhannisyan H, Gabaldon T (2019) Transcriptome sequencing approaches to elucidate host-microbe interactions in opportunistic human fungal pathogens. Curr Top Microbiol Immunol 422:193–235. https://doi.org/10.1007/82_2018_122

Hui AW, Lau HW, Chan TH, Tsui SK (2013) The human microbiota: a new direction in the investigation of thoracic diseases. J Thorac Dis 5(Suppl 2):S127–S131. https://doi.org/10.3978/j.issn.2072-1439.2013.07.41

Ichinohe T, Pang IK, Kumamoto Y, Peaper DR, Ho JH, Murray TS, Iwasaki A (2011) Microbiota regulates immune defense against respiratory tract influenza a virus infection. Proc Natl Acad Sci U S A 108(13):5354–5359. https://doi.org/10.1073/pnas.1019378108

Jiang D, Armour CR, Hu C, Mei M, Tian C, Sharpton TJ, Jiang Y (2019) Microbiome multi-Omics network analysis: statistical considerations, limitations, and opportunities. Front Genet 10:995. https://doi.org/10.3389/fgene.2019.00995

Johnson AJ, Zheng JJ, Kang JW, Saboe A, Knights D, Zivkovic AM (2020) A guide to diet-microbiome study design. Front Nutr 7:79. https://doi.org/10.3389/fnut.2020.00079

Keim NL, Martin RJ (2014) Dietary whole grain-microbiota interactions: insights into mechanisms for human health. Adv Nutr 5(5):556–557. https://doi.org/10.3945/an.114.006536

Kesh K, Mendez R, Abdelrahman L, Banerjee S, Banerjee S (2020) Type 2 diabetes induced microbiome dysbiosis is associated with therapy resistance in pancreatic adenocarcinoma. Microb Cell Factories 19(1):75. https://doi.org/10.1186/s12934-020-01330-3

Khatiwada S, Subedi A (2020) Lung microbiome and coronavirus disease 2019 (COVID-19): possible link and implications. Hum Microb J 17:100073. https://doi.org/10.1016/j.humic.2020.100073

Kim S, Covington A, Pamer EG (2017) The intestinal microbiota: antibiotics, colonization resistance, and enteric pathogens. Immunol Rev 279(1):90–105. https://doi.org/10.1111/imr.12563

Kirby TO, Ochoa-Reparaz J (2018) The gut microbiome in multiple sclerosis: a potential therapeutic avenue. Med Sci (Basel) 6(3). https://doi.org/10.3390/medsci6030069

Konstantinidis T, Tsigalou C, Karvelas A, Stavropoulou E, Voidarou C, Bezirtzoglou E (2020) Effects of antibiotics upon the gut microbiome: a review of the literature. Biomedicine 8(11). https://doi.org/10.3390/biomedicines8110502

Kyle MH, Glassman ME, Khan A, Fernandez CR, Hanft E, Emeruwa UN, Scripps T, Walzer L, Liao GV, Saslaw M, Rubenstein D, Hirsch DS, Keown MK, Stephens A, Mollicone I, Bence ML, Gupta A, Sultan S, Sibblies C, Whittier S, Abreu W, Akita F, Penn A, Orange JS, Saiman L, Welch MG, Gyamfi-Bannerman C, Stockwell MS, Dumitriu D (2020) A review of newborn outcomes during the COVID-19 pandemic. Semin Perinatol 44(7):151286. https://doi.org/10.1016/j.semperi.2020.151286

Langevin S, Pichon M, Smith E, Morrison J, Bent Z, Green R, Barker K, Solberg O, Gillet Y, Javouhey E, Lina B, Katze MG, Josset L (2017) Early nasopharyngeal microbial signature associated with severe influenza in children: a retrospective pilot study. J Gen Virol 98(10):2425–2437. https://doi.org/10.1099/jgv.0.000920

Lau K, Srivatsav V, Rizwan A, Nashed A, Liu R, Shen R, Akhtar M (2017) Bridging the gap between gut microbial Dysbiosis and cardiovascular diseases. Nutrients 9(8). https://doi.org/10.3390/nu9080859

Lebeer S, Spacova I (2019) Exploring human host-microbiome interactions in health and disease-how to not get lost in translation. Genome Biol 20(1):56. https://doi.org/10.1186/s13059-019-1669-4

Lee SR (2018) Critical role of zinc as either an antioxidant or a Prooxidant in cellular systems. Oxidative Med Cell Longev 2018:9156285. https://doi.org/10.1155/2018/9156285

Lee SY, Lee E, Park YM, Hong SJ (2018) Microbiome in the gut-skin Axis in atopic dermatitis. Allergy Asthma Immunol Res 10(4):354–362. https://doi.org/10.4168/aair.2018.10.4.354

Lee YB, Byun EJ, Kim HS (2019) Potential role of the microbiome in acne: a comprehensive review. J Clin Med 8(7). https://doi.org/10.3390/jcm8070987

Leis JA, Born KB, Theriault G, Ostrow O, Grill A, Johnston KB (2020) Using antibiotics wisely for respiratory tract infection in the era of covid-19. BMJ 371:m4125. https://doi.org/10.1136/bmj.m4125

Lemon KP, Klepac-Ceraj V, Schiffer HK, Brodie EL, Lynch SV, Kolter R (2010) Comparative analyses of the bacterial microbiota of the human nostril and oropharynx. mBio 1(3). https://doi.org/10.1128/mBio.00129-10

Lim S, Bae JH, Kwon HS, Nauck MA (2021) COVID-19 and diabetes mellitus: from pathophysiology to clinical management. Nat Rev Endocrinol 17(1):11–30. https://doi.org/10.1038/s41574-020-00435-4

Ling Z, Liu X, Luo Y, Yuan L, Nelson KE, Wang Y, Xiang C, Li L (2013) Pyrosequencing analysis of the human microbiota of healthy Chinese undergraduates. BMC Genomics 14:390. https://doi.org/10.1186/1471-2164-14-390

Magana M, Pushpanathan M, Santos AL, Leanse L, Fernandez M, Ioannidis A, Giulianotti MA, Apidianakis Y, Bradfute S, Ferguson AL, Cherkasov A, Seleem MN, Pinilla C, de la Fuente-Nunez C, Lazaridis T, Dai T, Houghten RA, Hancock REW, Tegos GP (2020) The value of antimicrobial peptides in the age of resistance. Lancet Infect Dis 20(9):e216–e230. https://doi.org/10.1016/S1473-3099(20)30327-3

Maleki Dana P, Kolahdooz F, Sadoughi F, Moazzami B, Chaichian S, Asemi Z (2020) COVID-19 and pregnancy: a review of current knowledge. Infez Med 28(suppl 1):46–51

Malinis M, McManus D, Davis M, Topal J (2020) An overview on the use of antivirals for the treatment of patients with COVID19 disease. Expert Opin Investig Drugs:1–15. https://doi.org/10.1080/13543784.2021.1847270

Malla MA, Dubey A, Kumar A, Yadav S, Hashem A, Abd Allah EF (2018) Exploring the human microbiome: the potential future role of next-generation sequencing in disease diagnosis and treatment. Front Immunol 9:2868. https://doi.org/10.3389/fimmu.2018.02868

Mansur JL (2020) Letter: low population mortality from COVID-19 in countries south of latitude 35 degrees north supports vitamin D as a factor determining severity. Aliment Pharmacol Ther 52(2):411–412. https://doi.org/10.1111/apt.15820

McCartney DM, Byrne DG (2020) Optimisation of vitamin D status for enhanced Immuno-protection against Covid-19. Ir Med J 113(4):58

McCullers JA, Rehg JE (2002) Lethal synergism between influenza virus and Streptococcus pneumoniae: characterization of a mouse model and the role of platelet-activating factor receptor. J Infect Dis 186(3):341–350. https://doi.org/10.1086/341462

Morais LH, HLT S, Mazmanian SK (2020) The gut microbiota-brain axis in behaviour and brain disorders. Nat Rev Microbiol. https://doi.org/10.1038/s41579-020-00460-0

Morris A, Beck JM, Schloss PD, Campbell TB, Crothers K, Curtis JL, Flores SC, Fontenot AP, Ghedin E, Huang L, Jablonski K, Kleerup E, Lynch SV, Sodergren E, Twigg H, Young VB, Bassis CM, Venkataraman A, Schmidt TM, Weinstock GM, Lung HIVMP (2013) Comparison of the respiratory microbiome in healthy nonsmokers and smokers. Am J Respir Crit Care Med 187(10):1067–1075. https://doi.org/10.1164/rccm.201210-1913OC

Neu J, Rushing J (2011) Cesarean versus vaginal delivery: long-term infant outcomes and the hygiene hypothesis. Clin Perinatol 38(2):321–331. https://doi.org/10.1016/j.clp.2011.03.008

Neuman H, Forsythe P, Uzan A, Avni O, Koren O (2018) Antibiotics in early life: dysbiosis and the damage done. FEMS Microbiol Rev 42(4):489–499. https://doi.org/10.1093/femsre/fuy018

Nuriel-Ohayon M, Neuman H, Koren O (2016) Microbial changes during pregnancy, birth, and infancy. Front Microbiol 7:1031. https://doi.org/10.3389/fmicb.2016.01031

O'Hara AM, Shanahan F (2006) The gut flora as a forgotten organ. EMBO Rep 7(7):688–693. https://doi.org/10.1038/sj.embor.7400731

Olaimat AN, Aolymat I, Al-Holy M, Ayyash M, Abu Ghoush M, Al-Nabulsi AA, Osaili T, Apostolopoulos V, Liu SQ, Shah NP (2020) The potential application of probiotics and prebiotics for the prevention and treatment of COVID-19. NPJ Sci Food 4:17. https://doi.org/10.1038/s41538-020-00078-9

Pascal M, Perez-Gordo M, Caballero T, Escribese MM, Lopez Longo MN, Luengo O, Manso L, Matheu V, Seoane E, Zamorano M, Labrador M, Mayorga C (2018) Microbiome and allergic diseases. Front Immunol 9:1584. https://doi.org/10.3389/fimmu.2018.01584

Pascarella G, Strumia A, Piliego C, Bruno F, Del Buono R, Costa F, Scarlata S, Agro FE (2020) COVID-19 diagnosis and management: a comprehensive review. J Intern Med 288(2):192–206. https://doi.org/10.1111/joim.13091

Peters HP, De Vries WR, Vanberge-Henegouwen GP, Akkermans LM (2001) Potential benefits and hazards of physical activity and exercise on the gastrointestinal tract. Gut 48(3):435–439. https://doi.org/10.1136/gut.48.3.435

Picca A, Fanelli F, Calvani R, Mule G, Pesce V, Sisto A, Pantanelli C, Bernabei R, Landi F, Marzetti E (2018) Gut Dysbiosis and muscle aging: searching for novel targets against sarcopenia. Mediat Inflamm 2018:7026198. https://doi.org/10.1155/2018/7026198

Pulikkan J, Maji A, Dhakan DB, Saxena R, Mohan B, Anto MM, Agarwal N, Grace T, Sharma VK (2018) Gut microbial Dysbiosis in Indian children with autism Spectrum disorders. Microb Ecol 76(4):1102–1114. https://doi.org/10.1007/s00248-018-1176-2

Quigley EM (2013) Gut bacteria in health and disease. Gastroenterol Hepatol (N Y) 9(9):560–569

Rajkumar RP (2020) COVID-19 and mental health: a review of the existing literature. Asian J Psychiatr 52:102066. https://doi.org/10.1016/j.ajp.2020.102066

Read SA, Obeid S, Ahlenstiel C, Ahlenstiel G (2019) The role of zinc in antiviral immunity. Adv Nutr 10(4):696–710. https://doi.org/10.1093/advances/nmz013

Riccio P, Rossano R (2018) Diet, gut microbiota, and vitamins D + a in multiple sclerosis. Neurotherapeutics 15(1):75–91. https://doi.org/10.1007/s13311-017-0581-4

Rishi P, Thakur K, Vij S, Rishi L, Singh A, Kaur IP, Patel SKS, Lee JK, Kalia VC (2020) Diet, gut microbiota and COVID-19. Indian J Microbiol:1–10. https://doi.org/10.1007/s12088-020-00908-0

Rodgers GP, Gibbons GH (2020) Obesity and hypertension in the time of COVID-19. JAMA 324(12):1163–1165. https://doi.org/10.1001/jama.2020.16753

Roy D, Tripathy S, Kar SK, Sharma N, Verma SK, Kaushal V (2020) Study of knowledge, attitude, anxiety & perceived mental healthcare need in Indian population during COVID-19 pandemic. Asian J Psychiatr 51:102083. https://doi.org/10.1016/j.ajp.2020.102083

Sadiq FA (2021) Is it time for microbiome-based therapies in viral infections? Virus Res 291:198203. https://doi.org/10.1016/j.virusres.2020.198203

Saffouri GB, Shields-Cutler RR, Chen J, Yang Y, Lekatz HR, Hale VL, Cho JM, Battaglioli EJ, Bhattarai Y, Thompson KJ, Kalari KK, Behera G, Berry JC, Peters SA, Patel R, Schuetz AN, Faith JJ, Camilleri M, Sonnenburg JL, Farrugia G, Swann JR, Grover M, Knights D, Kashyap PC (2019) Small intestinal microbial dysbiosis underlies symptoms associated with functional gastrointestinal disorders. Nat Commun 10(1):2012. https://doi.org/10.1038/s41467-019-09964-7

Sattar N, McInnes IB, McMurray JJV (2020) Obesity is a risk factor for severe COVID 19 infection: multiple potential mechanisms. Circulation 142(1):4–6. https://doi.org/10.1161/CIRCULATIONAHA.120.047659

Segal LN, Alekseyenko AV, Clemente JC, Kulkarni R, Wu B, Gao Z, Chen H, Berger KI, Goldring RM, Rom WN, Blaser MJ, Weiden MD (2013) Enrichment of lung microbiome with supraglottic taxa is associated with increased pulmonary inflammation. Microbiome 1(1):19. https://doi.org/10.1186/2049-2618-1-19

Shah VK, Firmal P, Alam A, Ganguly D, Chattopadhyay S (2020) Overview of immune response during SARS-CoV-2 infection: lessons from the past. Front Immunol 11:1949. https://doi.org/10.3389/fimmu.2020.01949

Sheflin AM, Whitney AK, Weir TL (2014) Cancer-promoting effects of microbial dysbiosis. Curr Oncol Rep 16(10):406. https://doi.org/10.1007/s11912-014-0406-0

Shen Z, Xiao Y, Kang L, Ma W, Shi L, Zhang L, Zhou Z, Yang J, Zhong J, Yang D, Guo L, Zhang G, Li H, Xu Y, Chen M, Gao Z, Wang J, Ren L, Li M (2020) Genomic diversity of severe acute respiratory syndrome-coronavirus 2 in patients with coronavirus disease 2019. Clin Infect Dis 71(15):713–720. https://doi.org/10.1093/cid/ciaa203

Shi Y, Wang G, Cai XP, Deng JW, Zheng L, Zhu HH, Zheng M, Yang B, Chen Z (2020) An overview of COVID-19. J Zhejiang Univ Sci B 21(5):343–360. https://doi.org/10.1631/jzus.B2000083

Shinu P, Morsy MA, Deb PK, Nair AB, Goyal M, Shah J, Kotta S (2020) SARS CoV-2 organotropism associated pathogenic relationship of gut-brain Axis and illness. Front Mol Biosci 7:606779. https://doi.org/10.3389/fmolb.2020.606779

Short KR, Habets MN, Hermans PW, Diavatopoulos DA (2012) Interactions between Streptococcus pneumoniae and influenza virus: a mutually beneficial relationship? Future Microbiol 7(5):609–624. https://doi.org/10.2217/fmb.12.29

Swiatecka D, Narbad A, Ridgway KP, Kostyra H (2011) The study on the impact of glycated pea proteins on human intestinal bacteria. Int J Food Microbiol 145(1):267–272. https://doi.org/10.1016/j.ijfoodmicro.2011.01.002

Thursby E, Juge N (2017) Introduction to the human gut microbiota. Biochem J 474(11):1823–1836. https://doi.org/10.1042/BCJ20160510

Vangay P, Ward T, Gerber JS, Knights D (2015) Antibiotics, pediatric dysbiosis, and disease. Cell Host Microbe 17(5):553–564. https://doi.org/10.1016/j.chom.2015.04.006

Varghese PM, Tsolaki AG, Yasmin H, Shastri A, Ferluga J, Vatish M, Madan T, Kishore U (2020) Host-pathogen interaction in COVID-19: pathogenesis, potential therapeutics and vaccination strategies. Immunobiology 225(6):152008. https://doi.org/10.1016/j.imbio.2020.152008

Venkataraman A, Bassis CM, Beck JM, Young VB, Curtis JL, Huffnagle GB, Schmidt TM (2015) Application of a neutral community model to assess structuring of the human lung microbiome. mBio 6(1). https://doi.org/10.1128/mBio.02284-14

Villapol S (2020) Gastrointestinal symptoms associated with COVID-19: impact on the gut microbiome. Transl Res 226:57–69. https://doi.org/10.1016/j.trsl.2020.08.004

West CE, Dzidic M, Prescott SL, Jenmalm MC (2017) Bugging allergy; role of pre-, pro- and synbiotics in allergy prevention. Allergol Int 66(4):529–538. https://doi.org/10.1016/j.alit.2017.08.001

Yang I, Corwin EJ, Brennan PA, Jordan S, Murphy JR, Dunlop A (2016) The infant microbiome: implications for infant health and neurocognitive development. Nurs Res 65(1):76–88. https://doi.org/10.1097/NNR.0000000000000133

Yang R, Mei H, Zheng T, Fu Q, Zhang Y, Buka S, Yao X, Tang Z, Zhang X, Qiu L, Zhang Y, Zhou J, Cao J, Wang Y, Zhou A (2020) Pregnant women with COVID-19 and risk of adverse birth outcomes and maternal-fetal vertical transmission: a population-based cohort study in Wuhan. China BMC Med 18(1):330. https://doi.org/10.1186/s12916-020-01798-1

Yang YJ, Sheu BS (2012) Probiotics-containing yogurts suppress helicobacter pylori load and modify immune response and intestinal microbiota in the helicobacter pylori-infected children. Helicobacter 17(4):297–304. https://doi.org/10.1111/j.1523-5378.2012.00941.x

Yi H, Yong D, Lee K, Cho YJ, Chun J (2014) Profiling bacterial community in upper respiratory tracts. BMC Infect Dis 14:583. https://doi.org/10.1186/s12879-014-0583-3

Young VB (2017) The role of the microbiome in human health and disease: an introduction for clinicians. BMJ 356:j831. https://doi.org/10.1136/bmj.j831

Zambrano LD, Ellington S, Strid P, Galang RR, Oduyebo T, Tong VT, Woodworth KR, Nahabedian JF 3rd, Azziz-Baumgartner E, Gilboa SM, Meaney-Delman D (2020) Pregnancy CC-R, infant linked outcomes T (2020) update: characteristics of symptomatic women of reproductive age with laboratory-confirmed SARS-CoV-2 infection by pregnancy status–United States, January 22-October 3. MMWR Morb Mortal Wkly Rep 69(44):1641–1647. https://doi.org/10.15585/mmwr.mm6944e3

Zuo T, Zhang F, Lui GCY, Yeoh YK, Li AYL, Zhan H, Wan Y, Chung ACK, Cheung CP, Chen N, Lai CKC, Chen Z, Tso EYK, Fung KSC, Chan V, Ling L, Joynt G, Hui DSC, Chan FKL, Chan PKS, Ng SC (2020) Alterations in gut microbiota of patients with COVID-19 during time of hospitalization. Gastroenterology 159(3):944–955. e948. https://doi.org/10.1053/j.gastro.2020.05.048

GPSR Compliance

The European Union's (EU) General Product Safety Regulation (GPSR) is a set of rules that requires consumer products to be safe and our obligations to ensure this.

If you have any concerns about our products, you can contact us on ProductSafety@springernature.com

In case Publisher is established outside the EU, the EU authorized representative is:

Springer Nature Customer Service Center GmbH
Europaplatz 3
69115 Heidelberg, Germany

Batch number: 10732044

Printed by Printforce, the Netherlands